INSECT FLIGHT

Previous Symposia of the Royal Entomological Society

NO. I. INSECT POLYMORPHISM edited by J.S.Kennedy
London: 1961

NO. 2. INSECT REPRODUCTION edited by K.C.Highnam
London: 1964

NO. 3. INSECT BEHAVIOUR edited by P.T. Haskell
London: 1966

NO. 4. INSECT ABUNDANCE edited by T.R.E.Southwood
Blackwell Scientific Publications, Oxford: 1968

NO. 5. INSECT ULTRASTRUCTURE edited by A.C.Neville
Blackwell Scientific Publications, Oxford: 1970

NO. 6. INSECT/PLANT RELATIONSHIPS edited by H.F.van Emden
Blackwell Scientific Publications, Oxford: 1973

SYMPOSIA OF THE ROYAL ENTOMOLOGICAL
SOCIETY OF LONDON: NUMBER SEVEN

Insect Flight

EDITED ON BEHALF OF THE SOCIETY BY

R.C. RAINEY, F.R.S.

A HALSTED PRESS BOOK

JOHN WILEY & SONS

NEW YORK

© The Royal Entomological Society 1976

Published by Blackwell Scientific Publications
Osney Mead, Oxford
85 Marylebone High Street, London W1
9 Forrest Road, Edinburgh
P.O. Box 9, North Balwyn, Victoria, Australia

Library of Congress Cataloging in Publication Data
Main entry under title:

Insect flight.

(Symposia of the Royal Entomological Society of London; 7)
1. Insects—Flight—Congresses. I. Rainey, Reginald Charles,
1913–   . II. Series: Royal Entomological Society of London.
Symposium; 7.

QL496.7.157   595.7'01'852   75-22091

SBN 0-470-70550-7

Printed in Great Britain

Foreword

The biennial symposia of the Royal Entomological Society were established to promote the coherence of entomology, and this, the seventh, was planned to bring to the notice of all who are interested in the flight of insects some notable recent advances by widely different approaches. Progress in elucidating the mechanisms of insect flight emphasizes the already recognized role of flight in moulding the morphology and dominating the physiology of insects; and it is hoped that entomologists concerned on the other hand with the behaviour, ecology, population dynamics and control of insects may be stimulated and challenged by striking new evidence on the scale and extent of flight in nature and on the relationships of flying insects to their atmospheric environment. It is hoped that this volume may find readers ranging in outlook from the naturalist to the aerodynamicist, the former following Darwin, Fitzroy and Wallace in recording the effects upon flying insects of the ceaseless drama of winds and weather, and the latter perhaps fascinated by an entirely new mechanism of lift-generation, the first to be discovered for more than half a century, and now found to be exploited by a minute wasp parasite of our familiar greenhouse whitefly.

The Society is greatly indebted to the speakers, especially for their co-operation in facing some formidable challenges of interdisciplinary communication, and in the range of new material received; to the chairmen and all who participated in the cogent and lively discussions; and to many others who contributed significantly to the occasion. My own appreciation is particularly due to Dr D.L.Gunn, C.B.E., for a quarter of a century of support and guidance in the study of insect flight, including his deputizing as convenor during my absence on field work immediately prior to this symposium; to Miss M.J.Haggis, for invaluable editorial assistance; and to the publishers, for their co-operation and patience during delays imposed by the demands of this rapidly developing subject.

R.C.RAINEY
*Symposium Convenor
and Editor*

August 1974

Contents

Foreword v

I. THE MECHANISMS OF INSECT FLIGHT

1   The muscles and sense organs involved in insect flight 3
J.W.S. PRINGLE

The rhythm of wing-beats   4
Mechanical sense organs in the thorax and wings   8
Other senses influencing flight   10
Conclusion   11
Discussion   11
References   14

2   Control of flight and related behaviour by the central nervous systems of insects
. . . . 16
BRIAN MULLONEY

Development of centrally generated motor patterns   19
   Genetic control of central pattern generators   19
Physiological mechanisms generating motor patterns   22
   Myogenic insects   23
   Neurogenic insects—mostly locusts and crickets   23
Discussion   26
References   29

3   Wing movements and the generation of aerodynamic forces by some medium-sized insects 31
WERNER NACHTIGALL

The aerodynamic forces on the wings of gliding butterflies   31
The aerodynamic forces on the elytra of beetles   34
The wing movements and aerodynamic forces of flies   36
The wing movements and aerodynamic forces of the honey-bee   41
Discussion   46
References   46

4   Energetics and aerodynamics of flapping flight: a synthesis 48
TORKEL WEIS-FOGH

Drag or lift?   49
Nature of aerodynamic lift   51
   Lift and circulation   51
   Bound and starting vortices   52
   Ordinary aerofoil action   53
Fast forward flight   54
   Horizontal flight of the Desert Locust   54

Aerodynamic power and speed in locusts   56
Normal hovering   60
   Coefficients of lift   60
   Aerodynamic power   62
   Dynamic efficiency   62
Novel aerodynamic mechanisms   63
   Flight of a tiny wasp and the 'fling' mechanism   64
   Flight of Syrphinae and Aeshnidae   67
General aspects of flapping flight   69
   A figure of merit   69
Acknowledgements   70
Discussion   70
References   72

II. THE FLYING INSECT AND ITS ENVIRONMENT

5   Flight behaviour and features of the atmospheric environment 75
   R.C.RAINEY

Summary   75
Introduction   75
Some dominating effects of behaviour   76
   Cohesion of travelling locust swarms   76
   Flight in persistent static populations   78
   Heights of nocturnal and diurnal flight   80
Down-wind displacement and its effects   82
   Rain and survival value   83
   New evidence on airborne insects in zones of wind-convergence   83
      Inter-Tropical Convergence Zone   86
      African Rift Convergence Zone   96
      Red Sea Convergence Zone   98
      Frontal systems of temperate latitudes   101
   Concentration of airborne insect populations by wind-convergence: some implications   104
      Mutual perception and behaviour   105
      Population dynamics   106
      Tactics and strategy of control   107
Conclusion   107
Acknowledgements   108
Discussion   108
References   110

6   Forecasting infestations of tropical migrant pests: the Desert Locust and the African
   Armyworm 113
   ELIZABETH BETTS

Desert Locust forecasting   114
   Methods   114
   Examples   116
      Major swarm redistribution, September to November 1968   116
      Problems of assessment from incomplete data   119
   Verification: assessment of forecasts   121
Armyworm forecasting   123
   Methods   123
   Forecasting the first infestations of the season   124
   Mid-season forecasting   124
   Verification and discussion   128
Conclusions   129
Acknowledgements   130
Discussion   131
References   132

7 Insect flight in relation to problems of pest control 135
   R.J.V.JOYCE

   The basis of present crop protection practice 135
   Pest outbreaks and adaptive dispersal 137
   Implications of adaptive dispersal for crop protection 140
   Insect control strategy in relation to insect flight 141
       Desert Locust 142
       Plague grasshopper in Sudan 142
       Spruce budworm 142
       Rice stemborers in Java 143
       Cotton pests in the Sudan Gezira 145
   Conclusions 150
   Acknowledgements 151
   Discussion 151
   References 152

8 Radar observations of insect flight 157
   G.W.SCHAEFER

   Introduction 157
   Elements of radar entomology: I Basic methods 158
       Equipment, performance and siting 158
       Radar detection and display 159
       Estimation of insect numbers and densities 159
       Measurement of insect ground-speed and track-direction; estimation of air-speed and orientation 160
   Elements of radar entomology: II Characteristics of radar echoes from insects 162
       Echo amplitude 162
       Orientation pattern 162
       Echo signatures 164
           Recording and analysis of echo signatures 164
           Wing-beat frequencies in the insect echo signature 167
           Modulation mechanism 168
           Use of wing-beat frequencies in insect identification 169
           Distinguishing echo signatures of insects from those of birds 169
           Echo signatures of Desert Locusts 170
           Echo signatures of Tree Locusts 171
           Echo signatures of Sudan grasshoppers 171
   Observations on take-off, height, density and duration of flight of insect populations 174
       Evening take-off 174
       Flying heights and densities after the take-off period 178
   Observations on concentration and transportation of airborne insects 181
       Concentration by convergent wind-fields 181
           Storm-outflow cold fronts 181
           Nocturnal wind-shift lines 183
           Inter-Tropical Front 184
           Bénard convection cells 186
       Echo layers 187
   Acknowledgements 189
   Appendix: some quantitative aspects of radar entomology 190
       Radar cross-section: definition 190
       Beam directivity pattern 190
       Radar cross-sections—theory 190
       Radar cross-sections—measurements 192
       Orientation pattern and detection aspects 193
   Discussion 195
   References 196

9   Foraging and homing flight of the honey-bee: some general problems of
    orientation 199
    MARTIN LINDAUER

    Teleorientation (long-range navigation) 199
       Estimation of distance 199
       Directional orientation 201
          Is the sun-compass orientation innate? 203
          Compensating for side-wind drift 203
    Orientation near the goal (short-range navigation) 206
       Olfactory orientation 206
       Pattern recognition 207
       Learning processes in approaching and identifying the goal 208
       Generalization and re-learning 212
    Discussion 213
    References 215

10 Lability of the flight system: a context for functional adaptation 217
    C.G.JOHNSON

    Differences in sustained flight between individuals of a species 217
    Changes in flight duration throughout life 218
    Some immediately post-emergence changes relating to flight duration 219
    Flight duration and muscular effort 221
       Fuel 221
       Comparisons between species 222
       Wing-loading 223
       Flight, ovary development and endocrines 223
    The degeneration and regeneration of flight muscles 225
    The collective expression of individual variation 227
    Conclusion 227
    Discussion 228
    References 230

11 The fossil record and insect flight 235
    R.J.WOOTTON

    Ephemeroptera 238
    Syntonopteridae 241
    Odonata and Protodonata 242
    Palaeodictyoptera, Megasecoptera, Diaphanopterodea and Archodonata 244
    Paoliidae and Protorthoptera 247
    Blattodea 248
    Palaeozoic nymphs and the origin of flight 249
    Conclusions 250
    Acknowledgements 251
    Discussion 251
    References 252

12 The evolution of insect flight 255
    V.B.WIGGLESWORTH

    Historical survey 255
    Re-examination of the gill theory 256
    The origin of flight 262
       Dispersal theory: physical considerations 262
       Physiological considerations 263
       Ecological considerations 264
       Climatic considerations 264
    Summary 266
    Discussion 266
    References 268

GENERAL DISCUSSION 271
CONTRIBUTORS TO DISCUSSIONS 277
INDEX 279

I · The mechanisms of insect flight

1 · The muscles and sense organs involved in insect flight

J.W.S.PRINGLE

Department of Zoology, University of Oxford

Nearly all problems connected with insect flight demand for their solution an understanding of the anatomy of the pterothorax, the physiology of the flight muscles and the organization of the sense organs and reflexes involved in flight. The last of these will be discussed by Professor Mulloney. This introductory paper reviews what is known about the flight muscles and the mechanical sense organs which monitor wing movement and forces; it also considers briefly the other senses which influence less directly the onset and orientation of flight.

It is now generally accepted that the ability to fly evolved in insects only once and that, in spite of considerable differences in present-day orders, there is a common pattern in the flight system of all insects. Perhaps because the segments lie closer to the centre of gravity, the meso- and meta-thorax of existing Pterygota contain most of the elements of the flight system, but, at least in the nervous system, there is some involvement of the prothorax [28], and sometimes (propodeum of Hymenoptera) the anatomy of the first abdominal segment is also significant. These features may be relics of an original, less differentiated arrangement.

The structure of a generalized pterothoracic segment is shown in Fig. 1.1. The wing articulates at two points on its dorsal fold to the scutum and at one point on its ventral fold to the pleuron, at the place where the strengthening pleural suture divides the episternum from the epimeron. At these articulations, the cuticle is folded and thickened in a complex manner, creating a number of discrete axillary sclerites at the base of the main wing veins. It is the relative movement of these sclerites that makes possible the complex patterns of wing twisting and changes of profile that must accompany the up and down movement in order to generate the correct aerodynamic forces.

At the pleural wing articulation and elsewhere in the flight system there occur pads of the elastic material, *resilin*, which store energy when deformed [45, 46]. These form one element of an elastic component in the flight motor which, combined with the inertia of the wings, generates a mechanical resonance constraining the wing-beats to occur at a particular frequency for any particular insect. Without this resonance, it would be impossible for the muscles to deliver sufficient power to maintain the wing movements. Because of it, small insects beat their wings at a higher frequency than large insects and this parameter, wing-beat frequency, is not able to be varied in order to control the speed or power of flight except over a narrow range.

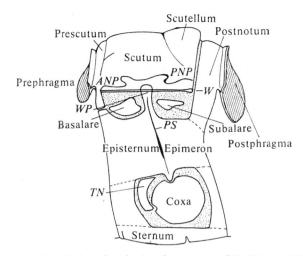

FIG. 1.1. **Diagrammatic lateral view of a wing-bearing segment of the thorax.** *ANP*, anterior notal articulation; *PNP*, posterior notal articulation; *WP*, pleural articulation; *PS*, pleural suture; *W*, wing (cut through); *TN*, trochantin of leg. (From Pringle [34].)

Ten pairs of muscles (or groups of muscles where they are anatomically divided) are involved in flight in each pterothoracic segment [35]. Their location in the insect is shown in Fig. 1.2. According to their mode of operation, they fall into three groups.

(1) *Indirect flight muscles*, causing the wings to move by distortion of the shape of the thoracic box—*dorsal longitudinal* (depressor), *oblique dorsal* and *dorsoventral* (elevator) muscles.

(2) *Direct wing muscles*, inserted at one end directly to the wing, axillary sclerites or moveable sclerites of the pleuron—*basalar* (depressor promotor), *subalar* (depressor supinotor), and *third axillary* (wing-folding) muscles.

(3) *Accessory indirect muscles*, attached at both ends to parts of the thorax and modifying the way in which the power-producing flight muscles cause the wings to move—*pleurosternal*, *anterior tergopleural*, *posterior tergopleural* and *intersegmental* muscles.

THE RHYTHM OF WING-BEATS

In the more primitive Orders of insects, wing movement is produced partly by muscles which either are, or clearly were leg muscles. Such bifunctional activity is still found in Orthoptera, Diptera and elsewhere. In the locust, the same sets of muscles produce wing or leg movement, depending on their pattern of co-ordination [48]. In muscid Diptera, most of the flight muscles are independent of those moving the legs, but one mesothoracic dorsoventral muscle, the *tergotrochanteral*, is used at the start of flight to produce both the leg depression for the jump and the first elevation of the wings [27]. In the original flight system, each contraction of the power-producing flight muscles is a twitch, as in fast leg muscles, and the command for it is given through the motor nerve as a single volley of impulses or a brief high-frequency burst [50]. This mechanism is found in many existing Orders. Because nervous and mechanical activity occur at the same frequency as the wing-beats, such insects are termed *synchronous* and the mechanism of the flight rhythm is *neurogenic*.

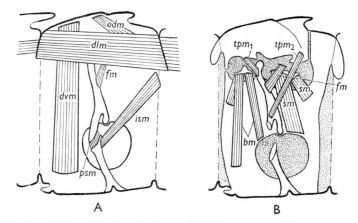

FIG. 1.2. **Diagrammatic internal view showing the location of the ten groups of muscles involved in flight.** A, indirect and some other muscles; B, lateral muscles. *bm*, basalar; *dlm*, dorsal longitudinal; *dvm*, dorsoventral; *fm*, wing-folding (third axillary); *ism*, intersegmental; *odm*, oblique dorsal; *psm*, pleurosternal; *sm*, subalar; *tpm₁*, *tpm₂*, anterior and posterior tergopleural. (From Pringle [34].)

By contrast, in certain higher Orders the rhythm of the wing-beats and of the causative contractions of the power-producing flight muscles is endogenous to the flight motor and results from the interaction of a peculiar property of these fibrillar muscles themselves with the mechanical resonance of the wing-thorax system already described. This mechanism has been termed *myogenic* [32, 33], but a better term is *asynchronous* [36], because the impulses in the motor nerves and the mechanical activity of the muscles do not occur at the same frequency. In these insects there is, in fact, no direct nervous control of the individual wing-beats and the central nervous system exercises control over the activity of the power-producing muscles in the same way as does the throttle control on an automobile; it adjusts the power which may be drawn from the flight motor, but does not control the individual cycles of motion.

The distribution of the two types of rhythmic mechanism is shown in Table 1.1. It is clear that the asynchronous mechanism of fibrillar muscle has evolved many times in the insects.

The way in which the rhythm of wing-beats is generated has a profound influence on all other aspects of flight. For example, the synchronous mechanism cannot produce a wing-beat frequency greater than about 100 per second (i.e. 100 Hz), because of the delays involved in conduction of the motor nerve impulses and the activation of the contractile system of the muscle fibres. Owing to the need to operate at mechanical resonance, this imposes a lower limit on the size of these insects. The smallest synchronous insects are probably the Micropterygidae, which are poor fliers, having to compensate for their relatively low wing-beat frequency by means of a very light wing-loading. By contrast, there is no limit to the wing-beat frequency in asynchronous insects, since the muscles are maintained in a continuous state of activation during flight. The highest wing-beat frequency recorded in an intact insect is about 1000 Hz in *Forcipomyia* sp. [41] and this can be doubled by experimentally reducing the wing area, which lowers the wing inertia, raises the resonant frequency and automatically leads to a higher frequency of contraction of the power-producing muscles. The evolution of very small insects required the prior evolution of the asynchronous mechanism.

TABLE 1.1. **Characteristics of some insect flight muscles** (From Pringle [34, 35]; Smart [39]; Daly [9]; Cullen [7], [8])

Name	Insertions	Structure in main wing segment				Function
		Tubular	Close-packed	Fibrillar	Absent	
A. INDIRECT MUSCLES 1. Dorsal longitudinal (*dlm*)	Prephragma to postphragma	Dictyoptera Odonata	Orthoptera Ephemeroptera Dermaptera Psocoptera (*Trogius*) Neuroptera Mecoptera Trichoptera Lepidoptera Homoptera (Cicadidae, Cercopidae, Membracidae, Flatidae, Cixiidae, Aleurodidae Coccidae) Hymenoptera (Symphyta, except *Xyela*)	Thysanoptera Psocoptera (*Psocus*) Homoptera (Jassidae, Aphididae Psyllidae) Heteroptera Diptera Hymenoptera (Apocrita, *Xyela*) Coleoptera		Power-producing depressor
2. Dorsoventral (*dvm*)	Scutum to sternum, coxa or trochanter	Diptera (tergotrochanteral muscle in many species)	*Similar to dlm*		Hymenoptera (Chalcidae *Leptomastix*)	Power-producing elevator
3. Oblique dorsal (*odm*)	Lateral scutum to postphragma	Orthoptera Dictyoptera	Phasmida *Otherwise similar to dlm*			Depressor in Phasmida Elevator in other Orders

B. DIRECT MUSCLES

4. Basalar (*bm*)	Basalare or episternum to sternum or coxa	Dictyoptera Diptera Hymenoptera (Aculeata)	Orthoptera Odonata Lepidoptera Homoptera (families as for *dlm*)	Homoptera (Jassidae, Aphididae, Psyllidae) Heteroptera (Belostomatidae, Naucoridae, Notonectidae) Hymenoptera (Ichneumonidae) Coleoptera	Heteroptera	Pronation Power-producing when close-packed or fibrillar
5. Subalar (*sm*)	Subalare or episternum to sternum or coxa	Dictyoptera Heteroptera (Geocorisae) Diptera (Brachycera, Cyclorrhapha) Hymenoptera (Apocrita)	Orthoptera Odonata	Diptera (Nematocera) Coleoptera	Heteroptera (Hydrocorisae)	Supination Power-producing when close-packed or fibrillar

Another result of the effective isolation of the rhythmic power-producing mechanism in these higher insects has been the independent evolution of the muscles and reflexes involved in the control of flight. Particularly in Hymenoptera and Diptera, the controlling muscles have become reduced in size and now determine the form of the wing movement through a servomechanism whose power is drawn from the flight motor. Tonic contraction of these muscles sets the position of the axillary sclerites in various ways, so that wing twisting takes place at different instants during the down-stroke when changes of orientation of the whole insect are required. The three-dimensional geometry of these control movements is difficult to describe in words and will be illustrated by an animated-drawing film ('The wing mechanism of the bee') later in the Symposium. The controlling muscles are homologous with those that perform the same function by twitch contractions at every stroke in synchronous insects and, in many cases, it is also clear that similar reflex mechanisms are involved. Study of the way in which the central nervous connections have become adapted to changes in the morphology and physiology of the flight motor is a fascinating and fertile field for comparative physiologists.

MECHANICAL SENSE ORGANS IN THE THORAX AND WINGS

Unlike vertebrate animals, insects do not normally have sensory endings in the muscles themselves in order to monitor position, movement and the force of contraction. Instead, position and movement are reported back to the central nervous system by external sense organs (hair plates at the joints [30]), or by means of chordotonal organs, internal sensory strands which are attached to the cuticle at two points on opposite sides of an articulation. The force produced by the muscles is monitored in the cuticle by means of campaniform sensilla, functioning as strain gauges [29].

The amplitude and frequency of the wing-beat are detected in locusts by a chordotonal stretch receptor which fires a brief volley of sensory impulses towards the end of the upstroke [13]. The reflex effect from this sense organ is to increase the frequency at which the ganglionic pacemaker generates the rhythmic patterns of motor nerve impulses to the flight muscles. The locust has a synchronous mechanism and it is therefore necessary for the centrally-generated rhythm to be matched to the mechanical resonance of the wing-thorax system. The reflex is relatively slow, modifying the discharge frequency only over several cycles [49], but a small phasic effect has recently been demonstrated [47]. The overall effect will be to stabilize the motor discharge frequency at the value which feeds maximum power into the wings.

The reactions of the aerodynamic forces on the moving wings are monitored by campaniform sensilla located on the basal wing veins [14, 15]. In the locust, the more important of these sense organs are situated on the lower surface of the hind-wings and their reflex effect is to change the angle of twisting of the fore-wings during the aerodynamically significant part of the down-stroke. The overall effect of this reflex on flight is twofold. It ensures, first, that the total lift force produced by the wings is maintained constant over a wide range of body angles, so that the locust can remain airborne in various attitudes. Secondly, it produces stability in the fore-and-aft plane, because the adjustable lift force from the fore-wings acts in front of the centre of gravity.

The campaniform sensilla on the fore-wings of the locust are arranged in two groups and reflexes from these endings have been shown to be important in the maintenance of stability in flight and to operate on a faster time scale than the lift-regulating reflex from

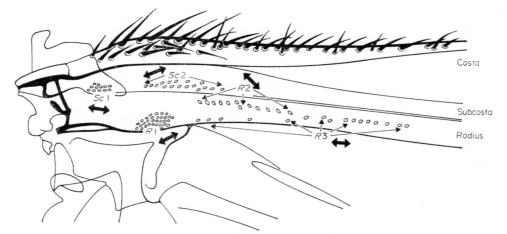

FIG. 1.3. **The wing-base** of *Empis tessalata* (Diptera), showing the location and orientation of groups of campaniform sensilla. R1, R2, R3, radial groups; Sc1, Sc2, subcostal groups; bold arrows indicate the orientation of the long axis of the sensilla, which monitor the aerodynamic forces resulting from the wing movements. (From Pringle [34].)

the sensilla on the hind-wings [15]. The detailed mechanisms, however, have not been elucidated.

Wing-beat frequency alters during reflex control adjustments in the way that would be expected in order to regulate the output power. Direct reflex modification of the motor excitation to the power-producing flight muscles is minimal in Hymenoptera and Diptera, but is important in Coleoptera, where the controlling basalar and subalar muscles, as well as the indirect flight muscles, are of the asynchronous type. In beetles, like locusts but unlike Hymenoptera and Diptera, turning movements are produced by differential excitation of the power-producing muscles on opposite sides of the body [3].

Monitoring the stroke amplitude is important and it is possible that this is always one role of the chordotonal organs at the wing-base, which are more numerous in higher insects than in the locust [34]. Control of stroke amplitude will affect both total power and balance in the fore-and-aft plane, two parameters which can be coupled in more primitive insects with high inherent stability but must be independently regulated in order to achieve maximum manœuvrability. The motor side of this reflex seems to involve the tergopleural complex, which becomes differentiated into several distinct muscles in Hymenoptera and Diptera [35]; the mechanism of control by two of these, the scutellar muscle and the muscle of the axillary lever in the honey-bee, is shown in the film. The campaniform sensilla on the wing veins are also much more differentiated in higher insects than in the locust and several functionally different groups of these sensilla can be recognized by their precise orientation and position (Fig. 1.3). There is probably separate reporting back to the central nervous system of the aerodynamic forces produced during different phases of the wing-stroke. The most elaborate and precise arrangement of campaniform sensilla is found in the halteres of Diptera (Fig. 1.4) and since these have evolved from hindwings it is reasonable to assume that the pattern of reflexes found in the halteres is also present from the wing sense organs.

In the halteres, aerodynamic forces produced by the reaction of the wings on the air have been replaced by inertial forces resulting from the acceleration of the mass of the terminal knob. In addition to the primary forces produced by the oscillation of the organ,

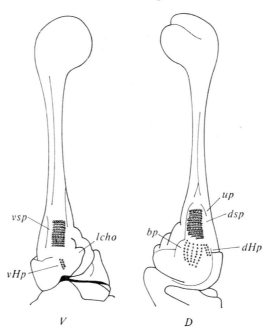

FIG. 1.4. **Ventral (V) and dorsal (D) aspects of the left haltere of** *Lucilia sericata,* showing the location of sensilla. Campaniform sensilla: *bp,* basal plate; *dsp,* dorsal scapal plate; *dHp,* dorsal Hicks papillae; *up,* undifferentiated papilla; *vsp,* ventral scapal plate; *vHp,* ventral Hicks papillae. Chordotonal organs: *lcho,* distal insertion of large chordotonal organ. (From Pringle [31].)

secondary gyroscopic forces are produced during rotations of the whole fly and the magnitude of these is selectively signalled by the sensilla of the basal plate [31]. Extirpation of the halteres results in instability in all three planes of space; in the intact insect pitching, rolling and yawing deviations are normally corrected by distinct patterns of wing twisting [12, 35]. The information from the haltere sense organs gives the fly a very precise control of its wing movements, and Diptera, alone among insects, can fly efficiently in complete darkness.

OTHER SENSES INFLUENCING FLIGHT

Apart from the sense organs on the wing or at the wing-base, flight is influenced and controlled by many other sensory systems; those which are best understood are mechanical sense organs giving indication of contact of the legs with the substratum, proprioceptors at the neck and other joints of the body, hairs on the front of the head detecting air-flow, the antennae and the eyes. Flight orientation may also be controlled by chemoreceptors, but the details of such regulation have not been worked out.

Loss of contact by the legs is a powerful stimulus for the initiation of flight in all insects except Coleoptera and some Heteroptera, where this stimulus may initiate running or swimming movements except when the insect is in the 'mood' for flight [11]. The exact sense organs involved in this reflex have not been determined but it is certainly not only or mainly the mechanoreceptors on the tarsi [37], since loss of the lower joints of the leg does not prevent the contact inhibition of flight. Wing movements can also be initiated by vertical bodily accelerations affecting the relative position of thorax and abdomen [10].

Although flight can be initiated by loss of leg contact, it does not usually continue for more than a short period unless one or more of the 'flight maintenance' reflexes is excited, and the relative importance of other sense organs is different in different insects. In locusts, frontal hairs and lift-detecting sense organs on the wings act synergically in this respect [44]. In *Drosophila*, *Musca* and *Muscina* flight will continue in tethered flies without an airflow [42, 26], but in the honey-bee backwards movement of the visual field is required [38].

With more careful study, it is becoming clear that these 'flight maintenance' reflexes are merely the overt sign of more sophisticated mechanisms for flight regulation. Thus, in the locust [18] and the blowfly [16] the antennae are actively moved by their own muscles to adjust the operating range of the mechanoreceptors that control flight; the antennae of the blowfly control both air-speed and direction [2]. In the locust the antennae and frontal hair plates, both reacting to air-flow, affect different parameters of the wing-stroke [19]. The frontal hairs can detect a lateral component in the air flow due to yaw [4], but the reflex effect on wing-stroke parameters is symmetrical [19] and the tendency for a pivoted tethered locust to turn into an air-stream [43] may be entirely mediated by postural changes in the legs and abdomen [5]. The effect of excitation of antennal and frontal mechanoreceptors on the velocity of forward flight is different, excitation of hairs on frons and vertex having an accelerating and excitation of the antennae a retarding effect. The latter may be part of the mechanism by which migrating locusts remain airborne for as long as possible, since slow flight is more economical of metabolic reserves [20].

Similarly with the visual reflexes: *Drosophila* and *Musca* flying in darkness develop double the thrust of that when the compound eyes are illuminated [1]. Backwards movement of the visual field has been shown both in the honey-bee [24] and in *Drosophila* [21] to have a quantitative effect on forward velocity. Lateral and rotatory movements of the visual field are important in control of direction and roll in many insects (locust [23]; honey-bee [25]; muscid flies [40]; *Drosophila* [21]; beetles [3]). Sudden reduction of the intensity of visual stimulation or specific movement patterns in the visual field may reduce the intensity of flight and elicit landing responses of the legs (locust [22]; *Oncopeltus* [6]). The eyes clearly play an important role in diurnal insects.

CONCLUSION

This short paper can do little more than give an outline of the complexity and refinement of the machinery of insect flight. Much remains to be learned, but it is perhaps consoling to remember that serious study has been in progress for less than one ten-millionth of the time that insects have taken to evolve the subject of our studies!

DISCUSSION

Professor T.Weis-Fogh (Chairman): A point about the campaniform sensilla on the wings of locusts, in comparison with those on the halteres of flies, is the difficulty, in using such strain gauges on the wings themselves, of differentiating between the aerodynamic forces and those caused by inertia, because they tend to run in the same direction. In locusts the campaniform sensilla sense the average lift produced, so that however the flying animal is tilted it tends to develop a constant vertical force. These sensilla are so placed that they are specifically sensitive to the aerodynamic forces; this can be shown by perforating the wing, so that it still has virtually the same mass but moves in almost a single plane, and it is thus possible to differentiate between the effects of aerodynamic and of inertial forces.

Pringle: This is extremely interesting because the gyroscopically-generated forces in the halteres are in the horizontal fore-and-aft plane and not in a vertical plane; the forces in the haltere base in the vertical plane are present all the time the haltere is oscillating, while the gyroscopic forces in the fore-and-aft plane at right angles to this can only be produced by the rotation of the whole insect. If, as Weis-Fogh says, the locust's campaniform sensilla are monitoring the forward component of the wing-stroke then it is the same component which is important for the halteres . . .

Weis-Fogh: . . . and the halteres are old wings . . .

Pringle: . . . and this is what the campaniform sensilla are monitoring at the bases of the halteres. The difficulty about the locust, on which most work has been done, is that its campaniform sensilla are undifferentiated and relatively unspecialized for discriminating between one force and another.

Dr R.J.Wootton: Professor Pringle said that asynchronous muscle systems arose in association with small size. Can he suggest a critical size of wings below which synchronous muscle-powered flight is impracticable?

Pringle: The critical frequency is about 100 wing-beats per second. The fastest synchronous mechanism known is probably in some small Lepidoptera. Wing-beat frequencies have been reported up to 90 or 95 in some Sphingidae, and could I think be just higher than that in some of the small Microlepidoptera, which as far as I know all have synchronous systems.

Wootton: Is this wing-beat frequency related to their size?

Pringle: Yes, because it is the resonance frequency of the thorax-wing system, it has to be. The smaller the insect, the faster will be its resonant frequency in general; some of the Sphingidae are quite large but are exceptional in this respect with high resonant frequency because the thorax is exceptionally rigid. Considering other groups for evidence of size at which the resonance would be at 100 cycles per second . . .

Wootton: Aleurodidae?

Pringle: Perhaps; their wings are large, in relation to the size of the body, and this low wing-loading keeps the frequency low.

Weis-Fogh: And the Aleurodidae are synchronous. This discussion has so far been conducted on the assumption that it is ordinary steady-state aerodynamics which is dominant. I do not think you can match the extreme mobility of a butterfly, and yet aerodynamically it is absurd unless it uses unsteady-state effects—which it does! To enable a small insect to take off easily, to hover and also to fly fast forwards, passing often from non-steady-state to steady-state flow and back again, higher wing-beat frequency and in turn the asynchronous muscle system will be advantageous.

Pringle: I think the reason the asynchronous mechanism has evolved so many times is because it is energetically more efficient. It can achieve nearly 50 per cent efficiency of conversion of adenosine triphosphate into mechanical energy (excluding energy losses in the initial conversion of fuel into ATP), whereas in ordinary muscle the comparable value would be only 25–30 per cent.

Professor W.Nachtigall (communicated subsequently): Asynchronous flight musculature is usually found in smaller insects, but there are exceptions: certain small moths generate a high wing-stroke frequency, up to about 100 per second, with synchronous musculature, while certain big bugs (Belostomatidae) generate a comparatively low frequency, up to about 50 per second, with asynchronous muscles. Is there other evidence to help to reconcile these facts with the generalization that small insects must have developed asynchronous muscles? The beating frequency of the hind-legs of the whirligig beetle *Gyrinus* is approximately 50 per second in fast swimming; these leg muscles are certainly of the synchronous type.

Pringle: Large bugs and beetles have probably evolved from smaller forms.

Nachtigall (communicated): Concerning halteres, which seem to be especially big in the genus *Tipula:* are there any further findings on the mechanical flight stabilization effects of the halteres, especially in this genus? Would you expect differences in the function of the halteres between a Tipulid and, for example, a Calliphorid fly?

Pringle: Not fundamentally, but the Tipulids obtain some stabilization from their long abdomens, which may have aided the transition of the hind-wing into the haltere.

Professor B.Hocking: On the significance of visual input: although visual stimuli are important in maintenance of flight in Hymenoptera and less so in Diptera, they have a strange negative importance in *Drosophila* and many other small flies. These insects will continue to fly indefinitely in the absence of visible patterns, but stop abruptly as soon as a conspicuous object appears in the visual field.

Pringle: So you would say there is potentially visual control?

Hocking: Yes, but in the opposite direction to that in Hymenoptera.

Pringle: However, visual input is not apparently essential to flies for balance in flight, in the way it appears to be in other insects; in locusts for example the indications are that most of the stability in the rolling plane is visual.

Weis-Fogh: But one must not forget they can fly in perfect darkness.

Pringle: How do you know that?

Weis-Fogh: I think Dr Schaefer* has some evidence on this point from radar.

Pringle: That is not complete darkness.

Weis-Fogh: No—it is not.

Dr R.C.Rainey: On this point, field experiments in Tanzania in 1955 with blinded locusts —locusts whose eyes had been completely painted over—showed that these were incapable of sustained steady flight. They flew readily, and, thrown into the air under sufficiently calm conditions, with dead still air in the early morning, initially succeeded in gaining some metres of additional height. Inevitably, however, they would get a wing down, and begin to fly in a wide circle, which became tighter and tighter; they would lose height, more and more rapidly, and always struck the ground apparently out of control. The speed with which control was thus lost, with which this instability developed, appeared to depend very much on the time of day—in the middle of the day, with more atmospheric turbulence, it happened more quickly; so that my guess is that they are not capable of stable flight in absolute darkness. As has been mentioned, I think that Dr Schaefer's observations, even in moonless conditions, may not have related to complete darkness in the sense of that experienced by an insect with its eyes painted over.

Mr M.V.Venkatesh (communicated): In many species of locusts and grasshoppers there is an asymmetrical arrangement of the bases of the tegmina, with either the right or the

* Dr Schaefer was not present at this discussion.

left tegminal base overlapping its counterpart while the insect is at rest. Contrary to earlier views, examination of tegminal bases in a couple of thousand locusts and grass-hoppers has shown no dominance of the tegmina of one side over the other, but the arrangement for an individual insect is fixed and cannot be reversed forcibly. The right-over-left or left-over-right arrangement has no relation to sex, place of origin, or season of hatching or fledging. Is there any information on the necessity or significance of this arrangement for flight, on whether it is reflected in the muscle arrangement, or on when the differentiation is brought about?

Pringle: I am afraid I have no evidence on this point.

REFERENCES

[1]   BUCHNER, E. (1972). Dark activation of the stationary flight of the fruitfly *Drosophila*. In Wehner, R. (ed.) *Information processing in the visual systems of Arthropods*, 141–146. Springer-Verlag, New York.

[2]   BURKHARDT, D. & GEWECKE, M. (1965). Mechanoreception in arthropoda: the chain from stimulus to behavioral pattern. *Cold Spring Harb. Symp. quant. Biol.*, 30 : 601–614.

[3]   BURTON, A.J. (1971). Directional change in a flying beetle. *J. exp. Biol.*, 54 : 574–585.

[4]   CAMHI, J.M. (1969). Locust wind receptors. I. Transducer mechanics and sensory response. *J. exp. Biol.*, 50 : 335–348.

[5]   CAMHI, J.M. (1970). Yaw-correcting postural changes in locusts. *J. exp. Biol.*, 52 : 519–531.

[6]   COGGSHALL, J. (1971). Sufficient stimuli for the landing response in *Oncopeltus fasciatus*. *Natur-wissenschaften*, 58 : 100–101.

[7]   CULLEN, M.J. (1971). *A comparative study of the anatomical basis of flight in Hemiptera*. D.Phil. Thesis, Oxford University.

[8]   CULLEN, M.J. (1974). The distribution of asynchronous muscle in insects with particular reference to the Hemiptera: an electron microscope study. *J. Ent. (A)*, 49 : 17–41.

[9]   DALY, H.V. (1963). Close-packed and fibrillar muscles in the Hymenoptera. *Ann. ent. Soc. Am.*, 56 : 295–306.

[10] DIAKONOFF, A. (1936). Contributions to the knowledge of the fly reflexes and the static sense of *Periplaneta americana*. *Archs néerl. Physiol.*, 21 : 104–129.

[11] DINGLE, H. (1961). Flight and swimming reflexes in giant water bugs. *Biol. Bull. mar. biol. Lab. Woods Hole*, 121 : 117–128.

[12] FAUST, R. (1952). Untersuchungen zum Halterenproblem. *Zool. Jb. (Allg. Zool.)*, 63 : 325–366.

[13] GETTRUP, E. (1962). Thoracic proprioceptors in the flight system of locusts. *Nature, Lond.*, 193 : 498–499.

[14] GETTRUP, E. (1965). Sensory mechanisms in locomotion: the campaniform sensilla of the insect wing and their function during flight. *Cold Spring Harb. Symp. quant. Biol.*, 30 : 615–622.

[15] GETTRUP, E. (1966). Sensory regulation of wing twisting in locusts. *J. exp. Biol.*, 44 : 1–16.

[16] GEWECKE, M. (1967). Die Wirkung von Luftströmung auf die Antennen und das Flugverhalten der Blauen Schmeissfliege (*Calliphora erythrocephala*). *Z. vergl. Physiol.*, 54 : 121–164.

[17] GEWECKE, M. (1970). Antennae: another wind-sensitive receptor in locusts. *Nature, Lond.*, 225 : 1263–1264.

[18] GEWECKE, M. (1972). Bewegungsmechanismus und Gelenkrezeptoren der Antennen von *Locusta migratoria* L. (Insecta, Orthoptera). *Z. Morph. Okol. Tiere*, 71 : 128–149.

[19] GEWECKE, M. (1972). Flight control in *Locusta migratoria* by antennae and hairs on the frons and vertex as air-current sense-organs. *J. comp. Physiol.*, 80 : 57–94.

[20] GEWECKE, M. (1972). Control of flying speed in locusts and its significance for their migrations. *Verh. dt. zool. Ges.*, 65 : 247–250.

[21] GOETZ, K. G. (1973). Processing of cues from the moving environment in the *Drosophila* navigation system. In Wehner, R. (ed.) *Information processing in the visual systems of Arthropods*, 255–263. Springer-Verlag, New York.

[22] GOODMAN, L.J. (1960). The landing responses of insects. I. The landing response of the fly, *Lucilia sericata*, and other Calliphorinae. *J. exp. Biol.*, 37 : 854–878.

[23] GOODMAN, L.J. (1965). The role of certain optomotor reactions in regulating stability in the rolling plane during flight in the Desert Locust, *Schistocerca gregaria*. *J. exp. Biol.*, **42** : 385–407.

[24] HERAN, H. (1959). Wahrnehmung und Regelung der Flugeigengeschwindigkeit bei *Apis mellifica* L. *Z. vergl. Physiol.*, **42** : 103–163.

[25] HERAN, H. & LINDAUER, M. (1963). Windkompensation und Flugeigengeschwindigkeit bei *Apis mellifica* L. *Z. vergl. Physiol.*, **47** : 39–55.

[26] HOLLICK, F.S.J. (1940). The flight of the dipterous fly *Muscina stabulans* Fallen. *Phil. Trans. R. Soc.* (B), **230** : 357–390.

[27] NACHTIGALL, W. (1968). Electrophysiological and kinematical investigations on start and stop of the dipteran flight motor. *Z. vergl. Physiol.*, **61** : 1–20.

[28] NUESCH, H. (1954). Segmentierung und Muskelinnervation bei *Telea polyphemus* (Lep.). *Rev. suisse Zool.*, **61** : 420–428.

[29] PRINGLE, J.W.S. (1938). Proprioception in insects. I. A new type of mechanical receptor from the palps of the cockroach. *J. exp. Biol.*, **15** : 101–113.

[30] PRINGLE, J.W.S. (1938). Proprioception in insects. III. The function of the hair sensilla at the joints. *J. exp. Biol.*, **15** : 467–473.

[31] PRINGLE, J.W.S. (1948). The gyroscopic mechanism of the halteres of Diptera. *Phil. Trans. R. Soc.* (B), **233** : 347–384.

[32] PRINGLE, J.W.S. (1954). The mechanism of the myogenic rhythm of certain insect striated muscles. *J. Physiol.*, **124** : 269–291.

[33] PRINGLE, J.W.S. (1957). Myogenic rhythms. In Scheer, B.T. (ed) *Recent advances in invertebrate physiology*, 99–115. Univ. of Oregon Press, Eugene.

[34] PRINGLE, J.W.S. (1957). *Insect flight*. Cambridge University Press. viii + 134 pp.

[35] PRINGLE, J.W.S. (1968). Comparative physiology of the flight motor. *Adv. Insect Physiol.*, **5** : 163–227.

[36] ROEDER, R.D. (1951). Movements of the thorax and potential changes in the thoracic muscles of insects during flight. *Biol. Bull. mar. biol. Lab. Woods Hole*, **100** : 95–106.

[37] RUNION, H.I. & USHERWOOD, P.N.R. (1968). Tarsal receptors and leg reflexes in the locust and grasshopper. *J. exp. Biol.*, **49** : 421–436.

[38] SCHALLER, F. (1960). Die optomotosische Komponente bei den Flugsteuerung den Insekten. *Zool Beitr.*, **5** : 483–496.

[39] SMART, J. (1959). Notes on the mesothoracic musculature of Diptera. *Smithson. misc. Collns*, **137** : 331–364.

[40] SMYTH, T. & YURKIEWICZ, W.J. (1966). Visual reflex control of indirect flight muscles in the sheep blowfly. *Comp. Biochem. Physiol.*, **17** : 1175–1180.

[41] SOTAVALTA, O. (1953). Recordings of high wing-stroke and thoracic vibration frequency in some midges. *Biol. Bull. mar. biol. Lab. Woods Hole*, **104** : 439–444.

[42] VOGEL, S. (1966). Flight in *Drosophila*. I. Flight performance of tethered flies. *J. exp. Biol.*, **44** : 567–578.

[43] WEIS-FOGH, T. (1950). An aerodynamic sense-organ in locusts. *8th Int. Congr. Ent., Stockholm* 1948 : 584–588.

[44] WEIS-FOGH, T. (1956). Biology and physics of locust flight. IV. Notes on sensory mechanisms in locust flight. *Phil. Trans. R. Soc.* (B), **239** : 553–584.

[45] WEIS-FOGH, T. (1960). A rubber-like protein in insect cuticle. *J. exp. Biol.*, **37** : 889–907.

[46] WEIS-FOGH, T. (1965). Elasticity and the wing movements of insects. *12th Int. Congr. Ent., London* 1964 : 186–188.

[47] WENDLER, G. (1972). Einfluss erwungener Flügelbewegung auf das motorische Flugmuster von Heuschrecken. *Naturwissenschaften*, **59** : 220–221.

[48] WILSON, D.M. (1962). Bifunctional muscles in the thorax of grasshoppers. *J. exp. Biol.*, **39** : 669–677.

[49] WILSON, D.M. & GETTRUP, E. (1963). A stretch reflex controlling wingbeat frequency in grasshoppers. *J. exp. Biol.*, **40** : 171–185.

[50] WILSON, D.M. (1968). The nervous control of insect flight and related behaviour. *Adv. Insect Physiol.*, **5** : 289–338.

2 · Control of flight and related behaviour by the central nervous systems of insects

BRIAN MULLONEY

Departments of Biology and Psychiatry, University of California, San Diego*

When insects fly, they move their wings in a highly stereotyped, cyclic series of movements. Two different mechanisms produce these movements in different groups of insects. Neurogenic fliers make co-ordinated wing-beats because their nervous systems drive their flight muscles in an ordered cycle of contractions [42]. Their flight muscles contract only when an action potential arrives in one of their motor neurones, so the nervous systems of neurogenic insects provide all the information which co-ordinates the movements of their wings. Myogenic fliers make their co-ordinated wing-beats because of the mechanical properties of their thoraces and the unique contractile properties of their fibrillar indirect flight muscles (pp 3–7 and [34]). Their flight muscles contract when stretched by the resonating thorax, and can do so much more often than action potentials normally occur in the motor neurones innervating them. Furthermore, the timing of action potentials in motor neurones innervating indirect flight muscles of myogenic insects is independent of wing position ([43], but see [1] and [12]). In both groups of insects, the same events occur at the neuromuscular junction. An action potential in a motor neurone innervating a flight muscle triggers the release from the neurone of a chemical transmitter which reacts with the muscle membrane, causing a transient increase in the free Ca^{++} inside the muscle fibre. In neurogenic insects, this increase in internal Ca^{++} causes the muscle to contract. In myogenic insects, the neurones innervating the indirect flight muscles regulate the levels of free Ca^{++} in the muscle fibres and thereby regulate the power produced by the contracting muscle.

During flight, the action potentials in the flight motor neurones of neurogenic and some myogenic insects occur in a temporally ordered pattern (Fig. 2.1). This pattern is characteristic of the behaviour, and stereotyped in the sense that it repeats with little variation many times during a flight. It is the pattern of action potentials which produces the co-ordinated wing-beats of neurogenic insects. Other patterns in the same neurones can cause different behaviours—warm-up [17, 26], stridulation [8], and, in neurones innervating bi-functional muscles [38] or leg and abdominal muscles, jumping [32]. Wilson [37] showed that, although abundant sensory information is available to the nervous systems of locusts, the flight pattern is produced by a mechanism in their central nervous system which can produce the complete, accurate flight pattern without any sensory information. Wilson's evidence for this conclusion was a series of experiments in which he showed that the three thoracic ganglia of a locust could themselves generate

* Now Department of Zoology, University of California, Davis 95616.

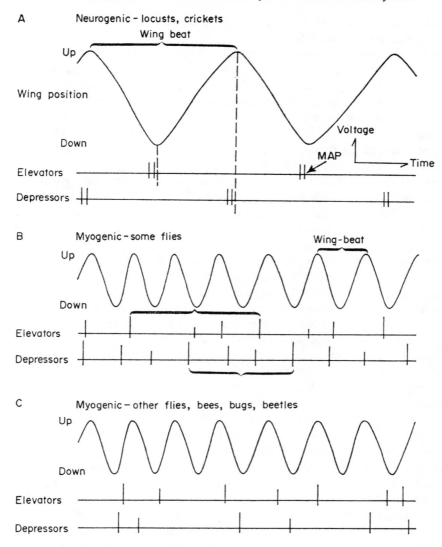

FIG. 2.1. **Diagrammatic illustration of the differences between neurogenic (synchronous) and myogenic (asynchronous) mechanisms in flying insects.** The MAPs (muscle action potentials), or spikes, are recorded electronically from the indicated muscles. Each MAP follows by a fixed delay an action potential in the neurone which innervates the muscle. Since MAPs occur only following nerve spikes, MAPs are convenient monitors of the firing of motor neurones. In neurogenic fliers, each flight motor neurone fires at a particular point in the cycle of wing movements (A). In myogenic fliers, each flight motor neurone fires without reference to the position of the wings (B and C). In some flies (B), each neurone maintains a particular phase relative to other neurones innervating the same muscle. In bees, bugs, beetles, and other flies, the indirect flight motor neurones fire at times independent both of wing position and of the firing of other flight motor neurones.

the flight pattern after their connections to the abdominal and cephalic ganglia had been cut, and after all the sensory nerves in the thorax had been cut. This means that the information needed to produce the flight pattern is somehow incorporated into the structure of the locust's nervous system. Wilson's singular observation has been extended to other animals and other behaviours, but in the context of this review the most interesting is singing, officially stridulation, in crickets.

Crickets fly with a neurogenic apparatus very similar to that of locusts. They sing by scraping their fore-wings together, and use the same muscles and motor neurones to make these stridulating movements as they use to fly. The firing patterns of the motor neurones are different during these two behaviours. Bentley [2, 3] showed that a cricket CNS (central nervous system), isolated except for the head and caudal cerci could generate correct flight patterns and, alternatively, stridulatory patterns in the appropriate motor neurones of the thorax. Bentley eliminated all the thoracic sensory information, but the cricket could still produce the motor patterns for flight and song. Kutsch and Huber [21] interfered with the sensory apparatus and normal motor performance of the cricket's thorax, and eliminated all sensory feed-back from the thorax, but still the crickets could sing. Finally, Kutsch and Otto [22] cut the connections between the suboesophageal ganglion and the prothoracic ganglion, and found that if enough time was allowed for the cricket to recover, its thoracic and abdominal nervous system alone could produce the motor patterns for stridulation. This disproved the earlier theory that stridulation was a behaviour whose motor patterns were generated in the brain and somehow transmitted through connections to the thoracic motor neurones. So, crickets can fly and stridulate with the same muscles and the same motor neurones; both behaviours are caused by motor patterns generated in the thoracic ganglia, and the ganglia do not need any particular sensory information to produce these behaviours.

Locusts and crickets, and probably other neurogenic insects, use very little of the available sensory information to regulate cycle-by-cycle their flight and stridulatory movements. Waldron [36] found that only flashes of light at nearly the wing-beat frequency could entrain the wing-beats of locusts. Flies, bees and milkweed bugs, and probably other myogenic insects do not regulate the time at which the motor neurones of their indirect flight muscles fire relative to the position of their wings (see [1] for an unresolved dispute with the author about certain neurones in bees). Therefore myogenic insects do not use phasic sensory information to generate the pattern in which their indirect flight motor neurones fire. These patterns are centrally generated, like those of locusts and crickets.

Heide [12] found that some of the neurones innervating direct flight muscles which control turning in Calliphorid flies are not only strongly influenced by visual input, but tend to fire at wing-beat frequency and with a preferred phase relative to the wing while they were actively trying to alter the animal's course. He demonstrated that sensory feed-back determines this preferred phase relation. It is not clear from his experiments if the sensory modality which contributes the timing information to the CNS is vision or the proprioceptors at the base of the wing. Perhaps the fly is seeing his wings out of the corner of his eyes, and Heide is seeing an entrainment like that of locusts by light flashes [36]. Whatever is going on, this synchrony of the basalar muscles and the wings during turns in flies is an outstanding exception to the rule that the firing patterns at flight motor neurones in myogenic insects is not influenced by phasic sensory input. This distinction between the direct and indirect flight muscles of Calliphorids supports Pringle's [33] idea of an anatomical separation of functions—power and steering—in myogenic insects.

Sensory information is critically important to insects and to us all if we are to adapt our behaviour to the vagaries of the real world [40], but in the rest of this review I will concentrate on centrally generated motor patterns and the questions they pose. Sensory integration of flight has been reviewed recently [41].

DEVELOPMENT OF CENTRALLY GENERATED MOTOR PATTERNS

Crickets have a hemimetabolous development, passing through 9–11 instars before reaching maturity. Only at the last moult do the cricket's wings appear; earlier they are immobile pads on the thorax. Only adult crickets, with wings, can fly or sing. Bentley and Hoy [6], working with the field cricket *Teleogryllus commodus*, have studied the post-embryonic development of stridulation and flight. Last-instar nymphs can, when subjected to Huberian brain cautery, produce perfect motor patterns for aggression song, calling song, and courtship song. The nymphal patterns are accurate adult song patterns in all respects, except no sound is heard because the wing-pads cannot move. These nymphs have the complete neural organization needed to sing before the mechanical apparatus is complete.

When these crickets fly, they hold their legs, abdomen and antennae in a character-istic posture. The motor activity which produces this posture is also part of the flight pattern, although most of us study only the activity of neurones innervating flight muscles. Bentley and Hoy found that nymphal crickets would assume this flight posture when suspended in a wind tunnel as early as the seventh instar—three or four moults before adulthood. They recorded the action potentials of antagonistic sets of flight muscles and found that last instar larvae could produce accurate, complete flight patterns. Nymphs four moults from adulthood produced only a few short bursts in hind-wing depressor motor units; nothing more. Older nymphs seemed to add elements of the adult pattern in a regular sequence: (i) short bursts in the hind-wing depressors, (ii) long trains of bursts, (iii) addition of hind-wing elevators and the fore-wing units, (iv) acceleration of the out-put to the adult frequency. All this goes on before the nymph would get any of the sensory stimulation associated with flight. Bentley [5] has followed the anatomical development of some flight motor neurones in crickets using the new methods discussed below.

Locusts have only five larval instars before their final moult. Kutsch [20] examined the development of the flight motor pattern in nymphal locusts. Unlike crickets, nymphal locusts cannot generate the flight motor pattern in neurones innervating flight muscles, although second through fifth-instar larvae do assume a flight posture similar to that of adults when suspended in an air-stream. Indeed, for the first week of adult life, the flight mechanism of locusts continues to mature; the wing-beat frequency increases, and the co-ordination of antagonistic muscles become more accurate. I find this difference between two seemingly similar insects surprising.

GENETIC CONTROL OF CENTRAL PATTERN GENERATORS

The presence of centrally generated motor patterns in immature animals implies that these patterns develop without any of the sensory information used by the adult to orient flight or to cue singing. Of itself, the demonstration of these patterns in juveniles is not proof that these pattern-generating mechanisms are genetically determined, but it is strong presumptive evidence for this hypothesis. Two groups have begun to analyse the genetics of locomotion in *Drosophila*. Hotta and Benzer [13] have used mosaic flies to locate the primary site of action of mutant genes affecting behaviour. Their methods allow them to sort out from the thousand or so known *Drosophila* mutants those which affect a given behaviour and have a primary focus in the CNS. Furthermore, they can localize the site of action to a comparatively small cluster of cells. As yet, the new under-standing of physiological mechanisms contributed by this study is very limited.

Brian Mulloney

Levine and Wyman [25] have extended earlier work (see below) on flight in flies to a study of a mutant in *Drosophila*. Levine and Hughes [24] developed a stereotaxic atlas of the flight muscles in *Drosophila* which they used to record from identified motor units during flight (Fig. 2.2). The mutation *stripe* (*sr*) causes no gross alteration of a fly's CNS, wings, or musculature, but flies with this mutation cannot fly. Their behaviour is

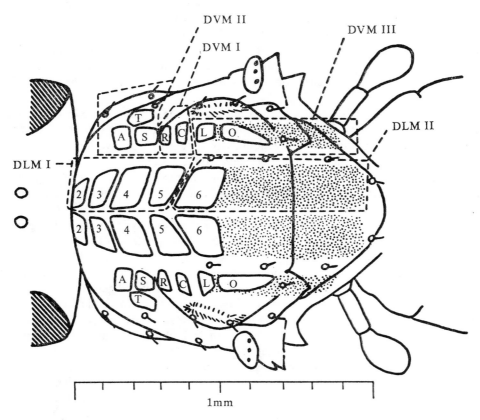

Fig. 2.2. Stereotaxic map of thorax of *Drosophila* [24]. Dorsal view, drawn as if transparent, to show relationship of superficial landmarks to subcuticular terminations of following flight muscle motor units, for which [24] also shows orientation of fibres. Dorsal longitudinal wing-depressors: Units 2–5—DLM I; Unit 6—DLM II. Dorso-ventral wing-elevators: A, T, S (anterior tergo-sternal)—DVM I; R, C (tergal remotor of coxa)—DVM II; L, O (lateral oblique dorsal)—DVM III.

otherwise normal. Levine and Wyman found that these flies have a disordered pattern of activity in their motor neurones. Their neurones burst at higher frequencies than those of wild-type flies, and then stop firing. The wings barely start to flutter before the burst is over. Levine and Wyman showed that the disorder is not caused by a loss of any known component of the pattern-generating mechanism—reciprocal inhibition and shared excitation of motor neurones belonging to a given pool—but by some unknown factor. Levine and Wyman are now extending this analysis to other mutants. Together with the work of Hotta and Benzer, it opens up an analysis of flight in *Drosophila* which may advance considerably our understanding of both the genetics and physiology of flight. An accurate statement about the relative contributions of genetics and interaction with an

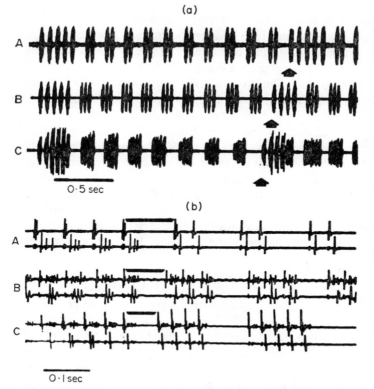

FIG. 2.3. **Genetic control of wing-movement patterns: neurone output in songs of hybrid crickets** [4].

(a) **Sound pulse patterns in calling songs.** Each record starts with one 5 to 6 pulse *chirp*, followed by a series of *trills* containing 2 to 4 sound pulses per trill. Together the chirp and trills form a *phrase* which repeats, starting again at the arrow.

(b) **Patterns of action potentials in motor units responsible for these calling songs.** Upper trace of each pair records spikes in a subalar (wing-opening) muscle, and lower trace in a promotor (wing-closing) muscle (with some cross-talk between the two traces); bar marker at transition from chirp to trills. Genetic information is precise enough to specify differences of one spike per trill between these hybrids.

A *Teleogryllus oceanicus:* wild type; 2 sound pulses per trill. B *T. oceanicus* × (*T. oceanicus* × *T. commodus*): back-cross; 3 pulses per trill. C *T. oceanicus* × *T. commodus*; 4 pulses per trill.

insect's environment to the maturation of its flight behaviour awaits an analysis as elegant as those of bird song [29] and of swimmerette-beating in crayfish [11].

A different approach to the genetics of centrally generated motor patterns involves hybridization of congeneric species of crickets with differentiable songs. Unlike *Drosophila*, the crickets have a little-known genetics. This is however compensated by advanced physiological experimentation and behavioural analysis on these insects; and stridulation in crickets is a centrally generated behaviour using some of the same neurones and muscles which are involved in flight. Those outriders of the physiological herd, Bentley and Hoy [4, 7] have crossed *Teleogryllus commodus* and *T. oceanicus*, respectively the Australian and Polynesian field crickets. The calling songs of these two species have the same qualitative structure but differ in the number of sound pulses per trill, the number of trills per phrase, and the duration of particular intervals (Fig. 2.3). Simultaneous recordings of sound pulses and action potentials in particular muscles show that each pulse is caused by a single action potential in each of these muscles. The hybrid crickets (F_1) sang songs which were intermediates between the parental songs. Careful back-crosses to both parental strains, taking note of the lack of a Y-chromosome in male crickets, showed that (i) calling song is a polygenic character with several linkage groups (this is true even for individual parameters of the song), (ii) some genes are sex-linked, others are autosomal, and (iii) song is under precise genetic control, and within wide limits is not affected by the cricket's environment. The quantitative parameters of the songs of hybrids and backcrosses form a striking series which correlates with the relative genetic contributions made by the two species. The precision of genetic control is enough to specify the number of action potentials which homologous neurones will fire in each trill, that is, to specify spike-by-spike the motor pattern which produces calling song. Is this also true of flight in these insects? Is it true of flight in *Drosophila?* What changes in a motor neurone or its input to alter so accurately the number of spikes per trill? Bentley and Hoy's work raises many questions about the organization of nervous systems, but none about the value and elegance of their work. It underscores the importance of an analysis of the pattern-generating mechanism.

PHYSIOLOGICAL MECHANISMS GENERATING MOTOR PATTERNS

What is it about the neurones innervating the flight muscles which makes them fire in the flight pattern, or in the stridulation pattern, but not in some other pattern? Two sets of theories exist about these mechanisms: (i) there is a neurone or set of neurones (oscillators) which determine repetition rate (wing-beat frequency), while the relative timing in each cycle of the pattern of action potentials in each neurone is determined by a network of synaptic connections. These oscillator neurones fire either spontaneous, endogenous bursts of spikes and are therefore called endogenous bursters, or fire one triggering spike per wing-beat. (ii) Both the repetition rate (wing-beat frequency) and the pattern within each cycle are fixed by a network of synaptic connections. There is no 'clock'; no neurones are endogenous bursters. Repetition rate and pattern are emergent properties of the synaptic network. These mechanisms could exist either among the motor neurones themselves or at some level in the nervous system which then drives the motor neurones.

The problems of testing these theories begin with the fact that there is no way to tell by inspecting a motor pattern at the periphery if it was generated by a network or by an endogenous burster mechanism. No known test which is applicable from the periphery of the CNS will distinguish these two classes of mechanisms. There is a large body of work analysing statistically the parameters of various flight patterns; in the end, all this work produces is plausible guesses about the organization of a nervous system. It can never prove or disprove a physiological hypothesis.

Another favoured procedure uses artificial stimulation of the axons of motor neurones out near the muscles—antidromic stimulation—to send action potentials back into the CNS. The idea is that if motor neurones make synaptic connections with one another, or with other neurones important for flight, then antidromic spikes inserted in a spontaneous pattern should affect subsequent events in predictable ways. Furthermore, if the neurone stimulated is an endogenous burster, antidromic spikes might reset the burst rhythm. The singular advantage of antidromic stimulation is that it allows experimental inter-ference with the CNS from the periphery; it avoids the problems of direct attack.

Unfortunately, it does not give clean yes or no answers. Mulloney and Selverston [28] showed that antidromic spikes do not necessarily reach the same regions of a neurone reached by spontaneous orthodromic spikes, or by orthodromic spikes generated by current injected into the soma of a neurone. In the example they presented, one motor neurone inhibited another directly. Whenever the presynaptic neurone fired, whether driven or spontaneously, the postsynaptic neurone showed synchronous inhibitory potential. Antidromic spikes in the presynaptic neurone reached the ganglion, but did not cause inhibitory postsynaptic potentials (IPSPs) in the postsynaptic neurone. In another set of motor neurones which are endogenous bursters, Mulloney and Selverston have shown that antidromic spikes, or bursts of spikes, do not reset the endogenous burst rhythm or reach the synapses made by these neurones. Together, these results mean that negative results of antidromic stimulation experiments mean nothing; they cannot be interpreted in the absence of other information. Positive results where the antidromic spike causes a clear-cut postsynaptic effect, can safely be taken as evidence for interaction or for endogenous bursting.

MYOGENIC INSECTS

Flies of the genera *Calliphora*, *Musca* and *Drosophila* produce a striking pattern of spikes in the motor neurones innervating each of their flight muscles [44, 45, 46]. Each neurone of a set fires at a particular phase relative to the firings of the other members of the set for many hundreds of spikes, despite wide fluctuations in the firing frequency of these neurones. Wyman advanced the hypothesis that this constant phase relation was caused by a reciprocal network of inhibitory synapses among the motor neurones or among neurones driving the motor neurones. Mulloney [27] confirmed this hypothesis by showing that antidromic action potentials in one neurone would reset the firing times, and sometimes the phases, of all other neurones innervating the same muscle. Antidromic action potentials did not affect the firing of neurones innervating other muscles, nor did they affect the firing of other neurones innervating the same muscle in another dipteran genus, *Eristalis*, which does not have a phase-constant motor pattern. Levine [23] has extended this work to *Drosophila*, and defined accurately the structures of the different functional pools of motor neurones in that insect.

This system is the best known pattern-generating mechanism, but is irritatingly off the main stream. Not all myogenic insects, not even all flies, have this phase-constant pattern. The pattern serves no known biological function; Wilson suggested that it is a decadent remnant of a more complex ancestral neurogenic pattern. In the context of this review, the *Calliphora* results are a striking example of a positive result from antidromic stimulation. If by chance antidromic spikes did not reach the inhibitory synapses, I would have been tempted to conclude they did not exist. That was the tentative conclusion I drew about *Eristalis*. Now all that can be said about *Eristalis* is that antidromic potentials do not affect the firing of other neurones.

NEUROGENIC INSECTS—MOSTLY LOCUSTS AND CRICKETS

After a decade of work describing the details of flight and stridulatory motor patterns, we do not know how any of these patterns are generated. Antidromic stimulation of various motor neurones in flying locusts does not reset the wing-beat phase or alter its frequency [39], and has very little affect on other neurones in the system [19], so we can say nothing from this source about interactions between motor neurones. Hypothetical models by the score have come from statistical analyses of motor patterns in various insects, but none has produced a proven mechanism.

Two new converging lines of attack have recently begun to yield results. The first is detailed study of the structures of identified neurones. By filling identified neurones with

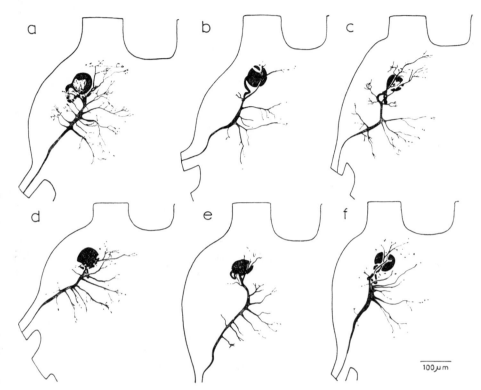

FIG. 2.4. **Flight motor neurone, number 113, in** *Chortoicetes terminifera*, from six different locusts. Each neurone was first identified physiologically, then filled with Co++ [31]. The Co++ was precipitated as CoS, the ganglion cleared, and the filled neurone drawn using a drawing tube. Only branches 1 μm and larger were drawn. Taken from Burrows [10].

Procion yellow [35] or with CoS [31], one can discover not only the location of cell bodies but the three-dimensional structure of a neurone (Fig. 2.4). Neurones can be filled either through microelectrodes in their somata or by axonal iontophoresis [16]. With these new methods, the neuropile is vulnerable to systematic exploration and structural analysis. The second line of attack is intracellular recording from identified neurones. The significance of intracellular recordings is that they often reveal the subthreshold, non-spiking events in the cell which are integrated to produce the final train of spikes recorded from the axon. The information obtained is more valuable than that from spike trains, but it is technically more difficult to obtain. As recently as 1970, intelligent opinion held no useful information could be recorded from the somata of insect motor neurones [14, 30]. The recent changes are due almost exclusively to improvements in microelectrodes, surgical procedures, and the persistance of investigators. It is worthwhile to review these developments. Initially, the procedures involved desheathing the ganglion, cutting the tracheae, and recording with KCl microelectrodes from unidentified regions of identified neurones [19, 30]. These methods recorded spikes but no postsynaptic potentials (PSPs), and preparations were obviously dying within half an hour. Better results were obtained when the sheath and tracheae were left largely intact, and the recordings were made with K^+acetate or K^+citrate electrodes [2, 3]. Preparations lasted many hours, and PSPs were recorded from neuropilar processes. Most recently, the sheath and air

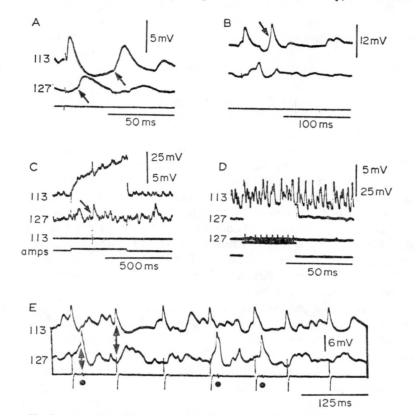

FIG. 2.5. **Simultaneous intracellular recordings from ipsilateral motor neurones 113 (a wing-elevator) and 127 (a wing-depressor).** A and B, antidromic stimulation of 113, and the resulting depolarization of 127; C, driving 113 by injecting current into its soma causes a depolarization of 127; D, driving 127 does not affect 113; E, responses of 113 and 127 to blowing on the locust's head. The arrows mark examples of the extracellularly recorded spikes of each neurone. Each of the tiny 127 spikes is marked by a dot. Taken from Burrows [9].

supply have been left completely intact; K^+acetate electrodes are simply punched through the sheath [15]. The results obtained are surprising when compared with the results yielded by the earlier procedures. Both attenuated spikes and subthreshold PSPs can reliably be recorded from the somata of identified neurones. Better yet, the animals respond to sensory input, and attempt to walk and fly. Preparations last many hours. The differences are probably due to two factors; desheathing the ganglia grossly disrupts the ionic environment of the neurones, and KCl electrodes shift the Cl^- equilibrium potential so that Cl^- conductance changes are masked [18]. The overall health of the preparations, their spontaneous co-ordinated activity, and the length of time they last are important measures of the worth of these various experimental procedures. The improved procedures for intracellular recordings and the new neuro-anatomical methods make possible a concerted attack on the mechanisms generating flight and singing, and tests of the two hypotheses discussed at the start of this section.

Burrows [9] has described the activity of two bilateral pairs of antagonistic flight motor neurones in the locust *Chortoicetes terminifera*. Using the procedures detailed in [15], he recorded subthreshold activity in these neurones and correlated it with behavioural acts and with activity in known neurones (Fig. 2.5). The particular power of this work comes from a combination of anatomical and physiological data, and the use of paired simultaneous intracellular recordings. The amount of useful, interpretable data obtained from paired recordings is vastly greater than from a series of single recordings. Burrows demonstrated subthreshold effects on one neurone of the pair of antidromic action potentials in the other, and of antidromic spikes in other neurones on both pairs on which he reports. These are not the same neurones on which Wilson [39] and Kendig [19] worked, and this is a different locust, but I think the important differences are the surgical procedure and microelectrodes. The results reported thus far are all recorded from somata, but the structural data on these neurones (Fig. 2.4) encourages the hope of repeatedly successful neuropile recordings, which might give different sorts of data, or confirm the soma recordings. The structure of these neurones is similar in different animals, and the diameters of some of the integrative regions are about 10 μm. Furthermore, the main integrative regions of these antagonistic neurones run parallel to each other for a long distance in the ganglion. Is this the region in which they receive common inputs? How many other flight motor and sensory neurones follow this same course, or send processes to this region? Functional stereotaxic atlases of the neuropile in ganglia of several species should take shape in the near future, and provide some insights into the evolution of neural organization in insects.

Burrows's work is a paradigm for studying the mechanisms which generate flight motor patterns. Paired recording from each of the possible pairs of flight motor neurones will yield, after considerable effort, a list of the direct interactions between these neurones. Furthermore, it will yield a description of the synaptic input common to various subsets of these neurones, and probably will yield information about interneurones and sensory neurones involved in flight. Accidental penetrations of neurones presynaptic to the motor neurones will inevitably occur during the attempts to hit motor neurones; Burrows has already found such neurones. Once filled with dye, their somata located, these interneurones will be regularly accessible to experiment. This research will be prolonged and tedious, but only by knowing the network of synaptic connections among the flight neurones, and examining them for endogenous bursting properties, can we settle the question of how flight patterns are generated.

DISCUSSION

Weis-Fogh (Chairman): Dr Mulloney's paper shows how entomological studies can lead directly into fundamental aspects of general biology and general physiology. Here we are dealing with highly sophisticated and extraordinarily accurate systems, and it is important to realize the basic difference between insects and many other invertebrates on the one hand, and on the other the vertebrates (with the special functions of the cortex) and other animals, with a greater capacity for learning. The reason why insects are so exceptionally useful for fundamental studies in this field is that they learn by the genetic mechanism, and not by experience during the lifetime of the individual.

Mulloney: The question of whether these insects can learn at all is the really interesting one, particularly in the light of the difference that Kutsch and Bentley find in the development of the thoracic nervous systems in locusts and crickets. I should very much like to see a detailed study of what the locusts are doing in that first week of adult life. Do they actually practise? If you prevent their getting a normal sensory input, say by immobilizing their wings, do you inhibit the development of motor patterns? I think not. I think that, as in the crickets, this is a 'hard-wired', genetically-determined system, simply on a different time-scale (and this is a view since supported by Kutsch, W. 1974. *J. comp. Physiol.*, **88** : 143–424), but it would be really interesting if it should turn out that the animals have to practise.

Weis-Fogh: May I comment on this question—though of course I have no answer to it. The cuticle of a newly-fledged locust needs to develop for at least another ten days [in the laboratory, or about a week in the field—Ed.] before it is thick and strong enough for sustained flight. If during the first one or two days the locust attempts to activate all its wing muscles in the normal way, it simply crumples; there may be a difference here in the developmental pattern of locusts and crickets.

Prof. J.W.S.Pringle: A possible experiment would be to treat a locust after the final moult with a new poison which has been found by Professor Sven Andersen in Copenhagen to prevent further deposition of cuticle. By treating a locust in this way, the whole mechanical build-up could be stopped and any effects on the development of the full flight-pattern could be studied. [See Post, L.C. & Vincent, W.R., 1973, *Naturwissenschaften* **9** : 431–432.]

Weis-Fogh: I believe the main effect is to prevent the formation of chitin?

Pringle: Yes—throughout the endocuticle.

Dr A.C.Neville: On what Professor Pringle has just said, has anyone tried to interfere with the sequence of development of flight motor output pattern by hormone experiments? Can it be retarded (with juvenile hormone) or accelerated (with ecdysone)?

Mulloney: To my knowledge, no; it is only quite recently that people have begun to think of the development of these systems as something which can be examined experimentally.

Professor M.Lindauer (communicated subsequently): Some learning process apparently takes place in the production of sound patterns by several Gomphocerine grasshoppers; since N.Elsner (1974, *J. comp. Physiol.*, **88** : 67–102) has found that these sound patterns differ between the left and the right hind-leg, and that the two hind-legs change their role from time to time—except when one is removed. The significance in communication of this 'duplicity' (*doppelzüngigkeit*) of language is not yet known. To the comments of Professor Weis-Fogh and Professor Mulloney I would add that Elsner has also detected in the thoracic ganglion of Gomphocerine grasshoppers bipolar motor neurons which can deliver two efferent action programmes, one commanding the flight muscles and the other the stridulatory muscles in the hind-leg. This represents an astonishingly economical command system, based apparently on simple, genetically fixed information units.

Dr C.Lewis: Dr Mulloney mentioned that there were some 90 motor neurones, perhaps nine or ten relating to a single muscle, and therefore a number of distinct muscle-units within a given muscle. Is this so? And does it confer a great deal of flexibility in the extent to which a particular muscle is employed?

Mulloney: While there are approximately 90 motor neurones altogether, the most usual number of neurones per muscle is less than nine or ten—characteristically there are one or two, and there are only a few muscles innervated by more than five motor neurones. There is indeed some recruitment of new activity in additional motor neurones innervating the same muscle as the power produced by a flying insect increases; there is also some indication that different motor neurones can operate when the same muscle is used in different types of behaviour. In the European grasshopper *Gomphocerippus* one of the leg muscles is innervated by two motor neurones, of which one is used in singing and one in walking when the animal is doing both. But there is not much in the way of graded responsiveness of muscles, by recruitment of further units, in the way to which we are accustomed in vertebrate muscles, because of the relatively small numbers of motor neurones involved in insects.

Weis-Fogh: I think this does not really do justice to the role of double-firing for example in the flight-muscles of locusts (Neville and Weis-Fogh, 1963, *J. exp. Biol.*, **40** : 87–104).

Professor W.Nachtigall (communicated): During steering movements there are differences in the kinematics of the left and right wings in locusts. Is it possible at the present time to correlate these differences to neural activity (thoracic ganglia), at least at the level of modelling?

Mulloney: Not as far as I know.

Dr R.C.Rainey (communicated): In expressing doubt as to whether insects learn at all during the lifetime of the individual, Dr Mulloney was of course referring to processes like the development of the motor patterns with which he has been concerned. Would he care to comment on any implications of the evidence of learning by experience during the lifetime of an individual bee, as outlined by Professor Lindauer? For example, in the course of a foraging life of only 10–15 days, a bee can learn to recognize its surroundings and to locate itself within an area of some 50 km². From the normal flying height of about 2–3 m in calm conditions (p.202), this would appear to imply familiarization with a total extent of visual field perhaps crudely comparable with that represented by a total area of 8 million km² (approaching that of the United States—!) from the point of view of a human pilot navigating visually in an aircraft at a height of 1000 m—though from the height of the foraging bee its navigation may well be facilitated by the number of landmarks towering above its horizon, in a manner rarely available to the human navigator.

Lindauer (communicated): I would like to underline Dr Rainey's point by illustrating how quickly bees learn to become familiar with the surroundings of their home. Inexperienced bees which never had left their hive before were allowed out for a single initial flight, from which they returned after four minutes. The bees were then caged and released from different compass directions and distances—they returned from a distance of several hundred metres. After a few more orientation flights, the successful release distance increased up to a few thousand metres; and, as every beekeeper knows, experienced forager bees will return to the home hive from distances of 6–8 km. Another remarkable orientation performance relates to the ability of the bees to learn to visit different species of flowers at the appropriate time of day for the nectar flow. We have been able to train a group of bees to collect from two, four and ten (!) different locations each at a fixed time of day. For example:

```
From  8– 9 hours food could be found at 200 m N
From 11–12 hours   „      „         „      200 m E
From 14–15 hours   „      „         „      200 m W
From 17–18 hours   „      „         „      200   S
```

On the test day the bees searched at the *right* time in the *right* place for food. Even ten different places at ten different times of day could be learned by the bees after a training time of 14 days.

Mulloney: Dr Rainey is right; I had in mind learning of motor skills, not learning navigation. The abilities of bees described by Drs Rainey and Lindauer are outside the range of things into which I have some physiological insight. Like our own thought processes, vocal abilities and sense of humour, they are realms physiologists have still to enter.

REFERENCES

[1] BASTIAN, J. & ESCH, H. (1970). The nervous control of the indirect flight muscles of the honeybee. *J. comp. Physiol.* **67**: 307–324.

[2] BENTLEY, D. (1969). Intracellular activity in cricket neurons during generation of song patterns. *Z. vergl. Physiol.*, **62** : 267–283.

[3] BENTLEY, D.R. (1969). Intracellular activity in cricket neurons during generation of behaviour patterns. *J. Insect Physiol.*, **15** : 677–701.

[4] BENTLEY, D.R. (1971). Genetic control of an insect neuronal network. *Science*, **174** : 1139–1141.

[5] BENTLEY, D.R. (1973). Postembryonic development of insect motor systems. In Young, D. (ed.) *Developmental neurobiology of arthropods.* Cambridge University Press.

[6] BENTLEY, D.R. & HOY, R.R. (1970). Postembryonic development of adult motor patterns in crickets: a neural analysis. *Science*, **170** : 1409–1411.

[7] BENTLEY, D.R. & HOY, R.R. (1972). Genetic control of the neuronal network generating cricket (*Teleogryllus, Gryllus*) song patterns. *Anim. Behav.*, **20** : 478–492.

[8] BENTLEY, D.R. & KUTSCH, W. (1966). The neuromuscular mechanism of stridulation in crickets (Orthoptera: Gryllidae). *J. exp. Biol.*, **45** : 151–165.

[9] BURROWS, M. (1973). The role of delayed excitation in the co-ordination of some metathoracic flight motoneurones of a locust. *J. comp. Physiol.*, **83** : 135–164.

[10] BURROWS, M. (1973). The morphology of an elevator and a depressor motoneuron of the hindwing of a locust. *J. comp. Physiol.*, **83** : 165–178.

[11] DAVIS, W.J. (1973). Development of locomotor patterns in the absence of peripheral sense organs and muscles. *Proc. Nat. Acad. Sci. U.S.*, **70** : 954–958.

[12] HEIDE, G. (1971). Die Funktion der nicht-fibrillären Flugmuskel von *Calliphora*. Teil I: Muskuläre Mechanismen der Flugsteuerung und ihre nervöse Kontrolle. *Zool. Jb. Abt. allg. Zool. u. Physiol.*, **76** : 99–137.

[13] HOTTA, Y. & BENZER, S. (1972). Mapping of behaviour in *Drosophila* mosaics. *Nature*, **240** : 527–535.

[14] HOYLE, G. (1970). Cellular mechanisms underlying behaviour–neuroethology. *Adv. Insect. Physiol.*, **7** : 349–444.

[15] HOYLE, G. & BURROWS, M. (1973). Neural mechanisms underlying behaviour in the locust *Schistocerca gregaria*. I. Physiology of identified motor neurons in the metathoracic ganglion. *J. Neurobiology*, **4** : 3–42.

[16] ILES, J.F. & MULLONEY, B. (1971). Procion yellow staining of cockroach motor neurones without the use of microelectrodes. *Brain Res.*, **30** : 397–400.

[17] KAMMER, A.E. (1970). A comparative study of motor patterns during preflight warm-up in hawk moths. *Z. vergl. Physiol.*, **70** : 45–56.

[18] KEHOE, JACSUE (1972). 1. Ionic mechanisms of a two-component cholinergic inhibition in *Aplysia* neurons. 2. Three Ach receptors in *Aplysia* neurons. 3. The physiological role of three Ach receptors in synaptic transmission in *Aplysia*. *J. Physiol.*, **225** : 85–114, 115–116, 147–172.

[19] KENDIG, J.J. (1968). Motor neurone coupling in locust flight. *J. exp. Biol.*, **48** : 389–404.

[20] KUTSCH, W. (1971). The development of the flight pattern in the desert locust *Schistocerca gregaria*. *Z. vergl. Physiol.*, **74** : 156–168.

[21] KUTSCH, W. & HUBER, F. (1970). Zentrale versus periphere Kontrolle des Gesanges von Grillen (*Gryllus campestris*). *Z. vergl. Physiol.*, **67** : 140–159.

[22] KUTSCH, W. & OTTO, D. (1972). Evidence for spontaneous song production independent of head ganglia in *Gryllus campestris*. *J. comp. Physiol.*, **81** : 115–119.

[23] LEVINE, J.D. (1973). Properties of the nervous system controlling flight in *Drosophila melanogaster*. *J. comp. Physiol.*, **84** : 129–166.

[24] LEVINE, J.D. & HUGHES, M. (1973). Stereotaxic map of the muscle fibers in the indirect flight muscles of *Drosophila melanogaster*. *J. Morphol.*, **140** : 153–158.

[25] LEVINE, J.D. & WYMAN, R.J. (1973). Neurophysiology of flight in wild-type and a mutant *Drosophila*. *Proc. Nat. Acad. Sci. U.S.*, **70** : 1050–1054.

[26] MULLONEY, B. (1970). Impulse patterns in the flight motor neurons of *Bombus californicus* and *Oncopeltus fasciatus*. *J. exp. Biol.*, **52** : 59–77.

[27] MULLONEY, B. (1970). Organization of flight motorneurons of Diptera. *J. Neurophysiol.*, **33** : 86–95.

[28] MULLONEY, B. & SELVERSTON, A. (1972). Antidromic action potentials fail to demonstrate known interactions between neurons. *Science*, **177** : 69–72.

[29] NOTTEBOHM, F. (1970). Ontogeny of bird song. *Science*, **167** : 950–956.

[30] PAGE, C. H. (1970). Unit responses in the metathoracic ganglion of the flying locust. *Comp. Biochem. Physiol.*, **37** : 565–571.

[31] PITMAN, R.M., TWEEDLE, C.D. & COHEN, M.J. (1972). Branching of central neurons: intracellular cobalt injection for light and electron microscopy. *Science*, **176** : 412–414.

[32] POND, C.M. (1972). The initiation of flight in unrestrained locusts *Schistocerca gregaria*. *J. comp. Physiol.*, **80** : 163–178.

[33] PRINGLE, J.W.S. (1960). The function of the direct flight muscles in the bee. *XI Internat. Kongr. f. Entomol. Wien.*, **1** : 660.

[34] PRINGLE, J.W.S. (1967). The contractile mechanism of insect fibrillar muscle. *Prog. Biophysic. & Mol. Biol.*, **17** : 3–60.

[35] STRETTON, A.O.W. & KRAVITZ, E.A. (1968). Neuronal geometry: determination with a technique of intracellular dye injection. *Science*, **162** : 132–135.

[36] WALDRON, I. (1968). The mechanism of coupling of the locust flight oscillator to oscillatory inputs. *Z. vergl. Physiol.*, **57** : 331–347.

[37] WILSON, D.M. (1961). The central nervous control of flight in a locust. *J. exp. Biol.*, **38** : 471–490.

[38] WILSON, D.M. (1962). Bifunctional muscles in the thorax of grasshoppers. *J. exp. Biol.*, **39** : 669–677.

[39] WILSON, D.M. (1964). Relative refractoriness and patterned discharge of locust flight motor neurons. *J. exp. Biol.*, **41** : 191–205.

[40] WILSON, D.M. (1968). Inherent asymmetry and reflex modulation of the locust flight motor pattern. *J. exp. Biol.*, **48** : 631–642.

[41] WILSON, D.M. (1972). Genetic and sensory mechanisms for locomotion and orientation in animals. *Am. Sci.*, **60** : 358–365.

[42] WILSON, D.M. & WEIS-FOGH, T. (1962). Patterned activity of coordinated motor units, studied in flying locusts. *J. exp. Biol.*, **39** : 643–667.

[43] WILSON, D.M. & WYMAN, R.J. (1963). Phasically unpatterned nervous control of Dipteran flight. *J. Ins. Physiol.*, **9** : 859–865.

[44] WYMAN, R.J. (1966). Multi-stable firing patterns among several neurons. *J. Neurophysiol.*, **29** : 807–833.

[45] WYMAN, R.J. (1969). Lateral inhibition in a motor output system. I. Reciprocal inhibition in the Dipteran flight motor system. *J. Neurophysiol.*, **32** : 297–306.

[46] WYMAN, R.J. (1969). Lateral inhibition in a motor output system. II. Diverse forms of patterning. *J. Neurophysiol.*, **32** : 307–314.

3 · Wing movements and the generation of aerodynamic forces by some medium-sized insects

WERNER NACHTIGALL

Fachbereich Biologie, Universität des Saarlandes

This paper is concerned with wing movements and with the aerodynamic forces of lift and drag which they generate, in insects of intermediate size, larger than *Drosophila* and thrips but smaller than grasshoppers and dragonflies. I have been asked in particular to present—without mathematics—some of my own work on the detailed wing movements of blowflies and on certain aspects of the flight of butterflies and beetles. I shall also be discussing some recent work on the wing movements of bees which may not be readily accessible in this country. Consideration is restricted to unaccelerated, straight and level flight; turning and other manœuvres are not discussed. On related topics, reference should be made to the critical review of basic principles, and subsequent work, by Professor Weis-Fogh and Martin Jensen ([11, 12] and pp. 48–72); to Professor Pringle's review of the comparative physiology of the flight motor ([9]; see also pp. 3–8); and to the recent work of Zarnack [15] on wing kinematics.

THE AERODYNAMIC FORCES ON THE WINGS OF GLIDING BUTTERFLIES

Many butterflies can glide, and, while the gliding phase of the Cabbage White (*Pieris brassicae*) and Peacock butterflies (*Inachis io*) lasts only a split second and is barely visible to the naked eye, the Swallowtail (*Papilio machaon*) and Scarce Swallowtail (*Iphiclides podalirius*) can glide for many seconds at a stretch. The gliding abilities of these insects depend largely upon the aerodynamic performance of the wings, and a good aerodynamic performance means that at a suitable inclination to the air-flow (the aerodynamic angle of attack—α) the wings produce a relatively small drag (d) opposing and parallel to the direction of flow, and a relatively greater force, the lift (l), perpendicular to the direction of flow. The standard aerodynamic characteristics of any wing are the curves of lift, drag, and of the lift/drag ratio, all as functions of the angle of attack, and the 'polar curve' of lift plotted against drag.

The graphs resulting from wind-tunnel measurements [7] at an air-speed of 2·3 metres per second on a single left fore-wing of an Underwing moth (*Catocala*) are given in Fig. 3.1. Both lift and drag rise with increasing angle of attack, but the lift becomes smaller again after reaching a maximum at $\alpha = 30°$. The maximal lift-to-drag ratio (l/d) amounts to 2·8 at $\alpha = +12°$; at this point, the highest percentage of aerodynamic force is transferred into lift (74 per cent of the total of the two force components). At greater angles of attack, the lift initially increases further, but the efficiency of utilization of the aerodymanic force decreases. At $\alpha = 40°$ the wing begins to stall. Therefore, the operational or physiological flight range of a flat, individual wing must remain within the range $0° < \alpha < +40°$.

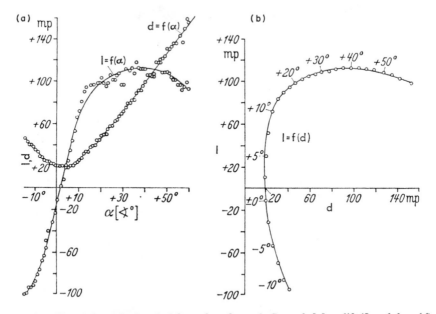

FIG. 3.1. **The lift and drag of a detached fore-wing of a moth** *Catocala* [7]. **a**: lift (*l*) and drag (*d*) as functions of the angle of attack (α) at which the wing strikes the air. 1 mp = 1 mgf. **b**: polar curve of relationship of lift to drag; angles of attack indicated on curve. Lift is at a maximum at an angle of attack of 30°, but the highest percentage of the total aerodynamic force is developed as lift at an angle of attack of 12° (at which the polar curve would be touched by a tangent through the origin).

The performance of the butterfly wing represents about the best that can be expected from a flat plate and from the properties of air at these scales of size and speed, for which the appropriate index to the flow regime is the Reynolds number *Re*. This is given by $Re = \rho Vc/\eta$, where ρ is the density of the air, η its viscosity, V the speed of the air relative to the wing, and c the chord of the wing (width in the fore-and-aft direction, used as a representative length); for the butterfly wing *Re* was about 3000. Relative to a flat plate, some improvement in lift could be expected from a camber of the wing, giving a wing section convex above and concave below; good gliders possess a slightly cambered wing.

Graphs of lift/drag ratio as a function of angle of attack have been found similar for a number of different species of Lepidoptera (Fig. 3.2). The lift/drag ratio increased to a maximum at some point between $\alpha = +5°$ and $\alpha = +15°$, followed by a slower decline; the highest *l/d* values measured were approximately 3·5. Measurements made with both fore- and hind-wings together have shown that larger aerodynamic forces are generated but their utilization is not necessarily improved very much. Measurements carried out on entire insects set in a gliding position have shown that different species gave quite similar polar curves.

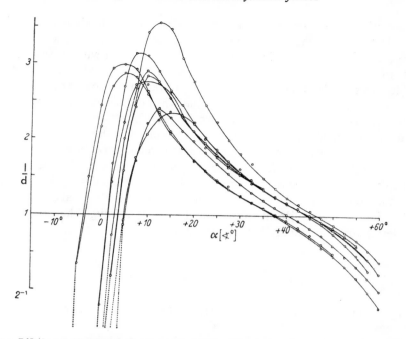

FIG. 3.2. **Lift/drag ratio l/d as a function of angle of attack** α. Curves for nine different species of butterflies and moths [7].

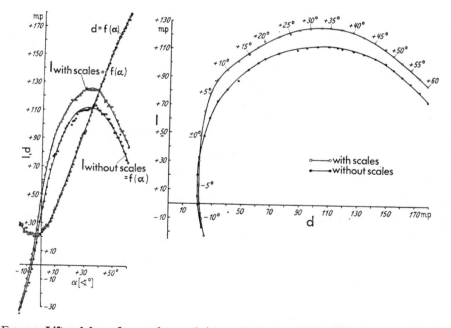

FIG. 3.3. **Lift and drag of a complete moth** (*Agrotis*) **with and without wing-scales** [7]. Lift *l* and drag *d* plotted against angle of attack α (*left*) and against each other as polar curves (*right*). Removal of the scales reduces the lift (without affecting the drag), probably by influencing the fine structure of the air-flow in the boundary layer in contact with the wing.

To determine the influence of the wing-scales, 3450 individual measurements of aerodynamic force were made with ten specimens of each of five species of butterflies and moths, first with the wings in normal condition and then without scales, and from $\alpha = -15°$ to $\alpha = +60°$ at a measuring interval of 1°. It was found that the scales had no measurable effect on the drag, and the minimum drag in particular remained constant. The lift, on the other hand, decreased somewhat when the scales were removed from the wings; the presence of the wing-scales increased the lift by an average of 15 per cent compared to the naked wings (Fig. 3.3). This occurred within the range of small to moderate angles of attack; from $\alpha = 45°$ upwards the removal of the scales ceased to have any effect. Thus the scales enable butterflies to glide over a longer distance from a given height or to remain in the air for a longer period of time when descending. This effect of the wing-scales on the lift must be because they project into the vital boundary layer of the air-flow immediately adjacent to the wing (see p. 71), though the detailed influence of the scales upon the boundary layer flow is still unknown.

THE AERODYNAMIC FORCES ON THE ELYTRA OF BEETLES

Using aerodynamic two-component balances in wind-tunnels, the aerodynamic forces on the elytra have been measured for the Rhinoceros beetle *Oryctes boas* [1] and the May beetle *Melolontha melolontha* [5]. The elytra of *Oryctes* beat in phase with the hind-wings, through a stroke angle of about 20°. The lift was measured, as a function of angle

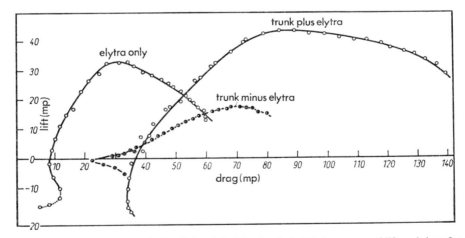

FIG. 3.4. **Effects of the elytra in the flight of the May beetle** [5]. Polar curves of lift and drag for elytra and body, separately and together.

of attack, at the upper and lower turning points and in a half-way position, with air-speeds of 1–5 m/s. Measurements were made on a dead specimen with elytra extended and rear wings removed; after these measurements the elytra were removed and the lift of the body alone was measured. The difference in lift, due to the elytra, was calculated as a percentage of the body weight. Maximum lift for all three positions of the elytra was achieved at an angle of attack of 25°. The average air-speed of the beetle is 4 m/s, at which the elytra support about 20 per cent of the body weight. In free flight the elytra not only beat in time with the rear wings but also rotate through a maximum of 14° around their longitudinal axis, with the angle of attack changing from 20° at the top of the stroke to 34° at the bottom. As with the wings of bees (p. 43) and flies a stroke oscillation is thus coupled with a rotating oscillation. The authors believe that the rotating oscillation is purely passive and comes into action due to the type of joint involved [1].

My own findings for the May beetle *Melolontha* are similar in principle. Wind-tunnel results with insects whose elytra were extended into a mid-stroke position are summarized in Figs. 3.4 and 3.5. Since the dynamic pressures of the moving air could not be measured

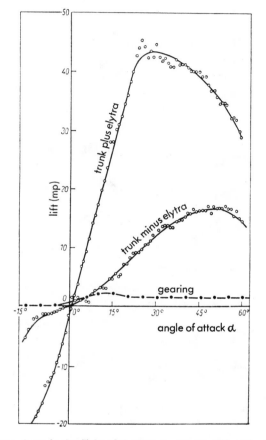

FIG. 3.5. **Effects of the elytra in the flight of the May beetle [5].** Lift of body with and without elytra, as a function of angle of attack. In rapid forward flight the elytra can support 5 to 10 per cent of the weight of the beetle.

accurately at such low air-speeds, the usual coefficients of the components of aerodynamic forces (which are standardized for air-speed) have not been used but rather the components themselves [see p. 46—Ed.]. While it can be seen that the extended elytra do act wings, increasing the body-lift in each position, the lift provided by the body with elytra and by the body without elytra was always smaller than the drag, though from the elytra only the lift was sometimes greater than the drag. At $\alpha = +5°$ the lift of the body with elytra was six times higher than without, and at $\alpha = 20°$ four and a half times, while around $\alpha = 27°$ the air-flow around the elytra became unsteady and they began to stall (Fig. 3.5). But the lift of the elytra is not large relative to the weight of the whole insect, since the elytral surface is small and the elytra are held in an inclined position (directed forwards and upwards). In rapid horizontal flight at 2·25 m/s and with a maximum angle of attack of 25° the elytra develop sufficient lift to support at most 10 per cent and at least 5 per cent of the animal's weight. Thus although the elytra themselves can act as relatively good wings they do not generate their maximal lift because of the position in which they are held.

THE WING MOVEMENTS AND AERODYNAMIC FORCES OF FLIES

For detailed kinematic analysis, the wing movements of the fly *Phormia regina* and the blowfly *Calliphora erythrocephala*, flying in steady-state conditions in front of a wind-tunnel and attached to an aerodynamic balance, were filmed simultaneously from three directions, at a maximum speed of 8000 frames per second [6]. Figure 3.6 shows the details of the wing-stroke as seen simultaneously in the three planes defined by the three

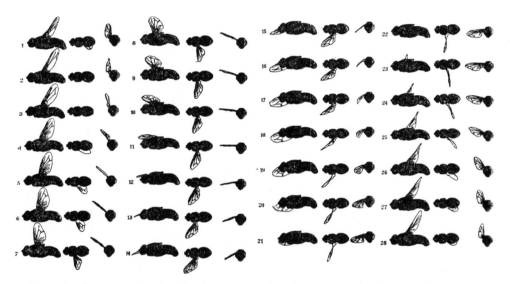

FIG. 3.6. **Wing movements of the fly** *Phormia regina* [6]. Photographed simultaneously in side, top and rear view with a time-interval of 1/3200 sec.

body-axes of the insect—the *x* or longitudinal axis, the *y* or transverse axis and the *z* or dorsoventral axis, defining the *x*, *y* plane (top view), the *x*, *z* plane (side view) and the *y*, *z* plane (rear view). From the photographs the movement of the longitudinal axis of the wing can be defined by three angles, between the projection of the longitudinal wing axis and appropriate axes in the three planes. These three angles are defined in Fig. 3.7 and the time-functions of the angles are shown in this figure. It is clear that the movement is not a simple harmonic oscillation.

The movement of the wing-tip in space can be reconstructed from these three projections and is, again, not a simple sinusoidal oscillation but contains higher harmonics. If the distance between wing-tip and wing-base is always the same, the wing-tip moves on a spherical surface of which the centre lies in the basal joint. The time-function of this movement can be constructed graphically and is shown at the bottom of

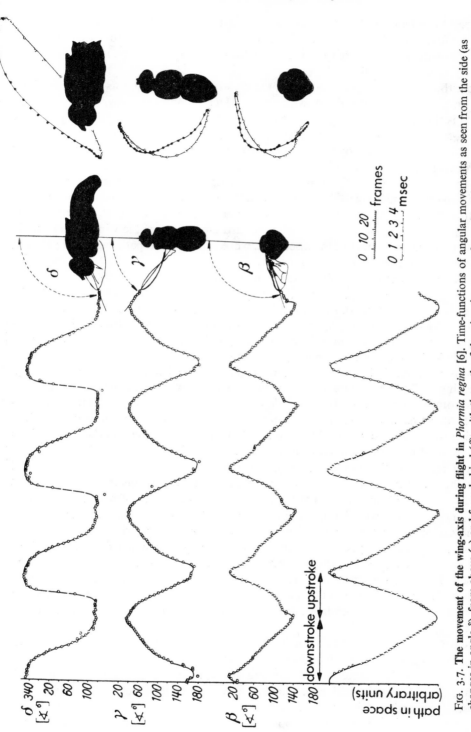

FIG. 3·7. **The movement of the wing-axis during flight in** *Phormia regina* [6]. Time-functions of angular movements as seen from the side (as changes in angle δ), from above (γ) and from behind (β), with the path of the wing-tip as seen from these three directions (*right*; up-stroke open dots, down-stroke solid dots), and the resultant time-function of the path of the tip in space (*below*).

Fig. 3.7. As can be seen, the up-stroke does not last as long as the down-stroke, in which the wing moves rapidly at first, slows, and then retains a constant velocity. Examples of the curves made by the path of the wing-tip on the three projections mentioned are shown to the right of the same figure.

From the side view it can be seen that the up-stroke path (open dots) lies well behind the down-stroke path (solid dots), with consequences which are shown in Fig. 3.8. To the right is the path of the wing-tip relative to the insect; it is similar to the projection of the wing-tip path on the x, z plane (Fig. 3.7, top right), but complicated by several factors (due to the type of projection and to a correction which was not taken into account) which distort the endings into loops but do not influence the aerodynamically essential central parts of the path. When allowance is made for the forward movement of the insect in space, the main closed loop opens out to a zig-zag path relative to the air (Fig. 3.8, left side). Since the up-stroke path lies behind the down-stroke path (relative to the insect) and since at up-stroke the initial backward movement of the wing is more rapid than the forward movement of the insect as a whole, the upward path, relative to the air, is directed first backward and then forward.

The kinematics of the flapping insect wing are not completely described by the path and velocity of the wing-tip. Even if the wing oscillates as a fixed plate (which may be true for its outer two-thirds), the wing still rotates about some axis in the course of the wing-beat. Assuming that this rotational axis is represented by the longitudinal axis of the wing, and analysing in more detail the three projections in Fig. 3.6 from this view-point, one finds that the fly's wings execute a rotating oscillation, combined with their beating oscillation, similar to the wing movements of the bee described below (p. 43). The beating oscillation is characterized by the path of the wing-tip, as has been discussed, and the rotational oscillation by the angle between a wing cross-section and an appropriate further system of co-ordinates. Zarnack [15] has described the movements of the fore-wing of *Locusta migratoria* by a more elaborate system, using Euler's angles, which should be used in further studies of insect flight kinematics.

For a representative cross-section of the wing the aerodynamically most essential part is the outer third and broadest part of the wing. The geometrical system [6] is accurately defined if one imagines around the wing-joint a sphere of which the radius is the distance from the wing-joint to the intersection of the longitudinal axis of the wing and the sample cross-section; this point always moves in a spherical surface and the cross-section of the wing is always tangential to this sphere. When the surface of the sphere is projected onto a plane, the wing is seen from every direction as a thin line because it is always observed along its longitudinal axis. To the left of Fig. 3.8 are the positions of the wing-tip and of the sample cross-section of the wing; the small triangular markers are placed at the leading edge of the wing and on its morphologically upper surface.

The wing section is seen to change its angle relative to the path of the wing-tip; this is the angle at which the wing strikes the air—the aerodynamic angle of attack α. Figure 3.8 shows that this angle is large at the beginning of the down-stroke, becomes smaller half-way through the stroke, and increases again towards the end. During the whole of the down-stroke the wing strikes the air with its morphologically under side. At the lowest turning point the stroke oscillation stops for a moment, and the rotational oscillation is at a maximum; the wing twists very rapidly. Because of the backwardly directed loop of the following up-stroke, the air is now struck by the morphologically upper side of the wing, at initially low angles of attack which increase and then decrease again. As the

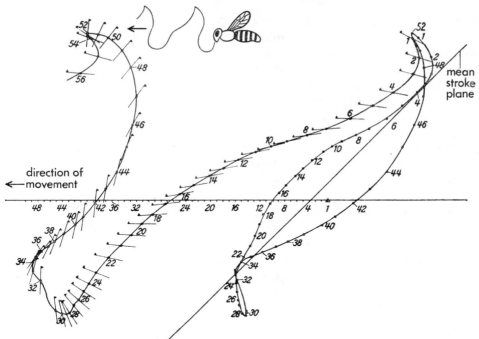

FIG. 3.8. **Movements of wing-tip and wing-surface during flight in** *Phormia regina* [6]. *Right*—movement of wing-tip relative to the fly; *Left*—movement of wing-tip and of a representative cross-section of the wing relative to the air; triangles mark the morphologically dorsal surface and anterior margin of the wing. The wing strikes the air with its morphologically lower surface during the down-stroke, producing most of the lift which carries the weight of the fly, and then twists sharply, to strike the air with its morphologically upper surface during the first part of the up-stroke, producing most of the thrust which propels the fly forwards.

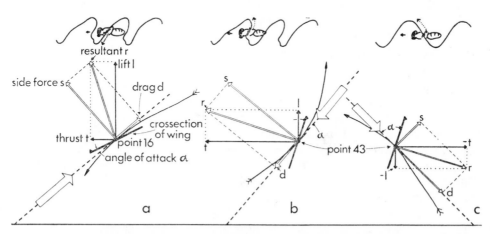

FIG. 3.9. **Aerodynamic forces developed by the wing of the fly** [6]. **a** during the down-stroke (point 16 in Fig. 3.8); **b** during the up-stroke (point 43); **c** during a theoretical up-stroke with imaginary simple sinusoidal wing movement. Large open arrow shows relative wind. Note how the main lift is produced during the down-stroke, and the main thrust during the up-stroke, with the latter dependent on the backwardly directed, non-sinusoidal path of the up-stroke.

up-stroke continues, the direction of the air-current changes and at the uppermost turn-ing point the wing is twisted back, so that the air will again be struck by the morpho-logically under side during the following down-stroke. The wing, with its stiffened frontal edge always leading, thus strikes the air from above during the down-stroke but from below at the beginning of the up-stroke. Since the wing is effectively a flat plate, neither cambered nor with a profile, this change of presentation of the morphologically upper and lower surfaces to the relative wind is functionally immaterial.

The questions now arise as to what forces are produced by the oscillating wing, and in particular what does the backward-running loop of the up-stroke mean?

Let us examine the wing just as it is passing through the horizontal x, y plane during the up- and the down-stroke; at these points conditions are exceptionally clear and easily presented (points 16 and 43 in Figs. 3.8 and 3.9). During the down-stroke (point 16) the wing, against which the air flows at a small positive aerodynamic angle of attack, can be expected to produce as usual an opposing drag d parallel to its path and a stronger side force s perpendicular to the path (Fig. 3.9a). Combining these two components would give a resultant aerodynamic force r, which could in turn be divided into a large upward lifting component, the lift l, and a smaller forward driving component, the thrust t. Examining in exactly the same way conditions in the first third of the upward stroke (point 43; Fig. 3.9b) implies that the resultant aerodynamic force r would now be directed strongly forwards, so that it could be divided into a stronger driving thrust t and a smaller lift l. Similar consideration of the remaining points indicates that both lift and thrust fluctuate rhythmically during the period of the stroke; at the same points of consecutive strokes the ratio of lift to drag is always the same. These kinematic data suggest that lift is at an overall maximum half-way through the down-stroke, while thrust is at a maximum in the first third of the up-stroke, when there is also a secondary maxi-mum of lift.

The fact that positive aerodynamic force components are also produced during the first part of the up-stroke is due to the backward-running loop of the up-stroke path. This can be shown by similar consideration of a simple theoretical sinusoidal wing-path, as a distortion of the actual path measured (Fig. 3.9c). At a position analogous to point 43 during the up-stroke in this hypothetical situation, the resultant aerodynamic force r would be directed downwards and backwards so that its components would now be backward drive $-t$ and downward drive $-l$, negative thrust and negative lift, which would be aerodynamically useless.

The actual forces so far envisaged may approximate only roughly to those developed during real wing-beat, though Wood's hot-wire air-flow measurements (p. 41) indicate [14] that they seem to be fundamentally correct. There are however three main points at which the assumptions and approximations so far made may well need refinement. Firstly, the outer two-thirds of the wing have been treated as a flat surface; this is not true, as the surface may be folded, the thin trailing edge can be bent, and the wing may perform twisting oscillations in the course of the stroke. Secondly, at each point of the stroke the wing was considered as if 'frozen', and the aerodynamic forces produced were envisaged as constant in time and the same as if the wing remained unchanged in position and configuration. However, since the establishment of a particular flow-condition takes time, and since the wing bends and rotates considerably, sometimes irregularly, and in any case very quickly, the wing may for example be capable of passing the dangerous period of high angles of attack so rapidly that the air-flow around the wing has no time to stall. The entire production of aerodynamic forces may well be altered under these non-steady conditions [see e.g. pp. 64-71—Ed.]. Finally the aerodynamic angles of attack which have been discussed here disregard the amount by which the wing rotates between two successive stroke positions considered, whereby the direction of the onflowing current will be altered. One can however be sure that during the aerodynamically important middle phases of the stroke, when the speed of stroke oscillation is maximal and the rotating oscillation minimal, the actual conditions will not differ much from those just considered.

Wood [14] has tested the same species of flies in front of a wind-tunnel, using hot-wire wind-measurement to determine the direction and speed of the induced down-wash of air behind the wings, which provides the actual lift. The beating and rotational movements of the wing found by him during the period of the stroke agree qualitatively with my own results. Furthermore, his hot-wire tests showed a marked peak of the downward vertical component in the air-current behind the wing approximately in the middle of the down-stroke, and a smaller peak for this component in the first part of the up-stroke; these results are entirely consistent with my earlier kinematic analyses and aerodynamic inferences. Further support for the earlier work was provided by his findings that during 'natural' flight in front of the wind-tunnel, the turbulence in the wake behind the wings was very small, and that, by superimposing the output of the hot-wire instrument for approximately 100 consecutive wing-beats, any random 'flutter' could be shown to be very small. Again, on turning the wind-tunnel off, I had found that the path of the wing altered drastically; the normally open O-shaped path of the wing tip was turned into a squashed figure-of-eight [6]. In these still-air conditions Wood measured very strong turbulence in the down-wash, which indicates stalling and implies that the wings may no longer have been acting as lift generators. He assumes that in this unphysiological case the aerodynamic angle of attack, especially during the down-stroke, is abnormally high and causes the wing to stall.

THE WING MOVEMENTS AND AERODYNAMIC FORCES OF THE HONEY-BEE

A bee can use its wings for three very different functions: firstly, flying; secondly, fanning (*fächeln*)—wing movements which produce air-currents to ventilate the hive; and, thirdly, wafting (*sterzeln*)—for generating a stream of air across an abdominal gland to transport scent molecules. Neuhaus [8] and Wohlgemuth [13] made preliminary studies of the different types of wing movement involved. In flight the bee moves its wings similarly to the fly, i.e. from above and behind to below and in front, and back again, but with the middle of the down- and up-stroke planes approximately coincident (Fig. 3.10*a*, *c*); the frequency is 200 wing-beats per second. The aerodynamic angle of attack amounts to 15° at the middle of the down-stroke (*c*, left), but is said to be almost zero during the up-stroke (*c*, right), so that the up-stroke (unlike that of the fly) does not apparently produce any useful aerodynamic forces. During wafting the bee holds tightly to the ground, slants its body and moves the wings in an almost vertical plane (Fig. 3.10*b*, *d*); during both up- and down-strokes, aerodynamic angles of attack are apparently high, so that during both strokes the wings induce a strong, almost horizontal current behind the insect. The angles of attack appear to be so high as just not to induce stalling, the thrust component is very large, and the muscles are heavily loaded. During fanning also the lift components must be strong, because the entire body of the bee oscillates up and down. There are smooth changes from fanning to wafting and to flight movements.

The biologist H.G.Herbst and the aerodynamicist K.Freund worked together on studying the wing structure of the honey-bee and wing movements during wafting, fanning and take-off [2, 3, 4]. The large fore-wing and smaller hind-wing of the bee are coupled together by a hook mechanism. The fore-wing is on average 9·9 mm long; total wing area, on one side, is $18·5+9·5 = 28$ mm^2, and weight is $0·109+0·032 = 0·141$ mp.* Compared to a bee's total weight of 80 mp, all four wings together weigh only 0·28 mp, i.e. 0·35 per cent. Maximum thickness of the veining amounts to 60 μm, and that of the wing membrane 2 μm. If one divides the wing area into parts by arbitrary lines and calculates the percentage of mass of the individual parts, one finds, because of the differences in the veining, an accumulation of mass on the frontal edge and towards the

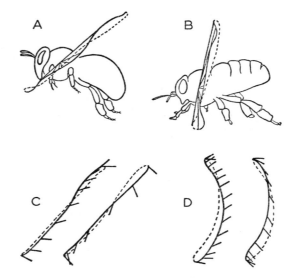

FIG. 3.10. **Differences in wing movements of honey-bee in flight** (at take-off; A and C) **and during stationary wafting** (B and D). Wing-stroke plane shown in A and B [13]; wing-beating and wing-twisting shown in C and D [8].

proximal end of the wing. The surface loading due to the weight of the wings themselves is 5·9 μp/mm^2 for the fore-wing; 3·4 μp/mm^2 for the hind-wing and 5·4 μp/mm^2 for both the wings together. In wafting there are 180 wing-beats per second and the wing-beat angle is 110°, so that the speed of wing-beating lies between 0 and $7·8 \times 10^4$ degrees per second. For one set of fore- and hind-wings, a centrifugal force of 72 mp can be calculated, and is nearly equal to total body weight. Such are the forces with which the swinging wing pulls at its roots. Apart from these inertial forces (and the additional effects of the clinging air mass, which can sometimes be significant [10], were not taken into account), elastic forces occur due to the deformation of the moving wings. The elasticity modulus of the wing was found to be 0·7 pond/mm^2. As was to be expected, tests carried out on wings in tension, and on various wing parts under load, gave rise to differing deflections by given forces. Downward loads can be divided into bending forces perpendicular to the wing surface and into forces along the length of the wing which tend to

* 1 millipond (mp) = 1 milligram-force.
  1 micropond (μp) = 1 microgram-force.

extend it. For the former forces, the results show a linear elastic angular deflection of about 8°/100 mp load.

The fundamental frequency of the elastic wing was shown by means of an oscillating-table to be 765/s, for the set of both wings of one side. There are, however, other resonance ranges. Living bees, under CO_2 narcosis with vibration stimulation from the thorax, showed resonance at a frequency of 230/s, which was extinguished when the wing-joint was blocked. Wing-beat frequency at take-off was measured to be 258/s, and is thus satisfactorily within the range of the fundamental frequencies of the thorax-wing system. The fundamental frequency of the wing itself is, as mentioned, approximately three times

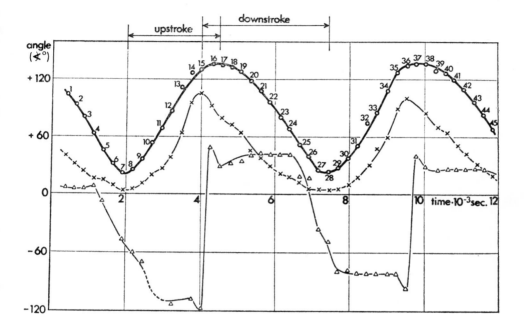

FIG. 3.11. **The wing movements of the honey-bee** [2]. *Circles*, projected path of wing-tip as seen in side view; *Crosses*, points in wing-beating oscillation (angle θ); *Triangles*, points in rotational (wing-twisting) oscillation (angle χ). Time-interval 1/3842 second.

as high. Another resonance at 415/s, which is largely independent of desiccation and joint blockage, can be traced to movements of parts of the thorax, perhaps the sterno-pleura, whose oscillations are transmitted via the pleural articulation to the wings.

The wing movements during wafting and the resulting air-currents were measured by means of wind-tunnels, high-frequency cameras and hot-wire instruments. To assist in establishing details of the wing movements a plexiglass model was built, with wings which were manœuvrable for visual comparison with individual film photos. On this model, it was possible to measure the angle of wing-beat and the angle of rotation [2].

As with the flies, it was shown that movements of the bee's wings consist of two coupled oscillations with a definite phase-shift. The wing beats downwards and forwards, and then up and back again, while rotating at the same time in a definite way around its longitudinal axis. The first component, the beating oscillation, is recorded by the angle of wing-beat; the second, rotational oscillation is recorded by the angle of rotation. The projection of the right wing-tip path on the x, z plane of the photography, together with the corresponding values of the angles of wing-beat and wing-twist as determined with the help of the model, are illustrated in Fig. 3.11. The changes of these angles with time, i.e. the angular velocities, are shown in Fig. 3.12.

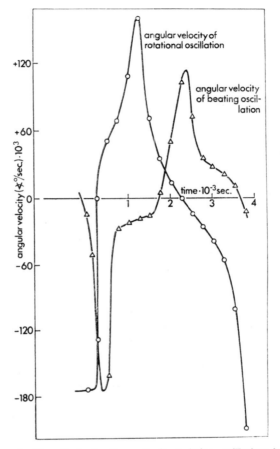

FIG. 3.12. **Angular velocities of wing-beating and wing-twisting oscillations in the honey-bee** [2]. See Fig. 3.11; angular velocity of beating oscillation

$$\omega_b = \frac{d\theta}{dt} \text{ and of rotational oscillation } \omega_r = \frac{d\chi}{dt}$$

The rotational oscillation at the upper turning point (the beginning of the down-stroke) can be seen to be very rapid; at the lower turning point it is slower. The up-stroke is more rapid than the down-stroke. The maximum speeds of the beating oscillation occur during the first third of the down-stroke and approximately half-way through the up-stroke. The maximum speeds of the rotary oscillation are near the turning points. There is a phase-shift between stroke and rotary oscillations, though as these are not simple harmonic oscillations a definite phase-angle cannot be given.

Measurements of the air-currents behind the buzzing wings (in flight) showed that the mean speed of the slipstream near the rear edge of the wings, determined by a pitot-tube, was 2·0–2·5 m/s, while the down-wash component was maximal at three torso lengths behind the wing-beat plane, where it reached 0·4–0·6 m/s.

During fanning the bee created a forward thrust which, at frequencies of 180–195 wing-beats per second, reached approximately 118 mp, i.e. four-thirds of the bee's weight. The bee must accordingly hold on with its feet so as not to blow itself away.

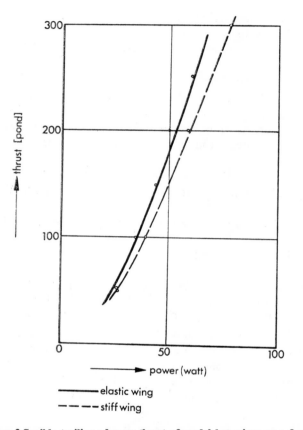

FIG. 3.13. **Effects of flexible trailing edge on thrust of model bee wings as a function of motor power** [2]. 1 pond = 1 g force.

An attempt was made to apply these measurements to model wing systems [2]. To reduce the speed, enlarged models were used in a water-tank, and a dye was injected into the water-stream from the frontal edge of the wing. High-speed film analysis was combined with measurements of static lift and thrust by two-component balances.

The wing model, which was about twenty times life-size, was set in motion by appropriate gearing to produce wing movements similar to those of real wings. It was shown that the thrust of model wings with elastic rear edges rose with increasing motor power more steeply (Fig. 3.13) than that of models which were more rigid. These measurements from the models definitely indicated that the contrast between the very soft and passively flexible rear edge of the bee's wing and the massively strengthened fore edge plays an important part in the generation of thrust.

DISCUSSION

Dr M.J. Samways: In the crickets (Gryllidae) and bush crickets (Tettigoniidae) that stridulate with their tegmina, one would think that selection has operated to increase efficiency of this sound-production apparatus. Does Professor Nachtigall think that, in those crickets and bush crickets that fly actively as well as stridulating, aerodynamic efficiency may have been decreased by the presence of the stridulatory apparatus? This could have ecological and evolutionary significance.

Nachtigall: I am afraid I do not know.

Pringle (Chairman): I do not think anyone is certain of the aerodynamic effects involved yet.

Dr C.J.C.Rees: Is it invariably true that the body of an insect contributes very little to its lift in forward flight? There are instances where body profiles look aerodynamically reasonable—though the body is admittedly not very wide.

Nachtigall: Yes, this can certainly be true of body profile, but preliminary calculations have shown the lift to amount to no more than a few per cent of the total lift needed to sustain the weight of the body.

Rees: Has this been true of all the insects you have looked at?

Nachtigall: Yes, we have looked at May beetles (*Melolontha*), higher Diptera and Lepidoptera.

Professor B.Hocking: In flies there is a considerably greater contribution of lift from the body of the insect.

Nachtigall: Has this been measured?

Hocking: Yes, in a wind-tunnel.

Nachtigall: On which insects? And do you know the percentage?

Hocking (partly communicated): Horseflies (Tabanidae) and *Muscina;* and up to 22 per cent (Hocking, B. (1953) *Trans. R. Ent. Soc. Lond.,* **104**: 269).

Sir James Lighthill: May I ask a question on Professor Nachtigall's point about aerodynamic coefficients not having been used in plotting his results for beetle elytra, because of the difficulties in measuring dynamic pressure and therefore in estimating air-speed? His results suggested remarkable differences in the lift-curve slopes with the elytra and without; and one has a slight anxiety (if it was not possible to measure the air-speed) as to whether these lift curves were truly at the same air-speed. This is a good reason for always plotting such results in terms of the aerodynamic coefficients, because they would not then be sensitive to these differences.

Nachtigall: That is completely true, though there are two points I should add. The first is that we used a special, constant-speed, inverted squirrel-cage induction motor*, in order to keep the air-speed constant. The second point is that if there had been a difference in air-speed it should have been seen clearly in the drag measurements, which were taken at the same time as the measurements of lift; there were large differences in the lift curves and little difference in the drag curves. In the plots of drag as a function of angle of attack, all points of measurement were close to a single curve.

REFERENCES

[1]   BURTON, A.J. & SANDEMAN, D.C. (1961). The lift provided by the elytra of the rhinoceros beetle, *Oryctes boas. S. Afr. J. Sci.,* **57** : 107–109.

* *Kurzschlussaussenläufer.*

[2]   FREUND, K. (1969). Untersuchungen am schwingenden Bienenflügel. 2. Teil. Biologie-Bionik-wissenschaftlich-technische Revolution, 36–46.

[3]   HERBST, H.G. (1969). Untersuchungen am schwingenden Bienenflügel. 1. Teil. Biologie-Bionik-wissenschaftlich-technische Revolution, 29–35.

[4]   HERBST, H.G. & FREUND, K. (1962). Kinematik der Flügel bei ventilierenden Honigbienen. *Deutsch. ent. Z.*, N.F. **9** : Hft. I/II, 1–29.

[5]   NACHTIGALL, W. (1964). Zur Aerodynamik des Coleopterenflugs: Wirken die Elytren als Tragflügel? *Verh. dt. zool. Ges.*, **52** : 319–326.

[6]   NACHTIGALL, W. (1966). Die Kinematik der Schagflügelbewegungen von Dipteren. Methodische und analytische Grundlagen zur Biophysik des Insektenflugs. *Z. vergl. Physiol.*, **52** : 155–211.

[7]   NACHTIGALL, W. (1967). Aerodynamische Messungen am Tragflügelsystem segelnder Schmetterlinge. *Z. vergl. Physiol.*, **54** : 210–231.

[8]   NEUHAUS, W. & WOHLGEMUTH, R. (1960). Über das Fächeln der Bienen und dessen Verhaltnis zum Fliegen. *Z. vergl. Physiol.*, **43** : 615–641.

[9]   PRINGLE, J.W.S. (1968). Comparative physiology of the flight motor. *Adv. Insect Physiol.*, **5** : 163–223.

[10]   VOGEL, S. (1962). A possible rôle of the boundary layer in insect flight. *Nature, Lond.*, **193** : 1201–1202.

[11]   WEIS-FOGH, T. (1956). Biology and physics of locust flight. II. Flight performance of the Desert Locust (*Schistocerca gregaria*). *Phil. Trans. R. Soc.*, (B), **239** : 459–510.

[12]   WEIS-FOGH, T. & JENSEN, MARTIN (1956). Biology and physics of locust flight. I. Basic principles in insect flight. A critical review. *Phil. Trans. R. Soc.*, (B), **239** : 415–458.

[13]   WOHLGEMUTH, R. (1962). Die Schlagform des Bienenflügels beim Sterzeln im Vergleich zur Bewegungsweise beim fliegen und Fächeln. *Z. vergl. Physiol.*, **45** : 581–589.

[14]   WOOD, J. (1970). A study of the instantaneous air velocities in a plane behind the wings of certain diptera flying in a wind tunnel. *J. exp. Biol.*, **52** : 17–25.

[15]   ZARNACK, W. (1972). Flugbiophysik der Wanderheuschrecke (*Locusta migratoria* L.) I. Die Bewegungen der Vorderflügel. *Z. vergl. Physiol.*, **78** : 356–395.

4 · Energetics and aerodynamics of flapping flight: a synthesis

TORKEL WEIS-FOGH

Department of Zoology, University of Cambridge

During calm summer days in August one often sees small flocks of black-headed gulls gliding and circling in narrow spirals near my cottage in Tibirke, Denmark. Suddenly, a bird will 'stop', flap briefly and catch one of the hovering beetles which, together with floating seeds of willow herb, have been carried up by a feeble invisible dust-devil. This little scene embodies the large range of performances exhibited by flying organisms, from passive floating and gliding to active flapping flight where the energy to remain airborne is provided by the rapidly contracting wing muscles. The main problem is whether the basic aerodynamic mechanisms are the same in all kinds of flight, or whether different principles are involved, and to what extent. Obviously, floating seeds and aerial plankton of wingless arthropods must make use of aerodynamic *drag* to delay their descent, and this may apply to very small Pterygota in between bursts of activity [10]. On the other hand, gliding and soaring in bats, birds and insects depend on the usual aerofoil action of their wings, i.e. on the aerodynamic *lift* which acts perpendicular to the direction of movement through the air, while the drag acts directly against the direction of that movement; and the aerofoil action of a gliding animal complies with the well understood *steady* flow patterns of textbook aerodynamics and aeronautics. However, when the wings flap, fluctuating or *unsteady* flow patterns must occur, and may become dominant features in some cases; but we are only beginning to understand the nature of the problem and how animals make use of such unsteady flow [26, 6].

The lift principle as applied to steady flow is most likely to dominate the performance when the animal is relatively large and flies at a high forward speed, i.e. during *fast forward flight*. This is the situation for instance in most birds and in large insects like locusts and migrating hawk moths. However, the majority of insects are small and employ *hovering* and *slow flight*. Under these conditions the wings sweep through a large angle and their tips attain air-speeds far in excess of that of the body. During true gliding, the wing-tip of course travels at the same air-speed as that of the wing-base and of the body, but during normal flapping flight in many birds (pigeon, pheasant, rook, partridge, gull) the air-speed at the tip is 1·3–1·4 times larger than at the base; it is twice as large in the Desert Locust *Schistocerca gregaria*, three times larger in the horsefly *Tabanus affinis*, and five times larger in the mosquito *Aëdes nearcticus* [27]. During true hovering, remaining over a fixed spot in still air, the body does not move forward at all and is exposed only to the vertical wind induced by the moving wings. As we shall see,

48

hovering is also the type of flight during which the wings must be twisted at the highest angular velocities when they are pronated and supinated. The small forward speed and the high rate of wing-twisting both tend to increase the relative importance of unsteady flow patterns. It is therefore of particular interest to investigate the aerodynamics and energetics of hovering. In addition, hovering and slow flight characterize the vast majority of flying animals, the Pterygota. Probably ever since late Devonian times, about 350 million years ago, this group has been so successful in exploring the new environments provided by the emergence of tall terrestrial plants that they represent three quarters, or 750,000, of all known animal species, fossil and living.

An attempt will be made to provide a simple synthesis of present knowledge. The timing is appropriate partly because some important aspects of the energetics of fast forward flight have been analyzed recently [13, 14, 15, 17] and partly because new studies of hovering flight have offered more insight and revealed novel aerodynamic mechanisms of general interest [25, 26, 6]. In order to avoid heavy technical language and mathematical expressions, only the most essential physical relationships are described in the text and readers with special interest must consult the original literature.

DRAG OR LIFT?

Any actively flying animal must accelerate air downwards in order to counteract its own weight. This could be done either by using the drag of the wings, as that of the oars is used in rowing, or by means of the lift mechanism which is much less expensive in energy, at least for large and medium sized animals. When we compare birds, bats and insects of different size, from a wing span of over 2 m to 1 mm, the plan forms are surprizingly similar. Since wing area and body weight will respectively be roughly proportional to the square and cube of linear dimensions, this means that the wings of small flying animals carry less weight per unit of wing area than do those of large ones; some examples are given in Table 4.1. A consequence of this is that below a certain size the wings may be used not as aerofoils but as 'oars' which depend only on drag. This arises in the following way.

When a volume of air is accelerated its mass gives rise to inertial forces, and at the same time internal shearing forces are caused by its viscosity. The flow pattern depends on the ratio between the inertial and the viscous forces in a particular flow situation, the Reynolds number *Re*. [see p. 32]. This is proportional both to the size of the wing, as measured by a characteristic length such as wing width or chord, and to the velocity of the wing relative to the undisturbed air. It therefore decreases with decreasing size and some examples are given in Table 4.2. This means that the viscous forces and the drag are increased, relative to the inertial forces and the lift, when the animal is small. The lift is usually the dominant force for Reynolds numbers exceeding 100, but at the small values reached by really tiny insects [4, 11] the drag will dominate and the usual steady-state lift becomes insignificant [16]. This and the decreased wing-loading both suggest that small or lightly loaded insects might use a drag mechanism for flight rather than the lift principle. However, two arguments speak against this.

Firstly, we have not yet observed a flapping animal which uses drag rather than lift, not even lightly-loaded plume moths (Pterophoridae and Alucitidae [11]) nor the small Chalcid wasp *Encarsia formosa* (wing length 0·6 mm; see p. 63–67) [26]. Secondly, the drag principle is difficult to apply when the wings remain totally immersed in the fluid, as

TABLE 4.1. **Wing-loadings in some flying animals** expressed as the weight in newton carried per square metre of sustaining wing area [26]

	N/m²
*SMALL BATS	10–20
BIRDS	
medium sized and large	30–170
small passerines	20–50
*hummingbirds	20–30
*swifts, swallows, bee-eaters	13–25
INSECTS	
*Coleoptera, large—Lamellicornia	12–40
*Hymenoptera, large—Vespoidea & Apoidea	8–44
*Diptera, large—Brachycera & Cyclorrhapha	5–20
*Lepidoptera, large—Sphingidae	4–12
*Lepidoptera, medium sized—Noctuidae	3–6
*Coleoptera, small	1–6
**Syrphinae—true hoverflies	3–11
**Odonata—dragonflies	1–6
**Drosophila virilis (wing length 3 mm)	3–4
**Encarsia formosa (wing length 0·6 mm)	1·2
**Lepidoptera Rhopalocera—butterflies	0·4–2

* Most species show normal (p. 60) hovering.
** Slow forward flight and hovering involve unusual aerodynamic mechanisms.
1 N/m² = 0·0102 gram-force per cm².

TABLE 4.2. **Calculated average lift coefficients in hovering flight** [26]

		Airborne weight 10^{-3}N	Reynolds number (Re)	Lift coefficient† ($\overline{C_L}$)
Bats:	*Plecotus auritus*	90	14000	1·3
Hummingbird:	*Amazilia fimbriata*	50	7500	2·0
Coleoptera:	*Melolontha vulgaris*	5·9	4700	0·6
	Amphimallon solstitialis	2·8	3000	0·7
	Heliocopris sp.	125	23000	0·5
Lepidoptera:	*Pieris napi*	0·4	1400	(2·2)
	Sphinx ligustri	15·7	6300	1·2
	Manduca sexta	20·8	6700	1·2
	Macroglossum stellatarum	2·8	2800	1·1
Hymenoptera:	*Vespa crabro*	5·9	4200	0·8
	Bombus terrestris	8·6	4500	1·2
	Apis mellifica	0·98	1900	0·8
	Encarsia formosa	$2·5 \times 10^{-5}$	15	(3·2)
Diptera:	*Tipula* sp.	0·28	770	0·8
	Aëdes aegypti	0·01	170	0·6
	Eristalis tenax	1·5	2000	0·9
	Drosophila virilis	0·02	210	1·0
	Syrphus spp.	0·2–0·3	500	(2 to 3)
Odonata:	*Aeshna grandis*	8·4	1750	(2 to 3)

† A lift coefficient can be envisaged as a measure of the lift produced per unit area of wing at a standard air-speed and a standard air density.

* Hovering procedure different from that defined as normal (p. 60).

Brackets indicate unusual lift mechanisms outside simple steady-state theory.
10^{-3}N = 0·102 gram-force.

with wings in air, but unlike rowing with oars where the effective stroke is in water and the return stroke in air (sculling a dinghy from the stern by means of a continuously submerged oar is in fact using the lift principle, in contrast with the use of drag in propelling a canoe by a continuously-immersed single-bladed paddle). In order to produce a net force the drag must be large during the effective stroke and reduced during the return stroke. This would be feasible if the extended wings could be folded or bent extensively during one half of the stroke, but even this mechanism would be of limited use at low Reynolds numbers, at which the drag of a stiff extended insect wing tends to become independent of its actual shape and of the angle of attack of the wing, and is determined mainly by its length. Moreover, we approach conditions in fluid dynamics where, in general, the flow becomes reversible.

The extremely small Ptiliidae (Coleoptera), Trichogrammatidae and Mymaridae (Hymenoptera) have wings which are only 0·07–0·2 mm long and consist of a 'stem' surrounded by a flat marginal brim of hairs [4]. As pointed out by Norberg [11], they must operate at *Re* about 1 and this virtually excludes any lift action, probably even by the newly discovered 'fling' mechanism (p. 64). As a new possibility I venture to suggest an acceleration of the air caused by a *twisting* movement which is *propagated* from base to tip of the wing as a consequence of 'delayed elasticity' (p. 68). If this turns out to be the case, these tiny insects 'swim' actively by means of a screw-like action reminiscent of that of an undulating membrane. However, we do not have any direct evidence for this as yet.

NATURE OF AERODYNAMIC LIFT

If a solid cylinder is placed horizontally in still air and spun round its axis, the surrounding air is set in uniform motion due to viscous effects. At the surface itself the air rotates with the same speed and direction as the solid surface and the speed decreases linearly with the distance from the cylinder. The result is a steady cylindrical *vortex* of air 'bound' to the cylinder—the *bound vortex*—but this does not in itself imply any other forces than a small drag against its rotation. If, however, a horizontal wind blows as in Fig. 4.1A, from left to right, the resulting air velocities will become increased on the upper side and decreased on the lower side. This causes the pressure to decrease above the cylinder and to increase beneath it due to the usual *Bernoulli effect*. The cylinder will now experience an aerodynamic force in the *transverse* vertical direction which we call the lift; with the spinning cylinder the phenomenon is known as the *Magnus effect*. Figure 4.1A shows the streamlines which result when the horizontal wind is superimposed upon the vortex. Note that the air behind the cylinder has been accelerated downwards.

LIFT AND CIRCULATION

The magnitude of the lift L is determined only by the translational wind velocity v_t and a property of the vortex which is called its circulation Γ and is given by the product of the surface speed and the circumference. If ω is the angular velocity of rotation and r is the radius of the cylinder, then $\Gamma = \omega r . 2\pi r = 2\pi \omega r^2$, and can be expressed in cm² / s.

In other cases it may be difficult to estimate Γ but, whether we are dealing with a cylinder or any other long body like a wing, the fundamental relationship holds:

$$L/\text{unit length} = \rho v_t \Gamma, \qquad \text{(i)}$$

where L is lift, ρ the mass density of the air and v_t the translational wind velocity. This applies to an infinitely long wing, and has to be modified in a real wing because of the formation of *tip vortices*, due to air moving around the tip from the high-pressure region below into the low-pressure region above the wing. This gives rise to a loss in usable flow and energy which manifests itself in less lift, higher drag and the so-called *induced power loss* (p. 59). Because the loss is related to the pressure differential, i.e. to the actual lift produced, it can sometimes be calculated with fair precision, as first done by Prandtl for a wing with the elliptical pressure distribution along its span for which induced drag is in fact at a minimum.

Provided a circulation is established in some way or other, a wing will thus experience lift when exposed to a wind, in accordance with equation (i). But how is the circulation set up in the first place? The answer is essential for an understanding of several aspects of insect flight.

BOUND AND STARTING VORTICES

We return to the infinitely long aerofoil seen in transverse section in Fig. 4.1, i.e. to the simplified case of two-dimensional flow. Let us first assume that the aerofoil is immersed in a fluid which has mass but no viscosity, i.e. in an ideal fluid. It is then possible to calculate the flow pattern, but since a wing cannot generate (or destroy) vorticity and circulation in an inviscid fluid, the calculated ideal flow in Fig. 4.1B (irrotational or potential flow) results in zero lift and drag. This was one of the dilemmas in classical fluid dynamics. It can be seen that the fluid is not directed downwards behind the wing and also that there are some discontinuities near the trailing edge, which would lead to large shearing forces in a real fluid like air or water. If, however, the pure rotational flow in Fig. 4.1C is superimposed on the irrotational flow in Fig. 4.1B, the result is the orderly 'streamlined' flow in Fig. 4.1D which corresponds to the *observed flow* round an aerofoil after it has been in uniform motion relative to the fluid for some time, i.e. after it has reached a steady state. The streamlines are now directed downwards behind the profile and lift is produced.

The problem of creating the *bound vortex* with circulation round the aerofoil was solved by Prandtl in 1912, the main point being that in a real fluid of even very small viscosity like air, the flow in the boundary layer near the solid surface must be influenced by viscous forces, particularly near to the trailing edge in Fig. 4.1B. When an aerofoil starts from rest, or when its angle of attack is initially so small that no lift is produced and the angle is then suddenly increased, these viscous forces will induce a vortex *behind* the profile, the *starting vortex*, which has the opposite sense to that seen in Fig. 4.1C. Now, it is a fundamental rule in fluid dynamics that no vortex can be created unless a vortex of the opposite sense and strength is set up simultaneously. In our case this is the *bound vortex* in Fig. 4.1C which causes the lift.

At the start of movement or when the lift changes we now have to consider two opposite vortices close to each other, the starting and the bound vortex, and they interact destructively in inverse proportion to their distance apart. Therefore at the start of this movement the net circulation round the aerofoil will be only about half as large as later on when the starting vortex has been left well behind. This interaction, causing delay in building up the lift, is called the *Wagner effect* and represents an unavoidable *unsteady* phase in the action of an ordinary aerofoil. Conversely, when a flapping wing stops at the end of one half-stroke in hovering flight, and the lift is reduced, the bound vortex must be shed, and becomes a 'free' vortex which will interact with the new starting and bound vortices which initiate the return stroke [25].

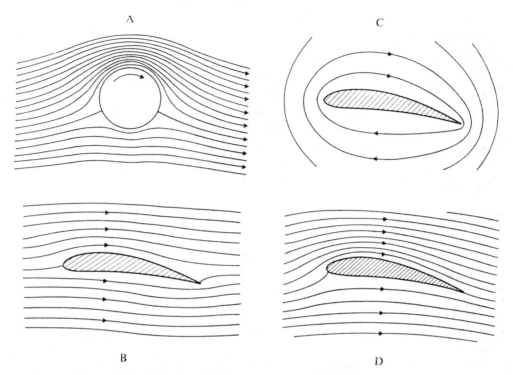

FIG. 4.1. **Circulation in ideal and real fluids.** A, a spinning cylinder in horizontally streaming air or water—the Magnus effect; B, the flow round an aerofoil in an ideal fluid (irrotational or potential flow); C, pure rotational flow; and D, real steady-state flow in a real fluid, where (B) and (C) are combined.

ORDINARY AEROFOIL ACTION

An ordinary wing or aerofoil is, then, a solid body which is shaped and placed relative to the streaming air in such a manner that viscous forces create a starting vortex near to its trailing edge and therefore a bound vortex of the opposite sense round its profile. The starting vortex is left behind and the steady-state bound vortex is maintained by the continuous shedding of small vortices; this gives rise to a small drag component. In addition, friction in the boundary layer close to the surface represents a skin drag; together the two are called the *profile drag*. Furthermore, a wing of limited span also experiences an *induced drag* caused by the tip vortices, as already explained.

After the flow has become steady it is customary to express the lift in terms of the *coefficient of lift* C_L rather than in terms of circulation. The two are related as follows,

$$L/\text{unit length} = \rho v_t \Gamma = \tfrac{1}{2}\rho v_t^2 c C_L, \tag{ii}$$

where c is the width or chord of the wing section in question. Because of the increased importance of viscous over inertial forces at low Reynolds numbers it can be understood intuitively that both the creation of a starting vortex and the maintenance of a bound vortex become expensive in energy and difficult to achieve. The result is that the lift decreases relative to the drag. The maximum C_L for a *Drosophila* wing at $Re = 200$ is about 0·9 [19], for the fore-wing of a locust at $Re = 2000$ it is 1·3 [5], while it may reach 2·0 or even more in bird wings operating at $Re = 5000$ or above. We may therefore use the calculated value of C_L as an indicator of whether ordinary steady flow is sufficient to explain a given flight performance or not [12, 27, 2, 25], though more detailed investigations are needed in order to estimate the true role of steady *versus* unsteady flow. This is particularly relevant to flapping flight where the lift changes rhythmically throughout the wing-stroke. The extreme is reached when the animal hovers because the lift then drops to zero at each end of the wing-path. In the case of ordinary aerofoil action, a hovering animal must experience the Wagner effect during an appreciable part of the wing-stroke. This need not be as serious a problem as one may think at first glance because stalling is also delayed and C_L can be increased above the usual stalling limit for brief periods of time so that the two effects tend to cancel each other [3, 25].

FAST FORWARD FLIGHT

Although the majority of insects and some small birds and bats use hovering and slow flight when feeding or courting, all birds and bats and some insects make extensive use of fast forward flight. The air-speed V of the body then corresponds to 50–300 body lengths per second, and analysis is complicated because the wind forces acting on the individual wing segments vary from base to tip as the square of the vector sum of the horizontal velocity V and the flapping velocity relative to the body. There is accordingly no exact theory for this type of flight but only approximations relating to certain aspects, including the generalized equations of Osborne [12] and the approximate expressions for the aerodynamic power components in birds derived by Pennycuick [13, 14] and recently revised by Tucker [17]. All theories applied so far rest on the assumption that the principles of steady flow dominate the pattern and this is indeed likely to be the case during forward flight, as discussed elsewhere [27]. However, we have only one complete *experimental* study of fast forward flight, namely Martin Jensen's analysis [5] of the Desert Locust, which offers direct and detailed insight into the aerodynamic processes. This and the recent studies of bird flight make it possible to extend his results and compare them with the power requirements of other animals in relation to the speed and cost of transport. The latter is particularly relevant to an understanding of the long-distance flights of migrating insects.

HORIZONTAL FLIGHT OF THE DESERT LOCUST

For an average *Schistocerca gregaria*, with a mass of 2 g, the ratio between the air-speed at the wing-tip and that at the wing-base is about 2, i.e. appreciably larger than in most birds (p. 48). In a detailed wind-tunnel study we first established the normal range of variation of the flight parameters [21] and then Martin Jensen [5] analyzed the kinematics, aerodynamics and energetics of a few selected examples on the basis of stroboscopic slow-motion films. Figure 4.2 shows the wing movements during the down-stroke and up-stroke when the insect was maintaining a vertical force equal to its weight (equivalent to horizontal flight) at an air-speed of 3·5 m/s. During the down-stroke the leading edge of both wings is twisted downwards, i.e. *pronated* and, in addition, the flap at the trailing edge of the fore-wing is tilted downwards when it passes the horizontal position. During the up-stroke, the leading edge is *supinated* (cf. frames 12 and 13) and the flap is bent upwards (frame 14, seen almost end on), creating the so-called Z-profile. The exact movements of the wings relative to the air were computed and Fig. 4.3 is an approximate graphical presentation of the results, A from the tip and B from the middle part of the fore-wing. In both cases (and in the hind-wings also) the individual wing profiles meet the air at small positive angles of attack during the down-stroke and almost edgewise during the more rapid up-stroke. This is consistent with an ordinary and orderly aerofoil

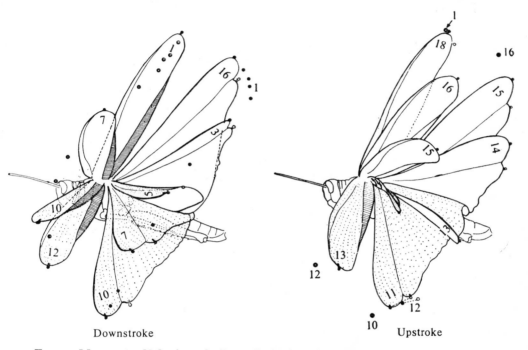

Downstroke   Upstroke

FIG. 4.2. **Movements of left wings of a Desert Locust in horizontal flight** at 17·5 wing-beats per second and air-speed of 3·5 m/s. Traces from numbered sequence of 18 equally spaced photographs, beginning from frame no. 1 with fore-wing at its top position; upper wing surfaces dotted and vannal flap of fore-wing line-shaded (from Weis-Fogh [26]).

action whereby aerodynamic lift is produced perpendicular to the tangent of the path of the wing through the air. The resulting force points upwards and provides thrust and a large vertical force during the down-stroke and a small vertical force and some negative thrust during the up-stroke. Indeed, this was the mechanism proposed by Otto Lilienthal [7] and illustrated in Fig. 4.4, redrawn from his famous monograph on flying storks which heralded the era of modern aeronautics (and his own death in a crash).

Martin Jensen then measured the lift and drag of detached locust wings in a special wind-tunnel in which there was a smooth velocity gradient from wing-base to tip similar to that observed during real flight, but of course under steady conditions of flow. On the assumptions that steady principles apply he could compute the fluctuating vertical and horizontal forces which the wings impart upon the body and compare them with the average forces measured in the aerodynamic balance during the cinephotography. Within a few per cent the two sets of results agree and it is therefore justified to conclude that fast forward flight of locusts depends on a succession of steady flow situations. He also took into account the mutual effect of the circulation round the two pairs of wings; it is small but significant. Another point was the effect of the new starting vortex which must be created between frames 18 and 1 in Fig. 4.2. Under the worst conceivable conditions the effect would amount to less than 10 per cent of the total, probably to 5 per cent reduction in lift, and this also applies to fast-flying hoverflies and mosquitoes.

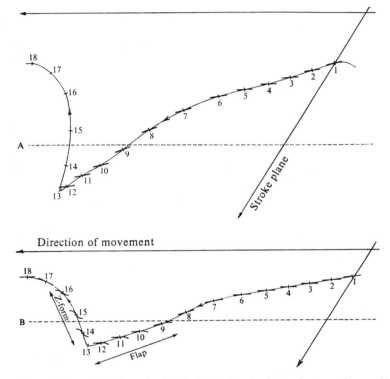

FIG. 4.3. **Movements of locust fore-wing in relation to the air**: shown for two wing profiles, near the tip (A) and half-way along the wing (B). The diagram is presented two-dimensionally, as on the unfolded surface of the elliptical cylinder upon which the mid-points of the two wing profiles appear to travel. (Redrawn from Jensen [5]). The numbers correspond to those in Fig. 4.2.

We are therefore able to proceed with the analysis of power requirements both in this and in other insects provided that the wings are in fact moved in such an orderly fashion. In his study of wing movements in the blowfly *Phormia regina*, presented on page 39 in a manner similar to that used for the locust data of Fig. 4.3, Nachtigall [9] has found a similar type of movement during the major part of the wing-stroke, but with an apparently disturbing difference at the top and at the bottom of the stroke. Here the rate of change in wing-twist is large and the angles of attack are very high. This may indicate important periods of unsteady flow (pp. 41, 68).

AERODYNAMIC POWER AND SPEED IN LOCUSTS

In the locust just mentioned [5] the power (energy per unit time) expended against aerodynamic forces, as determined directly from the wing movements, amounted to 0·86 W/N at the speed of 3·5 m/s in horizontal flight; this is the aerodynamic power in watts expended per unit (newton) of vertical force produced, or the *specific aerodynamic power P_a*. In another locust flying at 3·2 m/s and lifting 162 per cent of its body weight (equivalent to climbing flight), the specific power was 0·82 W/N [5], the average being 0·84 W/N at these speeds.

Comparing these results with the power calculated according to the procedure which Pennycuick [14] and Tucker [17] have applied to birds enables predictions to be made about migrating locusts and also shows the strength and shortcomings of the theory.

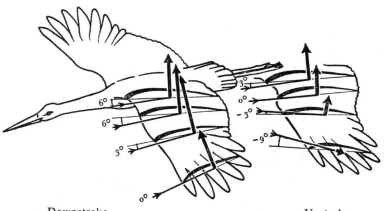

Downstroke   Upstroke

FIG. 4.4. **The main principle of fast forward flapping flight** in the stork (from Lilienthal [7]), and the Desert Locust (from Weis-Fogh [22]). Movement of air relative to each wing profile shown by finer arrow and angle of attack; direction and magnitude of resulting aerodynamic force shown by bolder arrow.

In insects, we need not incorporate the increased work done by the circulatory and respiratory systems during flight [17] because these components are negligible [24] compared with the work done against the air and against inertial forces due to the oscillating wing mass. As to the *inertial power* needed to oscillate the wings without doing any aerodynamic work, it is probably true that fast-flying birds make some use of the kinetic energy of the wings for aerodynamic work, as assumed implicitly in the theory [14, 17], although this does not apply to hovering hummingbirds [25]. In insects, the fast-flying locust also uses some kinetic energy in this way [5], but the major part is stored in an elastic system and paid back with high efficiency later during the wing-stroke [22, 25, 26]. We may therefore confine ourselves to the specific aerodynamic power P_a* which in a flying insect can be compared directly with the metabolic rate per unit of body weight. The ratio between the two is a measure of the mechanical efficiency of the flight system.

Since the power required equals drag times velocity, the method of Pennycuick [13, 14] is to split up the components of drag into three independent parts which are then summed. Let the air-speed of the animal be V. The drag D of the body and the appendages is the parasite drag and is proportional to V^2, similar to the lift in equation (ii). This means that the *parasite power* is DV^3. This is the uniformly increasing curve in Fig. 4.5. As already mentioned the flapping wings would exert some drag even if they were

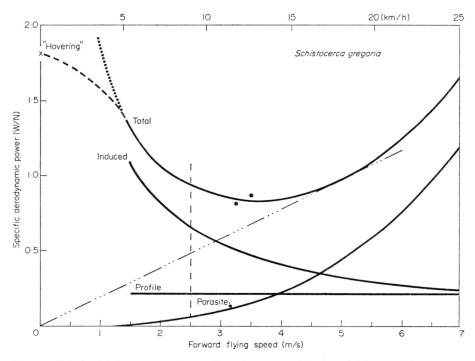

FIG. 4.5. **Relationship between specific aerodynamic power and horizontal flying speed in a Desert Locust.** The three curves for parasite power (needed to overcome the drag of the non-lifting points of the body), profile power (needed to overcome the drag of the wing system at zero lift) and induced power (needed to overcome the lift-dependent drag of the wing system; expended in accelerating air downwards to produce lift) are respectively estimated from measured parasite and profile drag [5] and calculated according to Pennycuick [14] with further correction factors for true flapping flight, as explained in the text. The tangent from the origin to the upper thick curve indicates the speed for minimum cost of transport. The two filled circles are values found independently by experiment (Jensen [5]).

not producing any lift. This profile drag is *not* a simple function of V and depends on the wing-stroke frequency and the stroke angle. However, detailed computations on the actual wing movements of a pigeon flying horizontally over a wide range of air-speeds showed that the *profile power* was roughly constant and independent of the speed [13] and this we shall assume to be the case also in locusts, indicated by the horizontal straight line in Fig. 4.5. In the present treatment, there is no need to introduce the modifications caused by Reynolds number [17] because we are dealing with a first-order approximation and the drag components were measured at an intermediate value of Reynolds number.

The third major power component is that expended in accelerating air downwards to produce lift, and represented by the tip vortices which are shed continuously and give rise to the induced drag already mentioned. During moderate to high forward speeds the *induced power* corresponding to this component has the form

$$\text{induced power} = 2G^2/(\pi \rho b^2 R' V), \tag{iii}$$

where G is the weight of the horizontally flying animal and equals the average vertical force, b is the wing span, and R' is a correction factor applied to the wing disc area, $\pi/4\, b^2$, through which air is accelerated downwards; note that the induced power required decreases as the speed increases. R' takes account for example of departures from the elliptical pressure distribution required for minimal induced drag with a fixed wing; in practice the correction factor usually varies from about 0·90 to 0·95 for conventional monoplanes. Tucker [17] reduced it to 0·7 in birds because of the width of the body relative to the total span. However, the authors on bird flight did not take into account (*a*) that the actual speed of the flapping wings is larger than V and so tends to reduce the induced power required; (*b*) that lift is produced mainly during the down-stroke in real birds and insects so that the effective G^2 is higher, which tends to increase R'; (*c*) that the down-stroke usually lasts longer than the up-stroke; and, most important, (*d*) that the aerodynamic lift is not directed vertically except when the wings happen to pass the horizontal position; the vortex sheet created by the tip vortices is therefore not horizontal. The last point is particularly important when the stroke angles are as large as in locusts and other insects. Without giving the details here, I have calculated the correction factor for a Desert Locust on the basis of all the known details of the wing-stroke [5, 21]. The result was $R' = 0·47$, or 0·5, and this is the value used for the steadily decreasing curve in Fig. 4.5. It should be noted that a real flying locust cannot carry its body weight at air-speeds lower than 2·5 m/s [21], indicated by the vertical line in Fig. 4.5, because its wings then begin to stall [27]. Also, the theoretical power needed for true hovering, if this were possible for the insect, would be too small because according to Pennycuick [14] it is estimated directly from the momentum theorem applied to an actuator disc of 100 per cent efficiency, and both the profile power and the induced power requirements are therefore not included. In the hummingbird this increases the figure by a factor of 2 and in *Drosophila* by 4 [25]. In Fig. 4.5, I have used the former value.

The summed power shown by the upper curve in Fig. 4.5 serves to bring the transport performance of fast-flying insects into a form comparable to that of birds and aeroplanes. The two values found by Martin Jensen from the direct analysis ([5] solid circles) are seen to be in remarkably good agreement with the indirect approach of Pennycuick provided that the correction factors are introduced as explained. This justifies the extrapolation to higher speeds. A number of interesting features then emerge. As pointed out by Pennycuick [14], the U-shaped curve is a consequence of the induced power being inversely proportional to the speed and the fact that the parasite power increases as V^3 so that it is negligible at small speeds. The minimum specific aerodynamic power for steady horizontal flight (i.e. for maximum endurance) in locusts (0·84 W/N) occurs exactly at the speed for steady prolonged flights in the laboratory, about 3·5 m/s [21]; this is also likely to be true of locusts flying freely in nature during long continuous migrations. In contrast, the speed for minimum cost of transport (i.e. for maximum range—in still air) is higher, about 4·5 to 5·5 m/s, as seen from the tangent drawn from the origin of the graph. This might explain why direct measurements of migrating *Schistocerca gregaria* in nature have resulted in values ranging mainly between 4 and 6 m/s [20] because most observations were obtained during intermittent flight when the locusts were disturbed[†] and probably tried to escape quickly (see discussion in [21] and [20]). However, from an aerodynamic and energetic point of view there is no reason why locusts should not exceed 7 m/s for *short* periods since they can double the metabolic rate relative to the rate needed for sustained flying at 3·5 m/s [23]. At the latter speed the metabolic rate is 7·7 W/N so that the overall metabolic efficiency during long migrations is 0·84/7·7 = 0·11.

If suitable methods for correcting the 'disc area' are introduced, there is therefore good reason to believe that Pennycuick's procedure can be applied to other migrating insects with advantage, for instance to armyworm moths (*Spodoptera* spp.) in eastern Africa. It is also of interest to compare the minimum aerodynamic power P_a^* in insects and birds. According to Tucker [17] it varies from 1·9 W/N in the budgerigar (mass 35 g) to 1·0 W/N in the laughing gull (322 g), 1·4 W/N being the average value calculated for non-passerine birds ranging in mass from 3 to 10,000 g. This is the same order of magnitude as found in locusts. It reflects what appears to be a general rule, namely that the aerodynamic power expenditure of flying animals tends to be directly proportional to the body weight throughout the animal kingdom. This also applies to hovering species (Table 4.3).

NORMAL HOVERING

A different approach based upon an analytical model of the moving wings has recently been proposed by Weis-Fogh [26] and has resulted in a reasonably complete theory of hovering. Essentially it consists of estimating the lift and drag forces at each instant, initially using steady-state lift/drag diagrams of known wings, and integrating the results over a complete wing-stroke. In this way the induced and the profile drag are incorporated simultaneously. One can then readily calculate the average lift coefficient $\bar{C}_L$ needed to hover over a stationary point and compare it with the maximum values to be expected from the type of wings and the Reynolds number involved. Only reliable flight data from freely flying animals were used. These showed that most animals hovered with the body inclined at a steep angle, the wings beating back and forth in a nearly horizontal plane, and the wing rotated at the start of each beat to enable the same leading edge to move forward at an angle of attack giving substantial lift. This is termed *normal hovering*, and is illustrated in Fig. 4.6.

COEFFICIENTS OF LIFT

All the animals listed in Table 4.2 perform normal hovering, as thus defined, with the exception of the true hoverflies (Syrphinae) and the dragonflies (Aeshnidae). The general result (assuming sinusoidal wing motion) is that the average coefficient of lift is sufficiently small for the flight to be explained on the basis of steady-state aerodynamics. This is true also of the heavily loaded Lamellicorn beetles hitherto considered exceptions [1, 2, 12] and of bees and wasps. However, there are some notable exceptions

† There was in fact no obvious indication of disturbance as the locusts passed overhead, above an observation point, but they may well have taken off (after temporary settling) only a few minutes before their air-speeds were recorded (photographically or by mirror)—Ed.

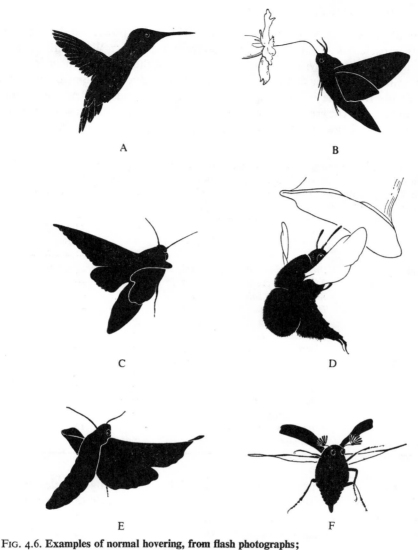

FIG. 4.6. **Examples of normal hovering, from flash photographs;**
A, hummingbird, *Archilochus colubris;*
B, hawk moth *Deilephila elpenor* (Sphingidae; from Nachtigall, 1969, *Naturwissenschaft &*
Medizin **6** (29), 9–20);
C, hawk moth *Manduca sexta* (Sphingidae);
D, bumble-bee *Bombus* about to land on a flower;
E, *Manduca sexta* during quick escape; note bending of wing axis;
F, cockchafer *Melolontha vulgaris* (Scarabaeidae) during vertical take-off.
(From Weis-Fogh [26])

for which the results are shown in brackets, namely butterflies (Rhopalocera) and the
tiny Chalcid wasp *Encarsia formosa*, as well as *Syrphus* and *Aeshna* species.

From these results it would be wrong to conclude that steady flow principles are the
only ones of importance during hovering but with some confidence we may assume that
they represent the major mechanism in most animals, even quite small ones. We shall
return to the exceptions shortly.

AERODYNAMIC POWER P_a*

The theory thus makes it possible realistically to calculate the power required for normal hovering in most cases (Table 4.3). It ranges from 1·4 to 4·7 W/N with an average of 2·5. Again, there is no obvious variation with size and the general conclusion must be that flying animals have adjusted their shape and performance to the mechanical power output available from the contracting wing muscles. This output depends mainly on muscle volume, i.e. on the cross-sectional area times the length of the muscles, because the contractile force F is determined by the area, and the shortening velocity U by the length.

TABLE 4.3. **The calculated aerodynamic power requirement per unit body weight lifted during normal hovering:** the specific power P_a*, together with the dynamic efficiency η, and the measured metabolic rate M, if known. Examples from Table 4.2. (From Weis-Fogh [26])

		P_a* (W/N)	η	M (W/N)
Hummingbird:	*Amazilia fimbriata*	2·6	0·51	24
Coleoptera:	*Melolontha vulgaris*	2·7	0·41	—
	Amphimallon solstitialis	2·3	0·47	—
	Heliocopris sp.	4·7	—	—
Lepidoptera:	*Sphinx ligustri*	1·8	0·49	—
	Manduca sexta	1·4	0·47	—
	Macroglossum stellatarum	1·7	0·33	—
Hymenoptera:	*Vespa crabro*	2·1	0·31	12
	Bombus terrestris	3·9	0·51	—
	Apis mellifica	2·1	0·30	35
Diptera:	*Tipula* sp.	2·1	0·39	—
	Aëdes aegypti	3·3	0·70	12
	Eristalis tenax	2·3	0·34	14
	Drosophila virilis	2·3	0·95	14

The power, $F \times U$, is then proportional to the weight of the wing muscles so that in similarly built animals the *specific power* is independent of size (although details may modify the relationship to some extent [14, 22]). Some animals may in fact not need as much power because of small wing-loadings, or because of gliding or soaring; they may then reduce the relative amount of muscle or fly in a more 'expensive' way in accordance with whatever selective advantage they may gain. It is in this light we should view the variations in shape and performance of flying animals.

DYNAMIC EFFICIENCY

A problem common to all hovering animals is that a major loss in energy must occur if the kinetic energy of the oscillating wings cannot be converted into stored elastic energy [25, 26]. How much this loss could amount to is expressed by the ratio η in Table 4.3 between the aerodynamic work and the aerodynamic + inertial work, as integrated over one complete wing-stroke. The results show that a hovering hummingbird, lacking an elastic system, must combust fuel at a much higher rate than needed for flight alone, whereas the potential loss in some insects like *Drosophila* is insignificant. However, the majority of insects have low efficiencies and could not fly at the relatively small metabolic rates shown in Table 4.3 *unless* they possessed an elastic storage system to decrease the loss and increase the true efficiency. The morphology of the insect pterothorax must be understood not only in terms of wing movements and muscle contractions but, equally important, in terms of the capacity to store elastic energy and release it effectively at high rates. This represents a large but difficult field of investigation for comparative morphologists. It has been suggested (Pennycuick, personal communication) that some birds may store useful energy when the primaries are bent towards the end of each half stroke but no such mechanism has so far been reported for insects or hummingbirds.

NOVEL AERODYNAMIC MECHANISMS

The four exceptions in Table 4.2 fall into two categories. The first consists of the butterfly *Pieris* and the small wasp *Encarsia* which appear to hover normally, with body nearly vertical and wings beating in a roughly horizontal plane, but at very high lift coefficients; and they 'clap' their wings together at one extreme wing position (*Encarsia*) or at both ends of the wing-stroke (*Pieris*) ([26] and unpublished). Such a clap may also occur in slowly flying *Drosophila* [18]. The other category consists of the true hoverflies (Syr-

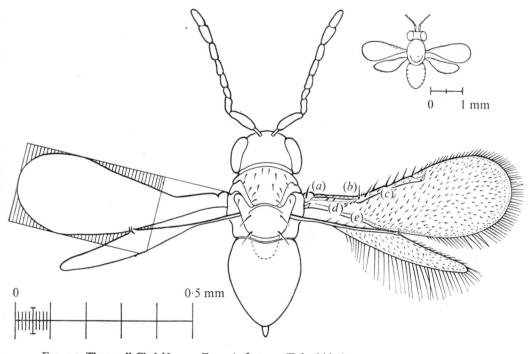

FIG. 4.7. **The small Chalcid wasp** *Encarsia formosa* (Eulophidae)
a, Tegula;
b–e, Elastic bending zone.
c, Marginal vein;
d, Submarginal vein.
Shaded rectangular area indicated in Fig. 4.9 and used in calculation of $\overline{C}_L$ (Table 4.2).
Inset: a larger related species, *Coccophagus spectabilis*, described by Compere, 1931, *Proc. U.S. natn. Mus.*, **78** art. 7, 1–132.
(From Weis-Fogh [26].)

phinae) and the large dragonflies (Aeshnidae) which hover with horizontal body and non-horizontal, obliquely beating wings. In addition, the stroke angles in the latter groups are so small that average steady-state lift coefficients would at the very least range between 2 and 3. All these forms must make use of unsteady-state aerodynamics. What is their nature?

The best understood case is the Chalcid wasp *Encarsia formosa* (Fig. 4.7) which is used for biological control of the greenhouse whitefly *Trialeurodes vaporariorum*.

FLIGHT OF A TINY WASP AND THE 'FLING' MECHANISM

By means of high-speed cinematography of freely flying *Encarsia* the following picture was obtained recently by Weis-Fogh [26]. Like many insects *Encarsia* can jump but the wing

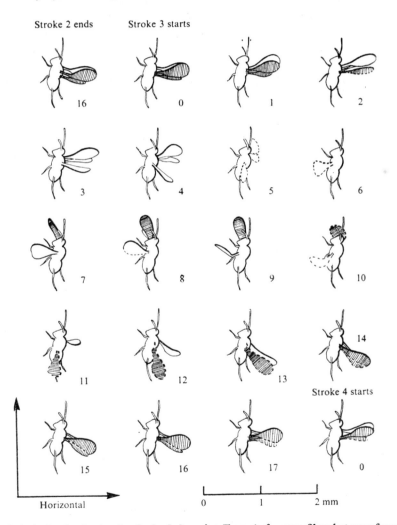

FIG. 4.8. **A single wing-beat cycle of a freely hovering** *Encarsia formosa*, filmed at 7150 frames per second, and successive frames numbered; leading (costal) edge of fore-wings shown by heavy line, and morphologically lower wing surface shaded. (From Weis-Fogh [26].)

movements seen in Fig. 4.8, at 403 wing-beats per second, relate to free unaided hovering with a slow climb; the long axis of the animal is nearly vertical but tilted by about 20° towards the camera. It is seen that the wings are moved essentially in a horizontal plane both during the morphological down-stroke (frames 0 to 7) and during the up-stroke (frames 7 to 14). This strongly indicates that the flight depends on a true lift mechanism.

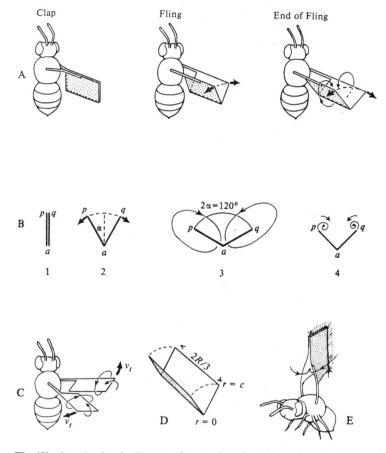

FIG. 4.9. **The 'fling' mechanism in** *Encarsia*, for creating circulation after the 'clap' and prior to the separation of the wings.

A, Start, middle and end of 'fling'.

B, Details of the flow, viewed parallel to long axis of wings.

C, Beginning of the horizontal, morphological down-stroke, with bound and tip vortices, after the wings have separated.

D, Geometry of the opening wings during the 'fling'.

E, Propagation of the cleft between the two wing pairs by 'delayed elasticity'. Air movements indicated by fine lines. (From Weis-Fogh [26].)

It is also seen that the sequence of movements contains three phases which are unusual in the sense that nobody seems to have noticed them or their significance before: (*a*) the *clap* preceeding the down-stroke where the two pairs of wings are brought together as a single vertical plate (frames 14 to 17); (*b*) a very rapid pronation where the wings are flung open in a manner reminiscent of the 'flinging open' of a book and *before* the two pairs of wings start the horizontal down-stroke; this I shall refer to simply as the *fling* (frame 1 at the beginning and frame 0 at the end); and (*c*) a very rapid supination before the up-stroke (between frames 7 and 9) which I shall call the *flip*, as in pancake-tossing. According to the direct analysis of the films these phases are always present in *Encarsia* during all kinds of flight. Moreover, the oscillations of the body in the vertical showed that the lift had equalled the body weight long before the wings in Fig. 4.8 reached maximum angular velocity, namely between frames 3 and 4, as if *circulation* had been built up prior to the wing-sweep itself and during the fling. This led to the idea illustrated in Fig. 4.9A. Towards the end of the clap, the wings are essentially at rest relative to the air.

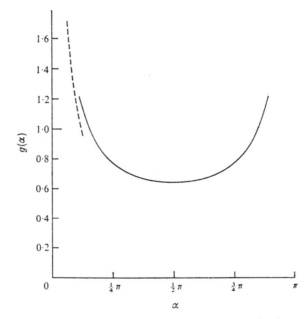

FIG. 4.10. Coefficient $g(\alpha)$—analogous to lift coefficient—in equation (iv) for the fling mechanism (after Lighthill [6]). α is the half-angle between the two wing chords; the wings from the two sides separate, and begin to swing 'downwards', at $\alpha \simeq \pi/3$, when $g(\alpha)$ is changing only very slowly. Broken line: approximate form for small α.

During the fling the air remains at rest along the 'hinge' represented by the hind margins of the hind-wings, but is sucked into the opening gap to fill the triangular void being created between the two upper wing surfaces, and this flow gives rise to two bound vortices of equal strength and opposite sign. When the wings break apart along the 'hinge' and start the horizontal down-stroke in Fig. 4.9C, they already have circulation and can produce lift in accordance with equation (i) and without any separate starting vortex, the two vortices both being bound vortices, so that there is no delay and no Wagner effect. Also, the creation of the two bound vortices of opposite sign and equal strength does not depend on viscous forces but can be understood in terms of irrotational flow applicable also to an ideal inviscid fluid. Moreover, the energy needed to start the vortices is higher in the beginning (between Fig. 4.9B—1 and 2) than later (Fig. 4.9B—3) and the major cost occurs before viscous forces create serious problems [26].

These results are surprising at first glance because they appear to violate established principles of aerodynamics but, in fact, they do not. They have been analyzed and amplified in a theoretical study by Lighthill [6] who was able to derive an exact expression for the circulation Γ round each wing. Combined with the general expression in equation (i) the unsteady lift caused by the fling takes the form

$$L/unit\ length = \rho v_t \Omega c^2 g(\alpha), \tag{iv}$$

where ρ is the mass density of air (as before), v_t is the flapping velocity indicated by the bolder arrows in Fig. 4.9C, Ω is the angular velocity with which the two sets of wings open during the fling (cf. Fig. 4.9B—1 and 2), c is the wing chord, and $g(\alpha)$ is a function analogous to the lift coefficient in steady flow and depending on the angle α through which each wing has rotated from the start of the fling (Fig. 4.9B—2), before they part at the start of the down-stroke. The function $g(\alpha)$ can be computed exactly (Fig. 4.10), and has the property that its value changes only very slowly once α has risen to the angle at which the separation of the wings is observed. This is significant not only because it indicates that the lift is insensitive to the exact angle at which the wings break apart, but also because it suggests that the pressure difference across the gap is negligible. At a small angle of 10° the coefficient is as high as 1·7, in conformity with the need for energy being greatest during the initial stages of the fling.

In the case of *Encarsia*, $\Gamma = 0·69\Omega c^2$ and this is equal to a circulation of 2·7 cm²/s. Using this value in equation (i), the flapping speed at which lift equals body weight is reached about frame 3 in Fig. 4.8, i.e. when *direct* film observations of the displacements of the body in the vertical showed that parity between lift and body weight was in fact achieved. Lighthill [6] also analyzed some aspects of the boundary layer (cf. Fig. 4.9B—4) and the three-dimensional flow; the reader should consult his paper on these important points.

The new mechanism of lift generation which I call the *fling mechanism* is, therefore, that during the rapid opening phase after a clap useful circulation is created round each wing profile independently of the usual Prandtl mechanism of aerofoil action and prior to the translation of the wing as a whole through the air. The clap-fling situation is particularly simple and amenable to theoretical (and experimental) analyses but it need only represent one extreme possibility out of many. As already mentioned, the clap appears to occur in *Drosophila* and I have recently observed it during vigorous hovering flight of *Pieris brassicae* both at the top and bottom of the horizontal stroke.

FLIGHT OF SYRPHINAE AND AESHNIDAE

In *Encarsia*, at the other extreme wing position, the flip phase (Fig. 4.8—frames 7–9) cannot be resolved for detailed analysis on the basis of existing material, but here again the results indicate that the building up of circulation prior to the succeeding wing movement through the air is also possible, though without a clap and fling. I have attempted to explain similarly the flight of hoverflies and dragonflies which hover without any clap and with the wings on the two sides far apart and independent of each other (26). The suggested *flip mechanism* is illustrated in Fig. 4.11. It rests partly on Nachtigall's [9] measurements (p. 39) of the twisting speeds during pronation [frames 50–56 in his Fig. 3.8] and supination [his frames 20–40] in *Phormia*, a fly of the same size and wing-stroke frequency as a large hoverfly; partly on the observation that the wings of Syrphinae and Aeshnidae are stiff and reinforced at their anterior leading edge and very soft and pliable at the trailing edge; and, finally, on rough calculations of the speeds with which a torsional deformation during pronation and supination can travel from the base of the wing to the tip in the stiff anterior region as compared with the soft posterior part. The main result is that the deformations cannot reach the trailing edge before they are completed at the anterior part ('delayed elasticity', see also Fig. 4.9E). The trailing edge therefore represents a stagnation line relative to the air so that the pronation in Fig. 4.11 (or a similar supination) would result in two opposite vortices, an anterior one bound to the wing and a posterior one which is free and situated mainly behind the hind margin. When the actual

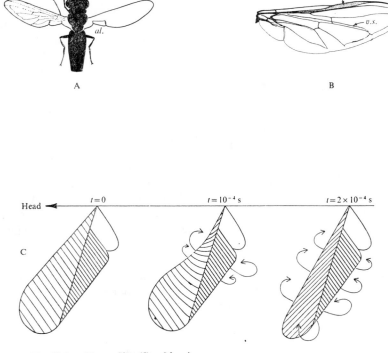

FIG. 4.11. **The flight of hoverflies (Syrphinae)**

A, *Platychirus peltatus;* alula (*al.*) can be bent upwards or downwards at right angles to the main wing surface, presumably by a pleuro-axillary muscle, and probably used for control and manœuvre.

B, A wing of *Syrphus balteatus* with stiff anterior part, including large pterostigma (*pt.*) and soft almost rubber-like posterior membrane (closely shaded in c), separated by vena spuria (*v.s.*), parallel to which the wing bends during pronation and supination.

C, Proposed 'flip' mechanism during pronation preceeding down-stroke, with suggested propagation of a twisting wave resulting in two opposite vortices, the trailing edge being a stagnation line; viewed obliquely from above. (From Weis-Fogh [26].)

translation starts during the down-stroke or the up-stroke, the bound vortex gives rise to lift from the very beginning of the stroke, although the resulting lift is somewhat reduced to begin with because of the Wagner effect which is unavoidable in this case. I suggest that this is the mechanism which enables Syrphinae and Aeshnidae to obtain lift coefficients at least as high as 2 or 3 during hovering flight. Further analysis showed that such a system could confer supreme manœuvrability, mainly by alterations of the wing-tip path [26], but much more direct evidence is needed before it is profitable to discuss this and other possible effects in detail.

GENERAL ASPECTS OF FLAPPING FLIGHT

There must be non-steady periods in any type of active flapping flight but the quantitative importance of non-steady aerodynamics is small during fast forward flight of birds, bats and most insects, and even when they hover. However, in very small insects and in some insect groups which have small wing-loadings (cf. Table 4.1), novel but definable non-steady-state principles based upon the creation of circulation, and thereby aerodynamic lift rather than drag, dominate the flow pattern. It is also significant that lift caused by the fling mechanism is independent of size and can operate at both small and high Reynolds numbers so that it could be used to good effect during vertical take-off and in emergency, and by birds. The wood pigeon (*Columba palumbus*), the rock dove (*Columba livia*) and its domestic descendants often start flight with one or two audible claps when disturbed, as already described in 19 B.C. by Virgil in the *Aeneid* (Book V, vv 213–217) and shown by Marey in 1890 [8] to be caused by the two fully stretched wings meeting dorsally in a real clap-fling. It should also be mentioned that the relatively small wing-loadings of terns and kestrels may make it possible for these birds to hover according to the flip principle. Only future studies can tell.

One point needs emphasis. As is the case with the revolving cylinder used to illustrate the Magnus effect, the creation of circulation of the air round the wing is the essential prerequisite for lift production and the actual shape of the wing is of less significance during the non-steady period. This may explain the otherwise abnormally high geometrical angles of attack (p. 39) found in *Phormia regina* towards each end of the stroke [9], and one should not use these observations to draw conclusions about forces based upon steady-state considerations alone. It may also explain some of the bizarre wing shapes seen particularly in many Neuroptera, Lepidoptera and Thysanoptera. Application of non-steady-state principles makes possible innumerable modifications of flight mechanisms and wing shapes and one must take this into account when discussing the evolution of winged insects.

A FIGURE OF MERIT

In order to estimate the relative importance of the unsteady fling mechanism as compared with the usual steady flow, I have recently derived a ratio between the circulation immediately after the fling and the circulation caused by the normal lift mechanism at the middle of a half-stroke. If we disregard the dying off of the unsteady flow in the course of the half-stroke, this figure of merit μ is unity when the two circulations are equal. It has the form

$$\mu = \frac{4\alpha c}{\pi \epsilon R} \times \frac{g(\alpha)}{\phi C_L}, \tag{v}$$

where 2α is the angle between the wing chords c just before the wings split apart in Fig. 4.9B—3 (usually about 2 rad), ϵ is the fraction of the stroke period occupied by pronation and supination (about 0·05 to 0·1), R is the wing length and ϕ the stroke angle. The interesting observation is that μ is independent of the absolute size and of the wing-stroke frequency. We also observe some interesting correlations between this expression and the morphology and flight of the animals now known to utilize unusual aerodynamics. For instance, animals with a small aspect ratio ($2R/c$) should benefit most from unsteady flow, butterflies being a good example. A small stroke angle also increases μ, in accordance with our observations on hoverflies and dragonflies. Finally, since $g(\alpha)$ does not vary much, it is seen that the animals with small wing-loadings and therefore small values of $\bar{C}_L$ should be able to take most advantage of the unsteady flow situations, exactly as observed and indicated by the double asterisks in Table 4.1.

For *Encarsia* and hovering butterflies, μ is larger than 5 and they are likely to depend almost entirely on unsteady flow. If there is a true effective clap in *Drosophila*, μ is about 3, but in these and similar cases one must remember that the fling circulation is likely to have been reduced due to viscosity at low Reynolds numbers when the wings have reached their full angular velocity. In hovering hummingbirds the figure would be 1·5 if they did use a clap (which they do not) but in a pigeon starting vertically from a perch with a real clap the figure of merit is 1 so that unsteady flow could contribute significantly during take-off. Virgil's observation that the rock dove starts with a few loud claps before it glides silently through the peaceful air probably indicates a real transition from unsteady to steady-state aerodynamics.

ACKNOWLEDGEMENTS

I am indebted to Professor Franz Blatt, the Carlsberg Foundation, Copenhagen, for checking the relevant passages in Virgil's *The Aeneid*, and to the Agricultural Research Council for support regarding high-speed cinematography.

DISCUSSION

Professor J.W.S.Pringle (Chairman): What were the Reynolds numbers for your observations on *Pieris brassicae?* Were there differences at different parts of the stroke?

Weis-Fogh: 400–500 in mid-stroke, with lower values at start and finish.

Mr P.Onyango-Odiyo: In locust flight, is there any coupling effect between the operation of the fore- and hind-wings?

Weis-Fogh: Yes, there are two types of coupling, the functional coupling between the fore-wings and the hind-wings based mainly on the firing sequences in the central nervous system, but in part also caused by the mechanisms of the pterothorax box. The effect of this is that the hind-wings move through a larger arc than the fore-wings and are out of phase with the fore-wings, actually leading the angular movement. The other coupling is of an aerodynamic nature and follows directly from the foregoing. Since each wing sets up a circulation corresponding to its instantaneous lift, the circulation around the fore-wings will interfere with the circulation around the hind-wings and *vice versa*. The effect is to decrease the circulations by some small amount, as calculated by Martin Jensen during level flight of the Desert Locust. The relative angular movements of the two pairs of wings are such that the effect of this interference is minimized.

(Remaining discussion communicated subsequently).

Professor W.Nachtigall: The role of unsteady-state aerodynamics in the flight of medium-sized insects like Calliphorid flies still remains to be worked out; I agree that it could be very important. For this reason discussion for example of the stalling behaviour of aerofoils at high angles of attack, on steady-state assumptions, may not now be very relevant. One can calculate the time-function of changing lift and thrust point-by-point in the frame-by-frame analysis of a high-speed film, assuming steady-state aerodynamics; but the only way to prove conclusively whether these calculations are valid or not would be the direct, phasic measurement of lift and thrust throughout the wing-stroke, by a sufficiently fast and sensitive two-component aerodynamic balance. This is not yet possible.

Sir James Lighthill: I should like to remark that the low Reynolds number involved in the flight of *Encarsia formosa* (about 30, in terms of wing chord and leading-edge velocity) may actually help it to achieve a high lift coefficient through the mechanism which Professor Weis-Fogh has discovered. This is because a high lift coefficient can be achieved with a sharp leading-edge only if the flow separating from the leading-edge can become re-attached to the upper surface of the wing to form a 'leading-edge separation bubble'. Fortunately, at low Reynolds numbers like 30 there is a powerful diffusion of momentum by viscous forces (which are then relatively strong), and this diffusion of momentum promotes such reattachment of the flow to the surface. By contrast, animals operating at Reynolds numbers of several thousand may need *turbulence*-promoting aids near the leading-edge to achieve the same effect; for example, the scales on butterfly wings mentioned by Professor Nachtigall as lift increasing devices may act in this way.

Weis-Fogh: Attention should be drawn to Professor Lighthill's detailed analysis of these phenomena [6], to which I have made only brief reference.

Nachtigall: Dragonflies sometimes make a specific noise when the wings touch each other at the end of the stroke.* Does Professor Weis-Fogh think this effect might have consequences for the circulation of the air-flow in the same way as he infers for the small Chalcid wasp? Or is this new establishment of circulation restricted to small Reynolds numbers?

Weis-Fogh: I do not know what the noise derives from and cannot answer this part of the question. However, direct observation shows that dragonflies do not use the fling mechanism, and calculations indicate that they do indeed use the flip mechanism. Neither of these mechanisms are confined to small Reynolds numbers but they are particularly useful for animals working at small wing-loadings.

Nachtigall: In a series of very fine papers Professor Weis-Fogh and his co-workers have included a detailed analysis of the kinematics of the fore-wing of *Schistocera gregaria*, in which they found small changes in wing camber such as the Z-profile shown during the up-stroke. It is apparently nearly impossible to include all these changes (small, but still visible to the naked eye in stroboscopic illumination) in a mathematical theory of the flapping aerofoil. A description of the kinematics of the fore-wing of *Locusta migratoria* has also been published by Zarnack (1972); but in his theoretical treatment he has assumed the fore-wing to be a stiff flat plate. Does Professor Weis-Fogh think this assumption represents a serious or a negligible departure from reality?

Weis-Fogh: It is clear from Martin Jensen's 1956 analysis (Fig. III, 13 in [5]) that the flap substantially increases the lift and the drag. Disregarding it would therefore represent a serious departure from reality.

---

* A possibly similar noise, showing the wing-beat frequency, is characteristic of locusts in flight soon after take-off, with the legs still extended, and is attributed to the wings striking the legs on the down-stroke (Haskell, P.T. 1957, *J. Insect Physiol.*, **1** : 52–75).—Ed.

REFERENCES

[1]   BENNETT, L. (1966). Insect aerodynamics: vertical sustaining force in near-hovering flight. *Science, N.Y.*, **152** : 1263–1266.

[2]   BENNETT, L. (1970). Insect flight: lift and rate of change of incidence. *Science, N.Y.*, **167** : 177–179.

[3]   HERTEL, H. (1966). *Structure, form, movement*. Reinholt, New York.

[4]   HORRIDGE, G.A. (1956). The flight of very small insects. *Nature, Lond.*, **178** : 1334–1335.

[5]   JENSEN, M. (1956). Biology and physics of locust flight. III. The aerodynamics of locust flight. *Phil. Trans. R. Soc.* (B), **239** : 511–552.

[6]   LIGHTHILL, M.J. (1973). On the Weis-Fogh mechanism of lift generation. *J. Fluid Mech.*, **60** : 1–17.

[7]   LILIENTHAL, O. (1889). *Der Vogelflug als Grundlage der Fliegekunst*. R. Oldenbourg, Berlin.

[8]   MAREY, E.-J. (1890). *Le vol des oiseaux*. G. Masson, Paris.

[9]   NACHTIGALL, W. (1966). Die Kinematik der Schlagflügelbewegungen von Dipteren. Methodische und analytische Grundlagen zur Biophysik des Insektenflugs. *Z. vergl. Physiol.*, **52** : 155–211.

[10] NORBERG, R.Å. (1972). Evolution of flight of insects. *Zoologica Scripta*, **1** : 247–250.

[11] NORBERG, R.Å. (1972). Flight characteristics of two plume moths, *Alucita pentadactyla* L. and *Orneodes hexadactyla* L. (Microlepidoptera). *Zoologica Scripta*, **1** : 241–246.

[12] OSBORNE, M.F.M. (1951). Aerodynamics of flapping flight with application to insects. *J. exp. Biol.*, **28** : 221–245.

[13] PENNYCUICK, C.J. (1968). Power requirements for horizontal flight in the pigeon *Columba livia*. *J. exp. Biol.*, **49** : 527–555.

[14] PENNYCUICK, C.J. (1969). The mechanism of bird migration. *Ibis*, **111** : 525–556.

[15] PENNYCUICK, C.J. (1972). *Animal Flight*. E. Arnold, London: 68 pp.

[16] THOM, A. & SWART, P. (1940). The forces on an aerofoil at very low speeds. *Jl. R. aeronaut. Soc.*, **44** : 761–770.

[17] TUCKER, V.A. (1973). Bird metabolism during flight: evaluation of a theory. *J. exp. Biol.*, **58** : 689–709.

[18] VOGEL, S. (1965). Studies on the flight performance and aerodynamics of *Drosophila*. Thesis, Harvard University.

[19] VOGEL, S. (1967). Flight in *Drosophila*. III. Aerodynamic characteristics of fly wings and wing models. *J. exp. Biol.*, **46** : 431–443.

[20] WALOFF, Z. (1972). Observations on the airspeeds of freely flying locusts. *Anim. Behav.*, **20** : 367–372.

[21] WEIS-FOGH, T. (1956). Biology and physics of locust flight. II. Flight performance of the Desert Locust (*Schistocerca gregaria*). *Phil. Trans. R. Soc.* (B), **239** : 459–510.

[22] WEIS-FOGH, T. (1961). Power in flapping flight. In Ramsay, J.A. & Wigglesworth, V.B. (eds.) *The cell and the organism*, 283–300. Cambridge University Press.

[23] WEIS-FOGH, T. (1964). Biology and physics of locust flight. VIII. Lift and metabolic rate of flying locusts. *J. exp. Biol.*, **41** : 257–271.

[24] WEIS-FOGH, T. (1967). Respiration and tracheal ventilation in locusts and other flying insects. *J. exp. Biol.*, **47** : 561–587.

[25] WEIS-FOGH, T. (1972). Energetics of hovering flight in hummingbirds and in *Drosophila*. *J. exp. Biol.*, **56** : 79–104.

[26] WEIS-FOGH, T. (1973). Quick estimates of flight fitness in hovering animals, including novel mechanisms for lift production. *J. exp. Biol.*, **59** : 169–230.

[27] WEIS-FOGH, T. & JENSEN, M. (1956). Biology and physics of locust flight. I. Basic principles in insect flight. A critical review. *Phil. Trans. R. Soc.* (B), **239** : 415–458.

II · The flying insect and its environment

5 · Flight behaviour and features of the atmospheric environment

R.C.RAINEY

Centre for Overseas Pest Research
(*formerly Anti-locust Research Centre*)
London

SUMMARY

A number of complex manifestations of insect flight behaviour in nature illustrate how consideration of the physical aspects of flight and meteorological aspects of the atmospheric environment can make possible the recognition of a dominating role of behaviour in these particular phenomena. These phenomena include the maintenance of cohesion of flying locust swarms travelling for hundreds of kilometres over periods of weeks in steady winds (probably the most clearcut of all manifestations of gregarious behaviour), the persistence of static and highly localized populations of other species that show considerable flight activity, and the sustained climb for some hundreds of metres after take-off by some night-flying grasshoppers. On the other hand, accumulating evidence of a dominant role of wind-systems in hourly, day-to-day and seasonal changes in distribution of insects of an increasing number of species is shown to have a series of biological and ecological implications. Attention is directed particularly to the concentration of airborne insects by wind-convergence to an extent which at times probably brings them to within range of mutual perception, and which is probably often significant in population dynamics and the incidence of crop-damage, and may at times provide potential targets for effective air-to-air spraying.

INTRODUCTION

Previous speakers have dealt with flight as studied in the laboratory; we now move into the outside world, and from controlled experiments to observations of flying insects in nature. Because time is limited, I have concentrated on aspects of behaviour, individual and collective, in which flight is central rather than peripheral, and have given less attention to aspects covered in the Society's previous Symposia [20] or recently reviewed elsewhere [22, 47, 48].

Without flight, the displacements of insects over the earth, in periods of hours and days, would indeed be very limited. But flight alone would not produce the effects that we observe—such as the geographical patterns of locust migration (e.g. in Fig. 6.2), or the range of swarm structures (Figs. 5.4, 5.23, 5.24), illustrated by the film sequences of flying locusts secured by John Sayer and his colleagues, and shown at this Symposium. Rather, it is movements of the atmosphere and features in its structure which dominate these visual effects in flying swarms, and atmospheric features on a larger scale which largely determine some of the most dramatic geographical and ecologically significant redistributions that we observe.

Such features of the atmospheric environment are illustrated by the now-familiar satellite pictures, in which the clouds show areas in which the air is rising and being replaced from below by air provided by converging winds at lower levels. The beautiful cloud patterns so formed reflect the circulation of the atmosphere, driven by the incoming energy from the sun and operating to transfer heat from the tropics towards the poles. This necessarily involves a return flow of cooler air towards the equator, organized in a manner which is basically different in higher and lower latitudes.

In higher latitudes, including our own, there is in general terms an alternation between successive surges of warmer air moving towards the pole and of cooler air moving equator-wards (as in Fig. 5.21), with their respective advances representing the characteristic warm and cold fronts of our familiar depressions. It is with these surges of warmer air that many insects, most of them enumerated in C.G.Johnson's massive review [22] and some subsequently [48], have been shown to move polewards in the spring, in North America, the Middle East and the Far East as well as in Europe, with their displacements sometimes rectified, in the terms of Kennedy's electrical analogy of long ago [25], by reduced flight activity in the cooler air following the cold front.

In lower, tropical latitudes on the other hand most of the poleward flow of warmed air takes place at high altitudes, with the return flow of effectively cooler air at lower levels providing the trade-wind and monsoon currents, characteristically so much more constant from day to day than the winds in our latitudes. The trade-winds and monsoons originating from the northern and southern hemispheres meet all around the globe at the almost permanent Inter-Tropical Convergence Zone (ITCZ). The seasonal movements of the ITCZ, following the sun, dominate not only the seasonal distribution of rains in the tropics but also the seasonal migrations and breeding seasons of locusts [43, 73] and of a number of other insects [5, 41 etc.].

SOME DOMINATING EFFECTS OF BEHAVIOUR

However, before considering some more recent findings on dominating effects of winds and weather-systems on airborne insects, it is instructive to examine some of the other phenomena presented by insect flight, in which behaviour can be dominant, and to see how in a number of such cases consideration of physical and meteorological factors has made possible a much fuller appreciation of these effects of behaviour.

COHESION OF TRAVELLING LOCUST SWARMS

While the major displacements (migrations) of Desert Locust swarms, as in Fig. 6.2, are now known to be very largely or wholly determined by the winds (see e.g. Figs. 5.1, 2, 3, 6, 7 and 15 and p. 82), what is very unlikely to be determined by wind is the continued cohesion of swarms in steady winds, as in Fig. 5.1.

The first of these swarms was found to travel at a ground-speed 40 per cent of the speed of the wind in which it was flying, and to maintain an area of 5 km² during the three days it remained under observation (roosting as usual each night), while the third swarm travelled at a speed 100 per cent* of that of its wind and maintained an area of

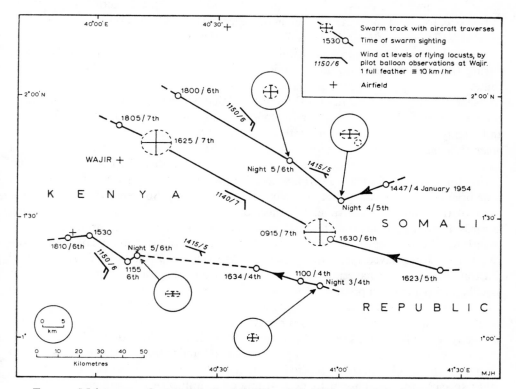

FIG. 5.1. **Maintenance of swarm-cohesion in steady winds.** While direction of swarm displacement is largely or wholly determined by wind, continued cohesion (at least of slower-moving swarms in steady winds, p. 109) appears to be maintained by orientation-behaviour as in Fig. 5.3.

Note how each of these Desert Locust swarms maintained its identity and even its plan-area (whether the swarm was settled or in flight [39] in a manner not yet understood), while their ground-speeds ranged from 40 to 100 per cent of the speed of the wind in which each was flying [43].

Systematic, progressive displacement, as shown by these swarm-tracks, is characteristically associated with steady winds, in contrast with the complex tracks followed in changing winds, as in Figs. 5.7 and 5.15.

150 km². We know of no type of wind-convergence system (p. 83) sufficiently persistent and with the appropriate range of mobility relative to the general wind-flow to account for such a range of performance [45, 74]. Morphometric differences had been established between neighbouring swarms similarly passing through this area at the same season of an earlier year [41], emphasizing their identity as discrete mobile populations (Fig. 5.24).

* For locust swarms (unlike the isolated locusts—and other individual flying insects—observed by radar), wind-speed appears to have represented an upper limit for ground-speed, in winds ranging from 6 to 15 km/h (Fig. 8.44); effectively randomly-orientated elements of the flying locusts have accordingly been postulated as pace-makers determining both speed and direction of displacement for such swarms [39, 40]. This is a point apparently not fully appreciated in recent work [74].

Under such conditions the potentially disruptive effects of the daytime turbulence associated with the powerful convection currents encountered (p. 80) are so strong that a cloud of passive airborne material like smoke, covering say an initial 50 km², could be expected to be dispersed in the horizontal at a rate which would double its diameter after about six hours and increase it by ten times after 70 hours [39]. In actual fact a swarm of this size, followed for some 370 km, across Kenya (from Garissa on the Tana river to the Rift Valley west of Nairobi), was found by a series of timed aircraft traverses to vary in area only by about 10 per cent (within the limits of reproducibility of such traverses) during the 70 hours the swarm spent in flight in the course of ten days along this route.

Nearly thirty years ago, Gunn [17] commented that 'when large numbers of individuals come together and remain together for long periods while moving through a complex and highly variable series of natural situations *with no discoverable common factor other than the crowd itself* [my italics—R.C.R.], one is left with the strong impression that the animals are bound together by reactions to one another'. In 1952 we found that during the passage of such a swarm, successive groups of flying locusts (like Fig. 5.2) were orientated in strikingly diverse directions, as in Fig. 5.3. So were locusts in different parts of the swarm at one time. We therefore postulated some reaction(s) deflecting inwards locusts reaching the perimeter of the swarm, to account for its continued cohesion. The photographic techniques of observation pioneered by Gunn [17], developed by John Sayer [58, 60], meticulously applied by Zena Waloff [71, 74], and illustrated by Fig. 5.3 have provided repeated and unambiguous evidence of just such an effect, with a marked predominance of inward orientations into each travelling swarm among the lower fliers at its leading, lateral and trailing edges [74].

Such a system, in which gregarious behaviour unmistakably operates against measurable physical effects, illustrates the manner in which consideration of atmospheric processes can make possible a fuller appreciation and understanding of the role of insect behaviour in the phenomenon studied. Thus flying insects of an increasing number of other species are now known to occur (under the influence of wind-convergence) at densities temporarily comparable with those of locust swarms (Table 5.3, etc.), so that it is the continued cohesion of flying swarms, rather than their initial formation, which now appears characteristic of flying locusts. Conversely, in interpreting an early radar sighting [37], one's original failure to appreciate the relative slowness of the turbulent dispersal to be expected even of passive airborne material, in a cloud as large as a big swarm, led to an initial over-emphasis of the possible role of auditory clues, relative to visual ones, in the maintenance of swarm-cohesion.

FLIGHT IN PERSISTENT STATIC POPULATIONS

Flight does not necessarily mean population displacement; even a flying locust swarm can sometimes remain above and roost on the same site for several successive days and nights; sometimes, as in Eritrea in June 1953, such a flying swarm can be held effectively stationary in a zone of persistent wind-convergence such as the ITCZ [43]. There are, however, other records of even more conspicuously immobile locust populations, showing considerable flight activity, for which such wind effects have clearly not been involved. In particular, localized populations of Red Locusts (*Nomadacris septemfasciata* Serville), studied in the long-grass habitats of the outbreak areas of this species in Tanzania and Zambia, repeatedly flew on disturbance by birds, game-animals and man, and

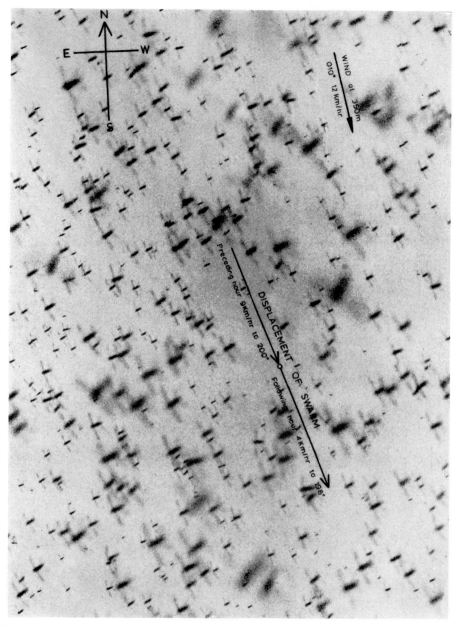

FIG. 5.2. **Group of uniformly-orientated flying locusts.** Viewed vertically upwards from ground, near Mtito Andei, Kenya, at 1625 on 10th February, 1955, during the passage of a swarm of which the track is shown in Fig. 5.3; the topmost locusts were recorded, by aircraft observation, at 720 m above the ground [43].

Despite the impression of purposefulness given by such a uniformly-orientated group, all locust swarms so far studied in this respect have been found to consist at any one time of large numbers of such groups, with the widest possible diversity of orientation between groups (see e.g. Fig. 5.3), resulting in a directly down-wind displacement of the swarm as a whole.

High-flying insects of a number of other species have been shown by radar (p. 162) to exhibit a uniformity of orientation comparable with that shown within this group.

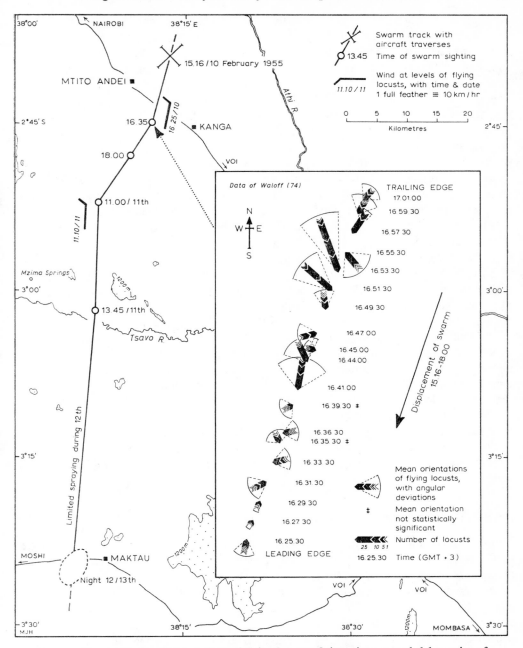

FIG. 5.3. **Swarm-track and orientations of flying locusts.** Orientations recorded by series of photographs similar to Fig. 5.2 taken at an adjacent fixed point at about 2-minute intervals throughout the passage of the swarm [74], and approximately equivalent to an instantaneous (Lagrangian) section through the lower levels of the swarm.

Note into-swarm orientations in both front and rear parts of swarm, considered to be of major importance in the maintenance of swarm-cohesion in steady winds (though see p. 109). Note again similarity of plan-area found for swarm in flight, on afternoon of 10th February, and when settled, on early morning of 13th; total locust numbers in this swarm estimated as 10^9, from densities of dead locusts found after drench-spraying runs with quick-acting insecticide (20 per cent DNC) made across swarm while completely settled, at Maktau [31, 38].

occasionally flew apparently spontaneously with a regular daily schedule of flight between tall roosts and feeding grounds [7, 9]. Yet some of these populations were observed to remain for periods of months within a few kilometres of the same position, despite winds so constant that even a few minutes of randomly orientated flight daily would have cleared the area completely in the time [55], and total numbers of locusts in such populations in a complete outbreak area of a few hundred square kilometres were likewise found effectively constant over similar periods [62]. The actual flying must necessarily have been orientated predominantly up-wind, and direct observation suggested visual perception of, and attraction to particular local features such as the conspicuous patches of tall grass among shorter grass characteristic of these habitats.

In such circumstances flight would also appear to have been confined to a biological 'boundary layer', as envisaged by L.R.Taylor [68], within which the air-speed of the insect exceeds the speed of the wind in which it flies and where each insect is accordingly able (with appropriate orientation behaviour) to remain within a limited ambit, perhaps of the order of hectares or square kilometres in extent. Such ambits were first demonstrated by C.H.N.Jackson for the tsetse fly *Glossina morsitans* Westw. [21]. It is suggested that the flight behaviour thus inferred ('keeping station' on landmarks, by flying low and mainly up-wind) may also account for the persistence of static localized populations of some other insects such as *Maniola jurtina* L. [12]. It is further suggested that flight behaviour of this kind, leading at times to the permanent occupation of localized ecological niches by individual flying insects and often requiring hovering flight (aerodynamically more exacting than fast forward flight, as Weis-Fogh points out—p. 49), may be more specialized and therefore perhaps a later evolutionary development than long-range seasonal displacements of wind-borne populations, now known to occur regularly and to be adaptively effective even for the spores of some fungi [47].

Best documented of all persistent static populations of actively flying insects is that of the beehive; and consideration not only of the strength but also of the turbulent structure of the winds encountered by foraging bees may perhaps assist further in the recognition and appreciation of the magnitude of their navigational achievements, with which Professor Lindauer deals (pp. 197–214).

HEIGHTS OF NOCTURNAL AND DIURNAL FLIGHT

Towards the end of the day, the locusts in swarms like those recorded in Figs 5.1 to 5.4 regularly flew lower and lower, as in Fig. 5.24, and by sunset all flight could be within a few metres of the ground, in striking contrast to flight up to heights of a kilometre or more regularly seen earlier in the day. Forty years ago Regnier in Morocco directed attention to probable effects on flying locusts of thermal convection currents, rising from the ground surface heated by the sun [56]; and twenty years ago it was found that the topmost locusts in the highest-flying swarms studied were at the upper limit of convection currents from the ground, with the decreasing height of flight in the evening associated with the weakening and final disappearance of such currents [39]. This upper limit of convection currents is determined by the upper air temperatures at the time, and in particular by the gradients of air temperature in the vertical. This is because air which has been warmed by contact with the ground, and is rising as a thermal, will continue to rise, despite its cooling by expansion, so long as it is warmer (and hence more buoyant) than the surrounding air through which it rises. This means that the upper limit of thermal up-currents can be readily estimated when surface and upper-air temperatures are available, with an accuracy, illustrated by gliding experience (Fig. 5.4), often better than 100 m [34].

Figure 5.4 strongly suggests that the ascent of these locusts was limited by the extent of the thermals, in just the same way that the climb of the sailplane was, and still higher-flying locusts, reported at some 2000 m above the ground over Somaliland and Uganda in the 1940's by early BOAC aircraft at the lower cruising levels of those days, were likewise recorded with vertical distributions of air temperature implying thermals rising from the ground over which the locusts were flying [39].

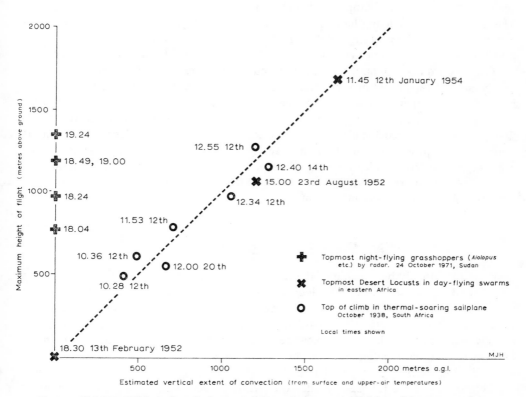

FIG. 5.4. **Heights of flight and vertical extent of thermal up-currents.** Heights of the topmost day-flying locusts appear to have been determined by the upper limit of thermal up-currents from the ground [39], exactly as the climb of the thermal-soaring sailplane undoubtedly was [34]; on the other hand, the heights attained by the night-flying grasshoppers (p. 178) were the result of active flapping flight in the complete absence of such up-currents [48].

As this appeared to be a satisfactorily complete story, it was mildly disturbing in 1968 to hear of Glen Schaefer's Niger radar observations of high-flying Desert Locusts by night (Fig. 8.45), when there is no ground heating to provide convection currents. This sense of uneasiness was accentuated in 1971 when, with guidance from Glen and his radar in the Sudan Gezira we caught night-flying grasshoppers, mainly *Aiolopus simulatrix* (Walk) up to 1200 m above the ground, by aircraft-trapping. Figure 5.4 includes an example of the steady rate of climb shown by the height of the topmost grasshoppers seen on the radar, here a rate of ascent of some 0.3 m/s up to a height of more than 1100 m. This must moreover be a minimum estimate of the rate of climb of the individual insects, and this was through air with a temperature distribution (an intensifying ground inversion) which inhibited completely the development of any thermal up-currents such as carried up the day-flying locusts. This air through which the grasshoppers climbed was in fact very smooth, in striking contrast to the turbulence encountered by the aircraft during day flights.

There was thus a radical difference in flight behaviour between the locusts in daytime, many in gliding flight, carried up by powerful up-currents commonly exceeding twice the sinking speed of a gliding locust, and the grasshoppers at night, climbing without any such assistance by active flapping flight shown by the modulation of their radar echoes. The flight of the day-flying locusts in Fig. 5.4 can thus be regarded as having more in common with the flight of the sailplane in the thermals than with that of the night-flying grasshoppers—reminiscent perhaps of Weis-Fogh's evidence (p. 69) that in other ways the fast forward flight of larger insects like locusts has more in common with the flight of aeroplanes than with that of very small insects like his tiny wasp (particularly in hovering flight), with their non-steady-state aerodynamics and their exploitation of novel lift principles. Again, this radical difference between the day-flying locusts and the night-flying grasshoppers became clear only when account was taken of relevant atmospheric structures and processes. An earlier finding for which study of the physical aspects of night flight assisted significantly in the interpretation of behaviour was the dorsal light reaction of the Desert Locust, first inferred from field observations [50] and later demonstrated by laboratory experimentation [15].

DOWN-WIND DISPLACEMENT AND ITS EFFECTS

Mention has already been made of the down-wind direction of displacement which has been found characteristic of flying locust swarms, and for which some of the evidence and implications may now be familiar. Recent evidence on wind-systems in relation to the distribution of insects other than locusts is extending this story to suggest implications which are in my view of sufficient general interest to justify presenting some initial findings as material for discussion at this Symposium.

Beginning by briefly recapitulating the part of the locust story which has already reached the textbooks [10, 22]: detailed observations such as those of Figs. 5.1, 3 and 7 on hour-to-hour movements of individual Desert Locust swarms and on the winds in which each was flying have demonstrated, for a total of 42 different occasions, a root-mean-square difference of only 15° between the direction of swarm displacement and that of the corresponding wind—a degree of agreement as close as could be expected from the known observational errors [43]. However, while our swarm tracks covered a wide range of wind-speeds, air temperatures and terrain, they related almost exclusively to sexually immature swarms and mainly to fair weather. Furthermore, the detailed biogeographical case-studies which have proved particularly rewarding in elucidating the relationships between locust population movements and weather systems are likewise vulnerable to a criticism of being perhaps inadequately representative as a basis for generalization. This point has been met by a comprehensive investigation of all the available data on both locusts and weather for a complete year of widespread and heavy infestation and for the entire invasion area. From this investigation, made possible both technically and financially by the World Meteorological Organization, it was concluded [51] that the movements and distribution of Desert Locusts, on scales from a hundred

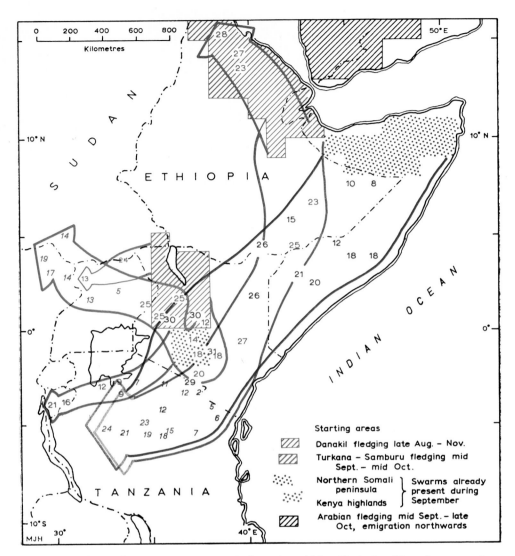

FIG. 5.5. **Desert Locust swarm movements in eastern Africa: September–November 1954.** A detail from the analysis summarized in Fig. 6.2, illustrating the identification of a series of separate population movements which was made possible by utilizing the known relationships between locusts, winds and weather to integrate and interpret the whole of the information available for this area and period on locusts (of all stages) and relevant meteorological factors. From this analysis each of the hundreds of swarm reports recorded can be assigned to one most probable source-area, up to 1700 km away and six weeks earlier, out of the four distinct source-areas which were involved in this situation. Numbers show date of first report in each area indicated, of swarm from starting-area of same colour; during Sept./Oct. from Turkana, during Oct. from Kenya highlands, and during Oct./Nov. from Danakil, northern Somali peninsula and Samburu. Such an analysis of current data—analogous to the synoptic analysis of the meteorologist—represents the first step in the production of a successful forecast (p. 116).

[*facing page* 83]

to thousands of kilometres, are not merely correlated with but are to a very large extent determined by the low-level wind-fields concerned, within the known range of limiting conditions for flight. The application of these findings has made possible the elucidation of series of population movements of a complexity illustrated by Fig. 5.5.

RAIN AND SURVIVAL VALUE

Down-wind displacement means on balance movements towards and with zones of low-level wind-convergence, defined as areas across whose boundaries surface winds show a net inflow, as indicated for example in Fig. 5.6 and demonstrated in Fig. 5.10. Horizontal wind-convergence, resulting in ascent of air, is a necessary (though not sufficient) condition for the production of rain, and so provides a mechanism for the association observed between the geographical pattern of Desert Locust migration and the distribution, in time and space, of the rains which are essential for successful locust breeding. Figure 5.6 shows an early example of the accumulation of swarms in the Inter-Tropical Convergence Zone already mentioned, which is associated with the seasonal 'short', 'long' and 'monsoon' rains of the tropics.

Flight has thus enabled insects to utilize kinetic energy from the atmospheric circulation for locating and exploiting the extensive but temporary energy sources provided by the ephemeral and seasonal vegetation of arid and semi-arid regions [48], in a manner which has been suggested as possibly relevant to the evolutionary origin of flight itself [36, 44, 76], and to which further consideration is given on pp. 262–271.

Again, on the same grounds, current weather data, particularly on winds and rain, have been used since 1950 for interpreting the current Desert Locust situation and for helping to forecast its development. Elizabeth Betts takes up this story and its extension (on the initiative of Eric Pearson and on the basis of the work of Eric Brown) to forecasting the incidence of another major migrant pest, the African armyworm, *Spodoptera exempta* Wlk. This species is similarly dependent on the vegetation produced by highly seasonal rains, and can, as she has shown [5], arrive in East Africa like the Desert Locust with the seasonal southward movement of the ITCZ.

NEW EVIDENCE ON AIRBORNE INSECTS IN ZONES OF WIND-CONVERGENCE

A further implication of down-wind population movement, and the one to which I particularly wish to draw attention, is that of concentration by convergent winds; concentration is to be expected if the incoming airborne insects are constrained against unlimited ascent, e.g. by effects of air temperature on flight activity. An early suggestion of such an effect was in Eritrea in 1950, when in late November scattered locusts were seen in dry conditions at a series of points extending for 93 km along the coastal plain, while on a second reconnaissance over the same route a fortnight later locusts were found, at higher densities, only within the 13 km stretch to which rains (totalling some 140 mm at Massawa, and providing evidence of vigorous wind-convergence) had been restricted in the meantime [36].

Early biogeographical analyses, of which Fig. 5.6 is an example, suggested that zones of wind-convergence were good places to search for swarms, and this was confirmed by the patterns of the tracks of individual swarms followed by aircraft in the vicinity of such zones (Fig. 5.7). John Sayer further noted [59], in a subsequent year in this area of the

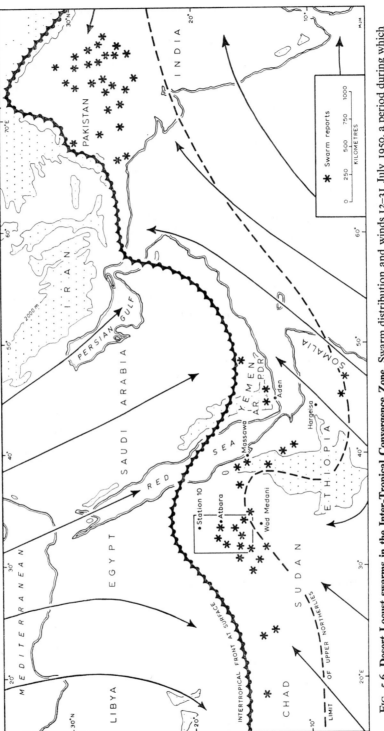

Fig. 5.6. **Desert Locust swarms in the Inter-Tropical Convergence Zone.** Swarm distribution and winds 12–31 July 1950, a period during which the ITCZ remained effectively stationary [36].

northern Somali Republic, that temporarily scattered locusts appeared to be reconcentrated into swarms by convergence at the discontinuity between opposing winds regularly found here at this time of year. The winds in a convergence zone can at times result in swarm-tracks forming closed loops, as in Fig. 5.15. Three years later, swarm movements in 'a vast circuit forming a series of loops' were similarly and independently established by air reconnaissance in west Africa, where it was concluded that . . . 'during the whole period of their stay in Senegal, the insects were held at the confluence of the winds . . . winds from eastern and western sectors predominated alternately, giving place progressively to the trade-winds. This alternation explains the curious loops effected by the swarms in the course of their movements' [32].

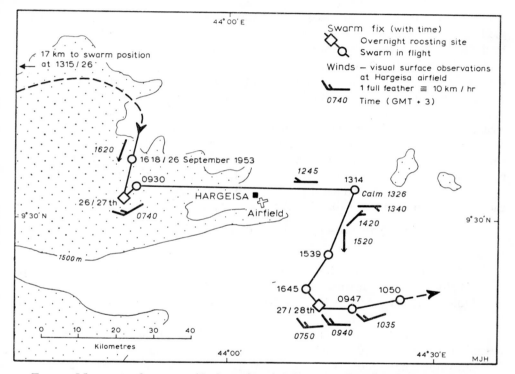

FIG. 5.7. **Movements of a swarm with alternating winds in the Inter-Tropical Convergence Zone.** Data of H.J.Sayer [59] and J.E.Allen

Since 1952 upper-wind observations, by pilot-balloon, had been found of value in directing air reconnaissance for swarms [42, 43, 53], but the logical extension of this approach, to the use of wind-finding equipment in reconnaissance aircraft, had proved impracticable at the time because of the difficulties of wind-finding by the standard visual drift-meter of those days in low flight under turbulent conditions. The advent of Doppler wind-finding radar for civil aircraft in the early 1960's was immediately recognized, by John Sayer, then Senior Scientist of the Desert Locust Control Organization for Eastern Africa, and his Director, Vernon Joyce, as a potential solution to this problem, and as offering the further facility of locating and searching zones of wind-convergence for possible concentrations of previously solitary-living locusts. The Doppler equipment senses the direction and speed of movement of the aircraft relative to the ground; the standard flying instruments of the aircraft sense its direction and speed of movement relative to the air; and the vectorial difference between the two is the wind.

Inter-Tropical Convergence Zone

It was not until 1970 that field trials with Doppler equipment became possible, thanks finally to support from industry but still on the initiative and under the direction of Vernon Joyce, to whom and to Ciba-Geigy Ltd the Centre for Overseas Pest Research has been very greatly indebted for opportunities to participate in these trials. Figure 5.8, which covers the area of the small rectangle in Fig. 5.6, shows our first Doppler traverse of the ITCZ. This traverse immediately demonstrated, within the ITCZ, an air-mass boundary (long recognized by the national meteorological services concerned as the Inter-Tropical Front—ITF) sufficiently sharply defined to be systematically explored by flight patterns such as those shown in Figs. 5.9, 11 and 14.

Before considering the insect catches made on these flights, the data of Fig. 5.9 can be used to illustrate the quantitative estimation of wind-convergence and in turn the rate of concentration of airborne insects to be expected at such a feature. If around any closed sampling route in a horizontal plane the wind experienced on one small element of the route, of length δl, has an inward component of velocity V (positive or negative), in a direction at right angles to the element, then the net rate of inflow across the whole route is $\Sigma V \delta l$. Expressing this in relation to the area A enclosed by the route gives the convergence, averaged over this area, as $\Sigma V \delta l / A$. For the near-rectangular flight pattern illustrated in Fig. 5.9, for which the mean wind components inward or outward across each side are shown in Fig. 5.10, this estimate of convergence simplifies to

$$[44(29{\cdot}7 + 1{\cdot}6) - 54(11{\cdot}9 + 0{\cdot}9)]/44 \times 54 = 0{\cdot}29 \text{ per hour.}$$

Beginning with the simplest of assumptions, airborne insects envisaged as constrained against ascent but otherwise without systematic movement of their own relative to the air, and initially uniformly distributed, would be concentrated by such convergence at an initial rate, averaged over the whole quadrilateral, of 29 per cent increase in area-density per hour.

This elementary treatment [11, 43] envisages only the beginning of the process and its operation relative to fixed ground positions; with most real areas of convergence, like Fig. 5.9, with some outflow as well as inflow, this initial rate of convergence would, as pointed out by Cochemé [11], be liable to decline as some of the insects, after initial concentration, would begin to leak away in the outflows at their increased density. In an area similarly fixed relative to the ground but without any horizontal outflow at all, our imaginary insects experiencing convergence at a constant rate c would be concentrated from their initially uniform area-density β_o to a mean density β_g within this area, after any time t, given as already indicated by: $\beta_g = \beta_o(1 + ct)$ which is an appropriate *initial* rate even when there is some outflow. If the reference area is envisaged instead as moving with the winds, and decreasing in extent because of their convergence, Cochemé has further shown that with steady convergence c, the density β_s after time t will be $\beta_s = \beta_o e^{ct}$. Considered in relation to a mobile reference area of this kind, steady convergence at the rate of 0·29 per hour would give concentration at a rate of 36 per cent per hour instead of the 29 per cent per hour already quoted for a fixed reference area. Finally, the rates of concentration thus far considered are those to be expected of randomly orientated flying insects; for a given intensity of wind-convergence, these rates of concentration would be

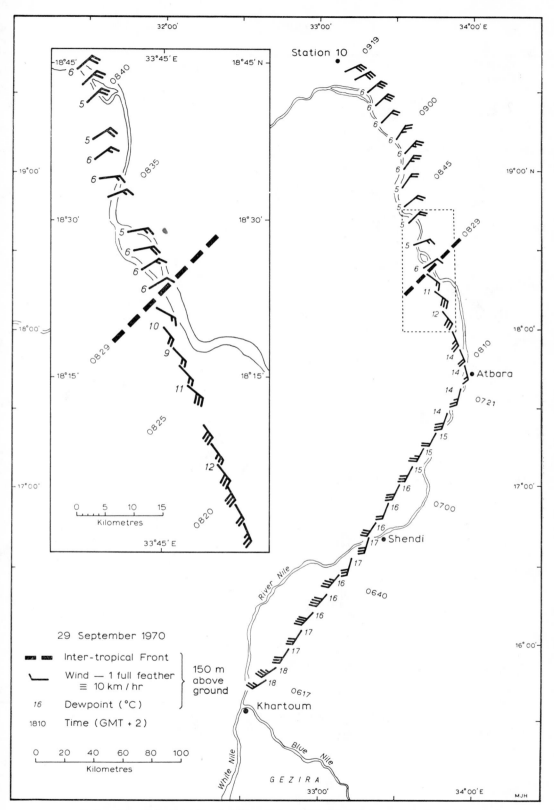

FIG. 5.8. **Location of the Inter-Tropical Front by aircraft.** Note characteristically abrupt wind-shift and contrast in humidity between NE trades and southerly monsoon.

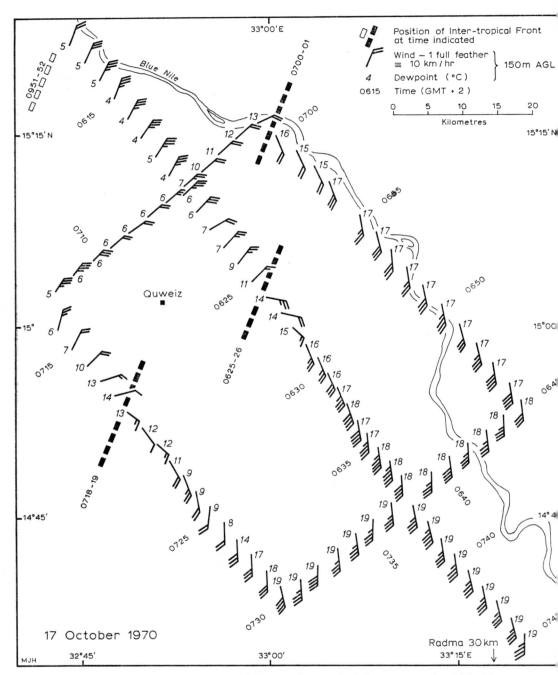

FIG. 5.9. **Exploration of the Inter-Tropical Front.** A zone of wind-convergence in which airborne insects of a number of taxa have been found at high densities.

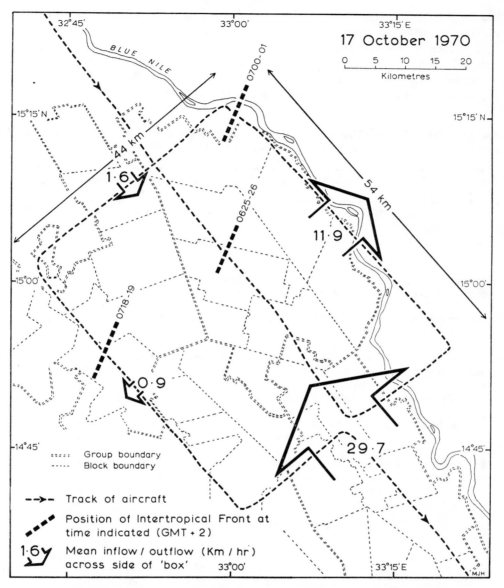

FIG. 5.10. **Quantitative estimation of wind-convergence.** The mean wind-components, averaged along each side of the rectangular aircraft-track of Fig. 5.9, clearly demonstrate net inflow, at a rate which on the simplest of assumptions (see text) could be expected to concentrate airborne insects at an initial rate (averaged over the whole rectangle) of 29 per cent increase in area-density per hour.

The figure also shows the boundaries of some of the 12 Groups and 103 Blocks which represent the operational units for current pest-control practice in the Gezira (p. 145); see also Fig. 5.13.

increased by any predominance of down-wind components of orientations, of the kind often indicated by radar observations (pp. 163–164).

Evidence on possible effects of the ITF on insect densities was at first sought by comparing the aircraft catches made during periods which included one or more traverses of the ITF, with those made during other trapping periods which did not, but were otherwise closely similar in height, route and time of day or night. Table 5.1 summarizes

TABLE 5.1. **Comparisons of airborne insect density in relation to the Inter-Tropical Front.** Sudan aircraf catches, October–November, 1970: enumerated and identified by Dr J.Bowden

Trapping period	Route relative to ITF	Insect density	Main taxa caught
A Sampled on night flights at 300 m above ground over route Khartoum – Wad Medani – Khartoum			
0201–0354/15/10	Well north of ITF, in dry northerly trades air	< 1 in 450,000 m³	Nil
0211–0353/17/10	Included 2 ITF traverses	1 in 7000 m³	Delphacidae Heteroptera
B Sampled on early-morning flights at 150 m above ground over route Khartoum – Wad Medani			
0630–0711/5/11	Well north of ITF, in dry northerly trades air	1 in 4500 m³	Hymenoptera
0603–0750/17/10	Included 3 ITF traverses (Fig. 5.9)	1 in 1700 m³	Delphacidae *Culicoides*
C Sampled on early-morning flights at 30 m above ground over route Khartoum – Wad Medani			
0609–0654/6/11	Well north of ITF, in dry northerly trades air	1 in 6000 m³	*Culicoides* Hymenoptera
0601–0650/27/10	Included 1 ITF traverse	1 in 1500 m³	*Culicoides* Delphacidae
0609–0648/20/10	Well south of ITF, in humid southerly monsoon air	1 in 940 m³	*Culicoides* Delphacidae
D Sampled on morning flight at 30 m above ground over route Wad Medani – Khartoum (Fig. 5.11)			
0913–0942*/27/10	North of ITF, in dry northerly trades air	1 in 900 m³	Hymenoptera
0755–0912/27/10	Included 3 ITF traverses	1 in 230 m³	Thysanoptera Hymenoptera
0734–0752/27/10	South of ITF, in humid westerly monsoon air	1 in 600 m³	Hymenoptera Thysanoptera

* Actual trapping period comprized 13 min at 30 m and a final 16 min at 300 m (on airport approach), treated as equivalent [52], with turbulent convective mixing, to a total of 15 min at 30 m.

the data for the seven catches which satisfied these criteria; we are greatly indebted to John Bowden for his examination of these catches. Interpretation was found to be complicated, e.g. in the third comparison because the densities of airborne insects found in the northerlies were, as might be expected from the predominantly desert track of these winds, substantially lower than in the south-westerlies. The most instructive of these comparisons was the fourth (details in Table 5.2), in which the difference in catching rate shown by the three successive nettings of Fig. 5.11 was statistically highly significant, with the highest density recorded during the part of the flight which included the three ITF traverses, though this is of course in itself not necessarily attributable to a process of concentration, since source-differences may also have been involved.

As the box pattern in Fig. 5.11 extended some 15 to 25 kilometres on either side of the ITF, the figure of one insect per 230 m³ (necessarily averaged over the whole box pattern) is likely (particularly in view of subsequent radar evidence such as Fig. 8.39) to have been a considerable under-estimate of the density in the more immediate vicinity of the front. Hourly catches at stationary suction-traps on a 15 m tower at Quweiz, made during passages of the ITF a few days earlier, are likely to have been somewhat more

TABLE 5.2. **Aircraft-trapping across the Inter-Tropical Front and in neighbouring air-masses.** Wad Medani–Khartoum 0734–0942/27/10/1970 at 30 m above ground. Fig. 5.11 and Table 5.1D

	Net 1 Outside ITF in monsoon air	Net 2 Including 3 ITF traverses	Net 3 Outside ITF in trades air
Thysanoptera	3	68	—
Hymenoptera	6	35	3
Psocoptera	—	13	—
Diptera—*Culicoides*	2	4	1
—other	—	8	—
Hemiptera—Delphacidae	—	4	—
—Aphididae	1	1	1
—other	—	1	1
Coleoptera	—	1	—
Araneae	—	—	1
Total airborne Arthropoda	12	135	7
Trapping period in minutes	18	77	15

Test for differences in catching-rate between netting periods:
$$\chi^2 = 23 \cdot 1, \ n = 2 \text{ and } P < 0 \cdot 001$$

representative of insect densities close to the front. Temporary spells of southerly winds, shown by the independent routine synoptic analyses of the Sudan Meteorological Department to have represented temporary northward movements of the ITF, gave peak insect catches (e.g. in Fig. 5.12) mainly of small grass- and cereal-feeding Cercopidae and Cicadellidae which reached a maximum volume-density of one insect in 30 m³ [4]. This density was, moreover, attained in the topmost trap, consistent with processes envisaged as operating above ground and crop levels.

Meanwhile, independent evidence of larger airborne insects at high densities at wind-shift lines had been provided by Glen Schaefer's radar observations of line echoes in Niger in 1968 [61] and in Australia in 1971 (p. 183). Under his direction, radar observations in the Sudan later in 1971 and again in 1973 similarly demonstrated line echoes at the ITF (Fig. 8.39), likewise attributable to concentrations of airborne insects; and the radar also provided guidance, already mentioned, which resulted in aircraft sampling of insect concentrations at greater heights and higher densities than in 1970.

Thus, during the evening of 16 October 1971, grasshoppers (*Aiolopus simulatrix* and *Catantops axillaris*) were taken by aircraft-trapping, with radar guidance, at 450 and 1200 m above the ground; a mean density of one grasshopper in 1000 m³ was recorded during a 50 km traverse at 450 m during which the estimated surface position of the ITF

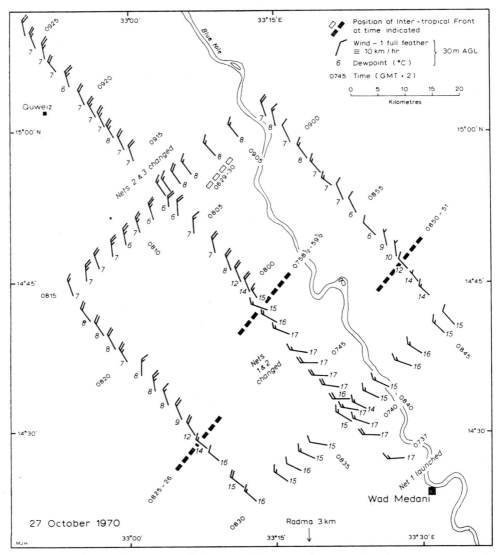

FIG. 5.11. **Airborne insects and the Inter-Tropical Front: aircraft-trapping.** A significantly higher insect density was given by net 2, flown during the three ITF traverses, than by nets 1 or 3, each flown wholly within a single air-mass (Table 5.2).

was overflown, in northerly winds which were markedly weakening downstream, from 34 to 10 km/hr, and thus suggesting active convergence as the ITF was approached at this level [49].

On one occasion (Fig. 5.13), when the aircraft was already airborne at the time a line of echoes appeared on the radar, it was found possible to make an appropriate wind-finding traverse at a height of 600 m. The line of echoes was found to correspond closely with a well marked wind-shift, between south-easterly and northerly winds, agreeing in position, orientation and direction of displacement with the ITF as independently indicated on the Khartoum synoptic analysis.

Before the 1971 Sudan trials, a series of traverses flown through the sea-breeze

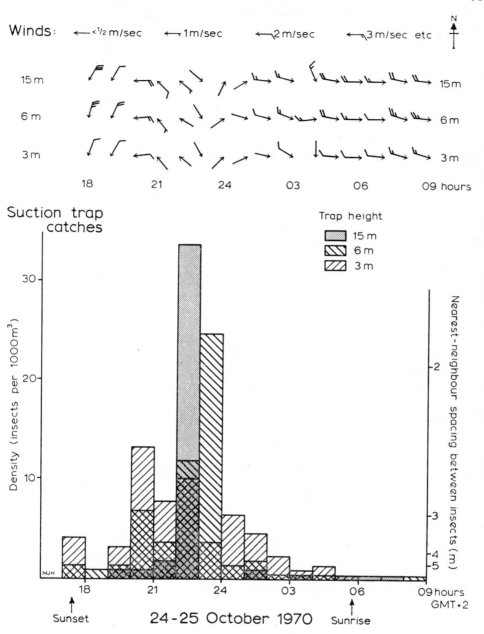

FIG. 5.12. **Airborne insects and the Inter-Tropical Front: suction-trap catches.** Data of Bowden and Gibbs [4]

front in southern England (Fig. 5.20) had enabled the pilot and observer concerned (K.H. and M.J.H.) to develop an improved operating procedure for the exploration of the immediate vicinity of a front. Figure 5.14 illustrates the subsequent application of this procedure to the ITF; a sample taken on this flight, over a period including four ITF traverses at a height of 300 m, gave a density exceeding one insect per 200 m³.

The winds found around the closed triangular area of some 120 km², shown on the

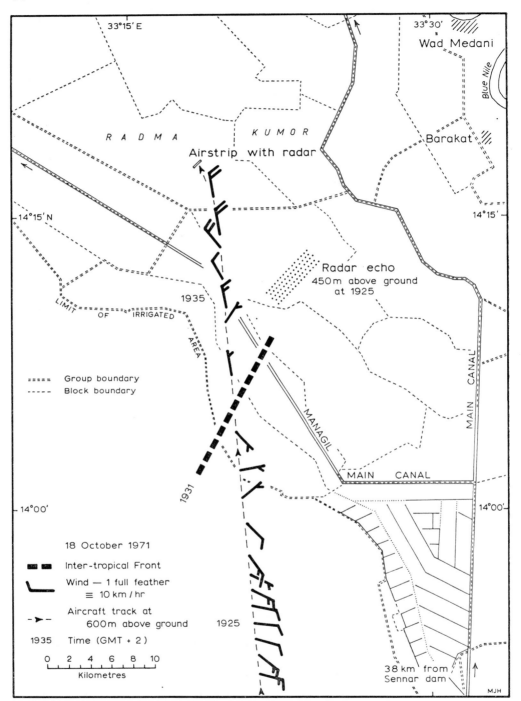

FIG. 5.13. **Radar line-echo and winds at the Inter-Tropical Front.** Radar echo largely or wholly due to airborne insects (p. 185); note agreement in position and orientation with wind-shift line found independently by aircraft (with orientation provided by other traverses further to SSW earlier in same flight).

Canal-network and drainage channels illustrated in a single Block; see also Fig. 5.10.

left of Fig. 5.14, provide an estimated rate of convergence subject to some uncertainty by reason of movement of the ITF but probably sufficient to give initially a four-fold increase per hour in the area-density of randomly-orientated airborne insects, envisaged as before and averaged over the 120 km². Substantially higher intensities of convergence occur at the squall-lines spreading out from cumulo-nimbus storm centres, and graphic radar evidence of accumulation of insects at such a squall-line in the Sudan is presented in Fig. 8.35. The probable genesis of a squall-line of this kind is illustrated by the uplifted Desert Locust swarm in Fig. 5.23, taken from one of John Sayer's air-to-air film sequences of flying swarms shown at the Symposium.

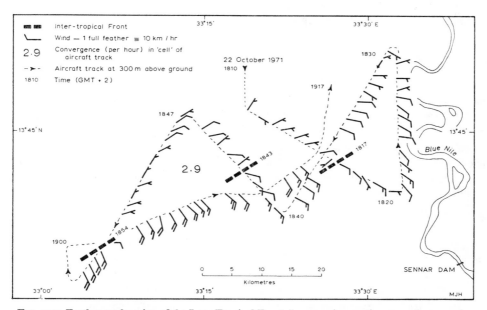

FIG. 5.14. **Further exploration of the Inter-Tropical Front.** Improved operating procedures made possible the exploration of the more immediate vicinity of the front, and recorded higher insect densities.

Particularly important evidence on effects of the ITF and other zones of convergence was provided by the suction-trap results obtained by N. Russel-Smith in October, 1971, when the traps on the tower were operated continuously at Radma over the period of the radar trials. High catches (i.e. with more than half the total catch for the day caught within a single two-hour period) in association with passages of the ITF were shown independently by each of the seven taxa which constituted the bulk of the catches and were separately recorded. These seven taxa, of which further details are given on p. 139, were whitefly, aphids and two genera of thrips, all flying predominantly by day; and Cicadellidae/Cixiidae/Delphacidae, Staphylinidae, and other Coleoptera (including *Gonocephalum simplex* (F.) and *Leptaleus sennarensis* (Pic)—cf. p. 98), all flying predominantly by night. Eight separate passages of the ITF across Radma (on 4, 12—twice, 13, 16, 19, 22 and 23 October) were each associated with high catches of 1–3 of these taxa; at least five of these occasions showed some increase in insect density with height (as had been found e.g. in Fig. 5.12), again as might perhaps be expected if the concentrations were formed by effects such as wind-fields operating well above crop level, rather than resulting from behaviour of the insects themselves at or before take-off. Two of the eight occasions were at times of day well removed from that of any regular diurnal peak of flight activity (and included the highest density noted at 15 m—thrips, mainly *Frankliniella* probably *dampfi* Pr., the cotton bud thrips, at one insect per 29 m³, 1100–1300 12th October).

Eight more high suction-trap catches were noted as associated with other wind-shifts well to the south of the ITF; most but not all of them were recorded on the Wad Medani anemograph as well as at Radma and accordingly represented wind-field features at least tens of kilometres in extent. Convergence zones between south-easterlies and south-westerlies, to the south of the ITF, are already recognized as sometimes significant in the Sudan [3], and may at times be analogous with the African Rift Convergence Zone (sometimes termed the Congo Air Boundary) further to the south.

African Rift Convergence Zone

Another semi-permanent zone of wind-convergence in which high densities of airborne insects have been found occurs in the vicinity of the Rift Valley [14] in eastern Africa (Fig. 5.16), where surface easterly winds from the Indian Ocean meet westerlies which sometimes extend all the way from the Atlantic. Eastward surges of this zone across East Africa, marked by temporary incursions of westerly winds, have occasionally extended far enough to produce exceptional eastward movements of Desert Locust swarms in the Coast Province of Kenya (Fig. 5.15). These particular swarm movements, in wind-fields showing very vigorous convergence, demonstrated by daily rainfall amounts up to 159 mm, brought these swarms considerably closer together, as shown, and two others, further to the north, appear in fact to have joined up at the same time. This rather special case provides the most direct evidence so far on record of an increase in the overall area-density of airborne insects established by the spatial distribution of particular populations before and after vigorous wind-convergence.

Incursions of westerly winds temporarily reaching the Nairobi area have repeatedly been found to be closely associated with large hourly light-trap catches of *Spodoptera exempta* moths, and those for the night before the situation of Fig. 5.16 are illustrated in Fig. 5.17 [18]. Low wind-speeds at the transition between the easterlies and westerlies may have augmented such light-trap catches by directly enhancing the collecting efficiency of the trap. However, this particular night (9th–10th March, 1970) also provided the largest suction-trap catch so far recorded for this species, with a corrected [69] density averaging one moth in 85 m³ over the 12 hour period from sunset to sunrise (Table 5.3). Several of these same westerly incursions proved to have been associated with eastward extensions of egglaying, producing important new armyworm infestations in and to the east of the Rift Valley, with which Elizabeth Betts deals (p. 126). A preliminary examination of further extensive light-trap data also secured by Eric Brown, and for which I am indebted to Margaret Haggis, has provided indications of a similar association of high light-trap catches with these incursions of westerly winds into the Nairobi area for another nine species of Noctuidae (*Amyna punctum* F., *Anomis sabulifera* Guen., *Cosmophila flava* Feld., *Eutelia discistriga* Walk., *Heliothis armigera* Hb. (for which N. Russell-Smith had already assembled East African evidence to the same effect), *Hypena* sp., *Plusia acuta* Walk., *P. limbirena* Guen., and *P. orichalcea* F.).

Figure 5.18 illustrates how sharply defined these wind-shifts can be, with more than 100 km of uniformly westerly and north-westerly winds meeting easterlies and south-easterlies with a fetch of at least 1000 km in a zone of transition near Narok, with succes-

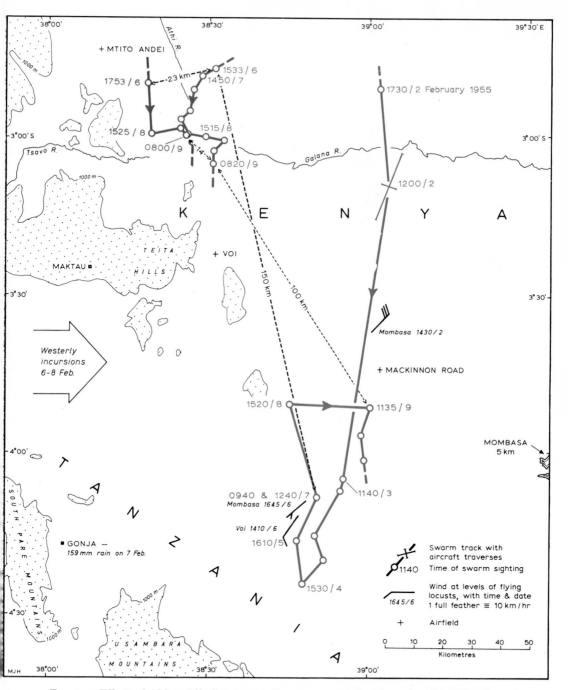

FIG. 5.15. **Effects of African Rift Convergence Zone on movements and spacing of neighbouring locust swarms.** All three swarms not only showed temporary eastward displacements with the temporary incursion of westerly winds but also were brought closer together (by a factor of roughly $\frac{2}{3}$ both on N/S and E/W axes) by the vigorous wind-convergence, which had thus roughly doubled the overall population density over an area of some 3000 km² in a total of some 12 hours of swarm movement.

[*facing page* 96]

sive wind observations of W 13 km/hr and ESE 13 km/hr recorded at points only 4 km apart. Insect collecting during the hour's flight including this traverse gave an overall mean density of one insect in 3300 m³, and the outward flight through the transition zone the same morning a density of one insect per 790 m³, at heights of 290–950 m above the ground (with Halticinae, probably *Longitarsus*, the most noticeably numerous

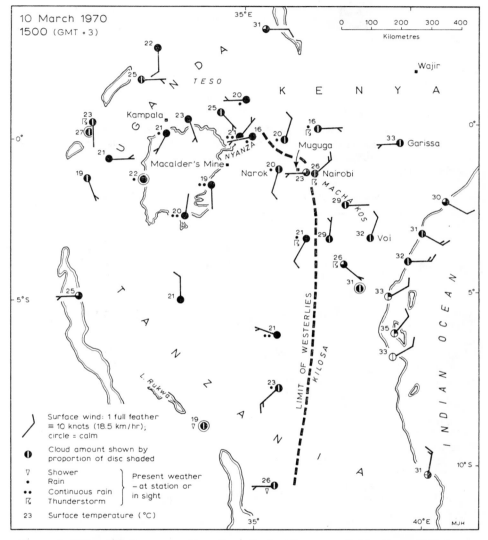

FIG. 5.16. **African Rift Convergence Zone.** Synoptic chart for 1500 10th March, 1970 [18] showing ARCZ displaced eastwards with incursion of westerly winds across Kenya highlands. This was associated with very high densities of armyworm and other moths (Table 5.3, Fig. 5.17 and p. 273), and followed by a spread of new armyworm infestations (Fig. 6.7).

constituent of both catches), in contrast with the density of only one insect in 15,000 m³ given by an intervening two hours' trapping at somewhat lower altitudes (130–600 m above ground) entirely within a uniform lake-breeze flow between the shore of Lake Victoria and 60 km inland near Macalder's Mine.

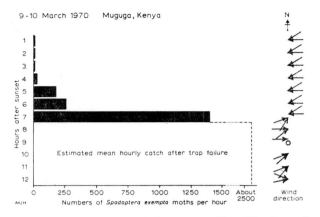

FIG. 5.17. **Armyworm moths and African Rift Convergence Zone.** Hourly catches in automatic light-trap [18, 63] recorded during passage of ARCZ (Fig. 5.16). Mechanism failed under overload after 7th hour; subsequent numbers estimated from total overnight catch at neighbouring standard light-trap.

TABLE 5.3. **Some high densities of airborne insects in zones of wind-convergence**

	Date and local time	Height above ground m	Mean volume per insect m³	Mean distance to nearest neighbour m	25 per cent with nearest neighbour closer than m
LEPIDOPTERA	March 1970:				
Spodoptera exempta p. 96; Fig. 5.17 [18]	1830/9–0630/10	2	85	2·4	1·8
HEMIPTERA	Oct. 1970:				
Cercopidae/Cicadellidae p. 91; Fig. 5.12 [4]	2200–2300/24	15	30	1·7	1·3
COLEOPTERA	Oct. 1971:				
Leptaleus sennarensis p. 273.	2158–2210/10	1600	190	3·2	2·4
THYSANOPTERA					
Frankliniella prob. *dampfi* p. 96.	1100–1300/12	15	29	1·7	1·3
ORTHOPTERA					
Aiolopus simulatrix p. 91; Fig. 8.7 [49]	1834–1851/16	450	1200	6·0	4·4

Nearest-neighbour distances derived from volume-densities on assumption of random spatial distribution within sample (see p. 105).
We are greatly indebted to colleagues at the British Museum (Natural History) for insect identifications.

Red Sea Convergence Zone

A further persistent zone of wind-convergence, long recognized as significant in relation to the distribution and breeding of the Desert Locust [36, 72], is to be found almost continuously from October to May within the Red Sea basin, associated with the winter rains characteristic of this area. In November 1970 a brief visit of the Doppler-equipped aircraft to Tokar and Port Sudan, for other purposes, made possible the observations of Fig. 5.19. This shows an air-mass boundary between cooler northerlies and warmer easterlies,

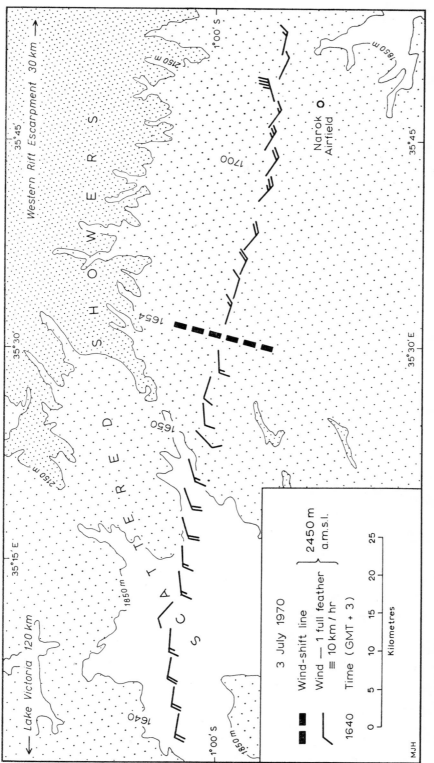

FIG. 5.18. African Rift Convergence Zone: wind-finding traverse. Flea-beetles (probably *Longitarsus*) were particularly noticeable in the catches taken on this occasion.

8

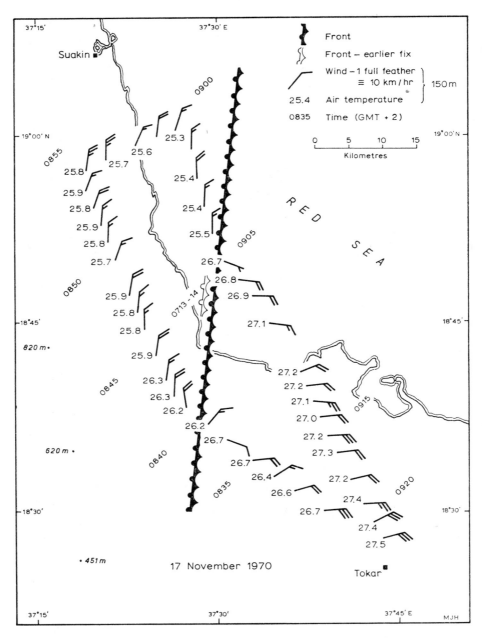

FIG. 5.19. **Exploration of Red Sea Convergence Zone.** The intensity of this wind-convergence was sufficient to concentrate airborne insects at initial rates corresponding (on the simplest of assumptions) to a 6 or 7-fold increase in area-density per hour over a belt 3–5 km wide.

orientated nearly north/south, and found to slope upwards towards the west, corresponding in all three respects with earlier inferences on the structure of the Red Sea Convergence Zone [33]. The wind-shift was characterized by a brief spell of turbulence, was particularly noticeable during the two traverses made in otherwise smooth air over the water, and was below the eastern edge of a belt of strato-cumulus cloud, with its western edge just short of Suakin, and extending away to the north over the sea. The wind sequences found on the two traverses were sufficiently similar to justify two-dimensional treatment, enabling convergence in the main transition zone, which was 3–5 km wide, to be estimated directly from the wind-components perpendicular to the orientation of the zone. The convergence so found was at rates of five per hour at 0905 and six per hour at 0713— which would concentrate airborne insects, envisaged as before, at initial rates corresponding to a six or seven-fold increase in area-density per hour. No insect netting was done on this flight, but these estimates of wind-convergence throw some light on the possibilities in particular of adult concentration of scattered Desert Locust populations, as envisaged in this same area many years ago, by Kennedy working in the Tokar delta [24].

Frontal systems of temperate latitudes

Compared with the semi-permanent nature of the tropical zones of convergence which have been considered, the zones of convergence in temperate latitudes, though often powerful and sometimes violent, are commonly more elusive and ephemeral phenomena, which may appear, develop, weaken and disappear within a matter of days and sometimes even of hours. Figure 5.20 presents observations made during a wind-finding flight undertaken in fair weather, as the training exercise already mentioned, through the coastal front of southern England. Here the summer sea-breeze from the Channel sometimes advances inland as a miniature cold front at which the warmer inland air is forced to rise for a thousand metres or so. This ascending air at the front provides lift which is now regularly exploited not only by birds but also by gliding pilots, whose observations have done much to elucidate the structure of this feature—particularly those made and collated by John Simpson [64, 65], whose briefing and co-operation, together with that of his colleagues of the Reading Department of Geophysics, made our own flight possible. He has drawn attention in particular to swifts (*Apus apus*) soaring in the belt of rising air at the front, repeatedly seen from gliders and chiefly at heights of about 270 m. The swifts have been present in sufficient numbers to render the front readily visible on the Marconi radar at Chelmsford, and have been 'apparently feeding on insects which have been concentrated in the convergence zone' [66].

Insects were caught in large numbers on our aircraft flight but, with the earlier type of trap net used (since substantially improved by John Spillman), in a condition which precluded accurate enumeration and identification; and we are indebted to Michael Way for recognizing among them many thrips, aphids, Diptera and parasitic Hymenoptera. Figure 5.20 shows the rates of convergence given by the winds found around each cell of this flight pattern, averaged over the total area of the cell. As already indicated in relation to insect collecting around the ITF, these average values of convergence will have been substantially exceeded in the immediate vicinity of the front itself; gliding observations in South Africa [34] and Germany [70] have shown that much of the ascending air resulting from convergence associated with a cold front can be confined to a narrow belt along the front ahead of the advancing cold air. The degree of localization of the effects of such a wind-field feature on the distribution and density of airborne insects will depend very much on how sharply defined is the frontal boundary. This sharpness of definition may accordingly be of biological significance; it has recently been emphasized [6] that the minimum thickness (δ) possible for a frontal zone between two air-masses, given the difference in wind velocity (ΔV) and temperature ($\Delta \theta$) across the

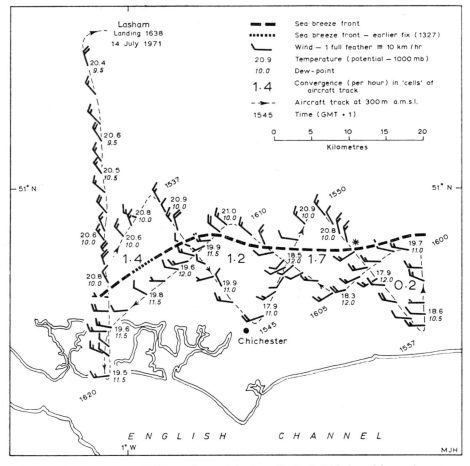

FIG. 5.20. **Exploration of sea-breeze front of southern England.** This is a fair-weather atmospheric feature which is frequented by swifts probably feeding on concentrations of small airborne insects in it. Note small but consistent air-mass difference in temperature and humidity between sea-breeze and warmer, drier air inland.

zone, must be limited by a balance between thermal stability (depending on the effective temperature-excess of the over-riding air-mass), tending to sharpen the transition, and turbulence generated by the difference in wind across the front, tending to make the transition more diffuse. This involves the Richardson number Ri, which is a criterion for the onset of turbulence and is given by:

$$Ri = \frac{g\Delta\theta\delta}{\theta(\Delta V)^2}.$$

The minimum thickness for the frontal zone corresponds with the Richardson number reaching a recognized critical value of about 0·25.

On the frontal traverse indicated by an asterisk in Fig. 5.20, a wind of 17 km/hr from 345° was recorded at 1551½ with an air temperature of 20·5°C, and a wind of 13 km/hr from 240° at 1552 with a temperature of 19·5° in the sea-breeze air. Taking these winds (with a vector difference of 6·7 m/s) and temperatures as representative of the two air-masses implies a transition layer with a minimum depth given by:

$$\delta = \frac{Ri\ \theta(\Delta V)^2}{g\Delta\theta} = \frac{0\cdot 25 \times 293 \times 6\cdot 7^2}{9\cdot 8 \times 1} = 340 \text{ m.}$$

High-speed recording of air temperature by a second aircraft, for which we are also indebted to the Reading Department of Geophysics and particularly to Trevor Lawson [28], showed that the actual transition at the front took place very largely within a zone

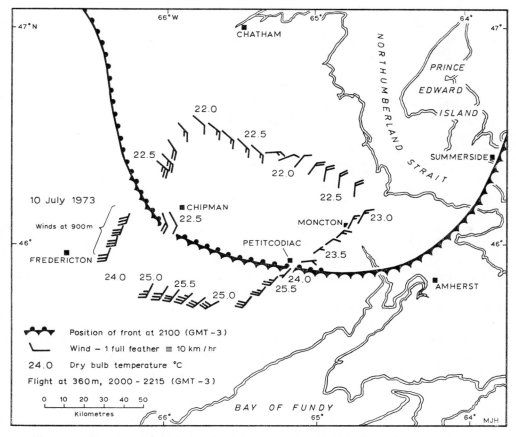

FIG. 5.21. **Convergent winds during spruce budworm moth flight.** This particular evening was recorded by visual, light-trap and radar observations at Chipman as one of noteworthy moth flight; Fig. 8.9B shows radar evidence of a layer of moths at the frontal interface.

some 500 m wide, which (with a probably steep frontal slope) is reasonably consistent with the estimate of minimum depth and within which most of the wind-convergence will have been concentrated—convergence accordingly probably an order of magnitude greater than the mean values shown in Fig. 5.20, and capable of giving an initial rate of at least a ten-fold increase in concentration per hour of insects envisaged as before as constrained against ascent. Computer studies of the more detailed redistribution of airborne insects to be expected from what is now known of the structure of sea-breeze fronts in this area are in progress at Reading. Coastal fronts of this kind can occur very regularly in lower latitudes [35], and have been found significant for Desert Locust swarm movements near the Gulf of Aden [54] and in western India [43].

The most important areas of wind-convergence in temperate latitudes are those of the travelling barometric depressions already mentioned, with, typically, alternating surges of warmer air towards the pole and of colder air towards the equator, and with these air-mass boundaries often in the form of sharply defined warm and cold fronts at which much of the low-level wind-convergence is localized. When this temperature change crosses a threshold for insect flight activity the effects of the alternating winds can as already indicated (p. 76) produce net poleward population movements of many species.

An example of such a frontal system, with evidence of effects on a major North American forest pest, is provided by Fig. 5.21, secured in the course of a research project of the Canadian Forestry Service, to whom and to David Greenbank as co-ordinator, COPR is greatly indebted for the opportunity to participate. Figure 5.21 shows the winds found with a Doppler-equipped DC3 aircraft during one of 14 successive evening flights around this route, undertaken to record conditions over this area of infestations of spruce budworm, *Choristoneura fumiferana* Clem. (Tortricidae), throughout the 1973 period of flight activity of these pest populations. On this particular evening, winds over the area were dominated by a front which as indicated was moving as a warm front to the north-west of Chipman and as a cold front to the south of Moncton, with the intervening sector temporarily stationary; we are greatly indebted to Professor R.B.B.Dickison for the detailed synoptic re-analysis of the relevant hourly meteorological observations at Chatham, Fredericton, Moncton, Summerside and Saint John, which convincingly elucidated this situation. The winds found from the aircraft give convergence which averaged 0·28 per hour (in fortuitously close agreement with the Sudan figure quoted above) over the whole 6800 km² enclosed by the quadrilateral, and probably something like an order of magnitude greater in the immediate vicinity of the front; no comparable convergence was recorded on any of the other evening flights*. Spruce budworm moths were observed striking the aircraft windscreen at a height of 360 m, and this particular evening was recorded by the research party at Chipman as one of noteworthy moth flight. It gave the highest light-trap catch of the season (with the highest percentage of female moths) although local emergence and exodus flights were almost over, high moth densities on Glen Schaefer's radar well after the peak time of local take-off, and an exceptional descent and deposition of moths in numbers in the Chipman area.

CONCENTRATION OF AIRBORNE INSECT POPULATIONS BY WIND-CONVERGENCE: SOME IMPLICATIONS

A given spatial distribution of airborne insects must represent the combined effects of their initial distribution at take-off, of their subsequent flight behaviour (particularly orientations and air-speeds), and of the transporting and concentrating or dispersing effects of the varying wind-systems subsequently experienced by them, integrated between limits set by the period over which the insects have been airborne and the areas and heights over which the population has occurred.

Furthermore, the potential concentrating effect of a given convergent wind-system must depend in a complex manner not only on the detailed structure of the wind-system but also on the way in which active flapping flight diminishes with height. Further

---

* Though it was, repeatedly, in similar observations during the 1974 season, with evidence of moth concentration.

laboratory data are needed on this point, which involves temperature thresholds (which can range e.g. from 6° for the winter moth *Operophtera brumata* [1] to 23° for the Desert Locust [75] in the absence of sunshine), together with the changing vertical components of velocity of the insects (relative to the air and its rate of ascent) as muscular activity declines and ceases, with the wings either remaining extended or finally closed.

The spatial distribution of airborne insects is accordingly unlikely to reflect exactly any instantaneous distribution of wind-convergence; and direct proof of the concentration of any particular airborne insect population by wind-convergence would require maintaining continuous contact with the same insects (as with the locust swarms in Fig. 5.15), as well as continuously monitoring the winds in which they fly, in a manner still well beyond the scope of the resources so far deployed. Nevertheless, the densities at which airborne insects of many taxa have now been found in zones of convergence, and the intensity of the convergence found on such occasions, provide sufficient circumstantial evidence on concentration of insects by wind-systems to justify examining some of the implications of this process.

Mutual perception and behaviour

In considering possibilities of mutual perception, in relation to sexual or gregarious behaviour, interest centres on distances between individuals, and particularly on the distance between each individual and its nearest neighbour. Suction-traps, aircraft nets and radar all provide estimates of the volume-density, expressed as number of insects per unit volume of air, and sometimes more conveniently envisaged as its reciprocal the volume V of air per insect. As a first step in considering distances between individuals, $\sqrt[3]{V}$ has been used for example in some recent studies of locust behaviour in flying swarms [74], and if each insect is envisaged as flying at the centre of a standard cube, $\sqrt[3]{V}$ can indeed be the distance from each insect to each of all six of its nearest neighbours. Less regular and correspondingly more plausible spatial arrangements will reduce the mean distance between each insect and its nearest neighbour to considerably less than $\sqrt[3]{V}$.

Within concentrations, random spatial distribution may often be a more realistic model than a uniform distribution, and has been shown [67] to give (with sufficiently large samples) a mean distance r between an individual insect and its nearest neighbour, where:

$$ r = \int_0^\infty r \left(\frac{4\pi r^3}{V} \right) e^{\frac{-4\pi r^3}{3V}} \, dr \quad = \quad \sqrt[3]{\frac{V}{36\pi}} \; \Gamma\tfrac{1}{3} = 0 \cdot 555 \sqrt[3]{V}. $$

Furthermore, 25 per cent of a large sample of randomly distributed insects would have their nearest neighbour closer than r_1, where $e^{\frac{-4\pi r_1^3}{3V}} = \tfrac{3}{4}$, that is $r_1 = 0 \cdot 41 \sqrt[3]{V}$. These factors are accordingly used in Table 5.3 for helping to envisage possibilities of mutual perception in a few of the denser airborne populations sampled.

Some quantitative evidence on the acuity of the sense organs concerned, which is necessary for the objective assessment of possibilities of mutual perception, is available for locusts, which have been found to respond visually to movement of a small light source over an angular distance of as little as 0·1° [8], equivalent to the length of a locust at about 30 m, and to show auditory responses to another locust in flight at a distance of 2 m or so [19]. Considered in relation to the development of gregarious behaviour, adult concentration, under the influence of convergent wind-fields, can operate over distances of many hundreds of kilometres. These distances are thus larger by perhaps four orders of magnitude than the range of mutual sensory perception of individual locusts, and adult concentration is perhaps to this kind of extent more significant than nymphal concentration in the process of gregarization [46].

No such data on sensory acuity appear to be available for the other species mentioned, though *Calliphora vomitoria* has recently been found capable of a roughly comparable visual resolution of angular movement [13]. It was however pointed out [20] at one of the Society's earlier Symposia that, even at a spacing of one insect per 1000 m^3, only $5\frac{1}{2}$ minutes of randomly orientated flight at an air-speed of 5 km/hr would involve 50 per cent of the insects in passing within 1 m of each other. It is accordingly suggested that other insects besides locusts may likewise be transported by the wind from long distances to within range of mutual perception; as already indicated (p. 78), flying locusts now appear to differ from other insects more in the extended periods over which high densities are maintained than in these densities themselves, which can be temporarily attained by species without gregarious behaviour. Considering the much lower volume-densities at which flying insects are often found, it was noted [20] at that earlier Symposium that for species with very short reproductive lives 'some process of concentration seems necessary if a reasonable proportion of the population is to mate at all'. It may now be suggested that wind-convergence may at times provide such a mechanism, and that for some insects concentrating effects of convergent wind-systems may perhaps represent a result of flight which is biologically more important than long-range transport. It now seems likely that concentration by wind-convergence may have been a factor in the high density of airborne insects which is characteristic of many records of insect migration [e.g. 77], and it is further suggested that the puzzling uniformity of orientation often recorded on such occasions may perhaps have developed, e.g. by optomotor reactions, in response to mutual perception resulting from concentration by wind-convergence.

Population dynamics

It now appears necessary to envisage insect populations as subject from time to time to wind-systems (both temperate and tropical) which can concentrate them at rates that may double area-densities of airborne insects over hundreds of square kilometres in periods of less than an hour. It would accordingly seem essential to begin to take into quantitative account such potential effects of wind-systems in considering the population densities of any species showing significant flight activity. Spatial changes in distribution are in any case likely to dominate the changes in numbers seen in a sample area which is small compared with the range of movement of an individual insect, though such changes appear often to have received less attention than changes with time in the numbers recorded; attempts to model the dynamics of mobile populations in terms only of scalar quantities must necessarily be unrealistic.

In meteorological studies of mobile atmospheric features such as turbulence eddies, it has been found necessary to distinguish and use both the Eulerian approach which considers the features from the point of view of a stationary observer, and the Lagrangian approach, which studies the features from the point of view of an observer travelling with them. It is suggested that realistic dynamics of insect populations showing active flight must likewise involve both approaches, as illustrated by Fig. 5.3, and by the assessment of the 1954 Desert Locust invasion of East Africa [42] in which the successive

FIG. 5.22. **Early spraying operation.** Near Korogwe, Tanzania, 1530, 25th February, 1954. This was one of four light spraying aircraft (provided by FAO) which also helped to provide the air reconnaissance data illustrated by Figs. 5.1 to 5.4, 5.7 and 5.15.

FIG. 5.23. **Swarm and downpour from tropical rain-storm.** Near Hargeisa, Somali Republic; 1630, 10th August, 1957; filmed by H.J.Sayer. The swarm, indicated by arrows, is being lifted from the ground, in a manner so far recorded only in the vicinity of such rain-storms [39], probably by an outflow of air cooled by the rain in the way which produces the kind of squall-line found by radar to concentrate airborne insects in the Sudan (Fig. 8.35).

FIG. 5.24. **Low-flying swarm in late afternoon.** Near Wajir, Kenya; 1715, 13th January, 1953; photo H.J.Sayer. Swarm covering 1 km² with locusts flying mainly below 20 m.

swarms seen were initially located relative to the ground and subsequently envisaged relative to each other.

Wind-formed concentrations of airborne pests may contribute to the degree of localization and severity of crop damage (p. 141); such concentrations [5] of armyworm moths (*S. exempta*) for example may explain the characteristically sudden and heavy attacks by larvae of this species—and its common name of 'mystery worm' in some areas.

Tactics and strategy of control

Flying insects at high density, in the form of locust swarms (Figs. 5.2 and 5.22–24), have been found (as envisaged many years ago by Kennedy, Toms and Gunn [16, 26]) to provide particularly efficient three-dimensional targets for aircraft spraying. A substantial proportion of the insecticide applied can be picked up directly on the flying locusts without reaching the ground, and field assessments have demonstrated locusts killed in numbers representing a surprisingly large proportion (up to 6 per cent) of the maximum theoretically possible from the quantity of insecticide applied [29, 38]. Vernon Joyce suggested some years ago [23] that other pests beside locusts might at times be concentrated in flight by wind-fields to an extent sufficient to provide targets for spraying operations, possibly more efficient and involving less environmental contamination than more conventional control tactics, and subsequent experience has supported the feasibility of this idea [49].

Thus for example the catch of *Aiolopus simulatrix* within the ITCZ, recorded in Table 5.3, demonstrated a mean volume-density, averaged along a 50 km traverse at 450 m above ground, of one grasshopper of this species per 1200 m^3—a density reaching the range which has been recorded photographically (and considerably nearer the ground) within coherent Desert Locust swarms [73]. A second pest which has provided evidence of possible air-to-air spray targets is spruce budworm (p. 104). For this species it is understood that there is already evidence that against moths (even settled) lower dosages of insecticide than those needed against larvae can be effective; and, even apart from concentration by convergence, there is radar evidence of moth densities of 1 in 300 m^3 up to a height of 200 m at the regular evening peak take-off time (p. 179). Theoretical models of established effectiveness are already available for the simulation of aircraft spraying against flying insects [30, 40, 53, 57], and direct attention for example to the increased pick-up of insecticide and correspondingly enhanced efficiency made possible by the greater vertical extent of the *Aiolopus* concentration relative to that of the locust swarm targets so far assessed, helping to offset the relatively lower density of the grasshopper target.

On control strategy, Vernon Joyce points out (p. 150) the need for synoptic survey and the scope for synchronous control provided by the mobility of many airborne pests; and the occurrence at high density of airborne insects of a number of taxa in semi-permanent zones of wind-convergence such as the ITCZ and the ARCZ (p. 96 and Table 5.3) suggest possibilities of shifting some of the control operations against other pests besides locusts away from the crop immediately threatened.

CONCLUSION

Circumstantial evidence suggesting concentration of airborne insects by convergent wind-systems is provided by many observations in the literature, of which time permits citing only two. The first relates to the rice-hoppers *Sogata furcifera* (Horv.) and *Nilaparvata lugens* (Stal.), caught at sea in many hundreds in the vicinity of frontal systems at points up to 500 km off the Japanese coast during June–July in four successive years [2, 27]. On this and other evidence the annual appearance of these pest species on rice in Japan at this time of year is now being attributed mainly to long-distance migration [27].

The final example, quoted by C.B. Williams in his 1930 classic [77], is provided by the vivid observations of Darwin and his captain, Fitzroy, on 4th December, 1832. A great flight of yellow butterflies, in all probability *Colias lesbia*, was seen from the 'Beagle' at sea off the Argentine coast, like a snowstorm, in a line many miles long, a mile wide and 600 feet high, which came with a storm from the northwest—suggesting the kind of concentration by convergence at a wind-shift line to which the radar and other evidence are now directing attention. This early record of what can now be suggested as a relevant weather feature may perhaps have owed something to the nature of the observation platform provided by a sailing ship of only 235 tons, on which the observers would have been very directly aware of the wind-systems being encountered by the insects. Perhaps the comparably direct involvement with features of the atmospheric environment, which is now made possible by aircraft and their instrumentation, may have a continuing value in extending the limited sensory experience with which we attempt to interpret the behaviour of flying insects. In conclusion, I would like to suggest that just as Weis-Fogh's beautiful study of the flight of his tiny Chalcid wasp demanded appreciation of the special properties and behaviour of air movements only fractions of a millimetre in extent, so the systems of winds and weather considered in the present paper, ranging from hundreds of metres to tens of thousands of kilometres in extent, represent features of the atmosphere environment which similarly need to be taken into account in the study of insect flight in nature.

ACKNOWLEDGEMENTS

The development of this approach has been made possible by the co-operation and stimulation provided by colleagues, past and present, of the Desert Locust Survey, the Anti-Locust Research Centre, the Centre for Overseas Pest Research and of the many organizations—governmental and industrial, national and international—with which we have been privileged to work. In particular, much has depended on the support, encouragement and initiative of R.J.V.Joyce, successively as Sudan Government Entomologist in the 1950's, first Director of the international Desert Locust Control Organization for Eastern Africa in the 1960's, and subsequently as Director of the Agricultural Aviation Research Unit of Ciba-Geigy Ltd.

DISCUSSION

After asking for further information on the process of concentration of airborne insects by wind-convergence (p. 86), which had been presented very briefly in the spoken text, the Chairman (**Dr D.L.Gunn**) opened the discussion by inviting comments from Dr Greenbank on the recent Canadian work.

Dr D.O.Greenbank: We have now sampled the subsequent egg-laying at our usual thousand points over the area of Dr Rainey's Fig. 5.21, but this particular weather situation occurred late in the adult season of the spruce budworm; the moths involved in the convergent winds were almost all spent, having already completed their laying; and we found no clear evidence of any greater number of eggs within the area affected by this convergence zone than elsewhere.

Gunn: I take it that in New Brunswick convergence is one of those frequent and erratic phenomena that we are familiar with in English weather, and is not such a regular phenomenon as the standard swaying back-and-forth of the Inter-Tropical Convergence Zone?

Greenbank: That is true.

Professor R.S.Scorer: When Dr Rainey was, as it were, apologizing to the biologists for introducing so much meteorology, and saying that there was a biological factor in the coherence of locust swarms because the locusts tend to fly inwards at the edge of the swarm, I think he was in danger of requiring the biological mechanisms to do more than is necessary. Now suppose that you had particles which remained at a fixed height above the ground, say 30–100 metres, and they travelled with the wind, I think they would in fact gather into swarms. That is to say that if the locusts were to fly at fixed height and not move relative to the air, then they would be found to occur in swarms, because this is what the air and its thermal up-currents would do to them. Now I am saying this not to make a take-over bid for the life history of the locust, but to say that the biologists do not have to do as much as they think they have got to do to keep the swarms together. I think the only thing that the inward flight at the edge of the swarm has got to do is to prevent diffusion of the swarm due to the random flight of the locusts, and it's nothing to do with the random motion of the air—which will itself produce swarms.

Gunn: But is not your thesis contrary to the Porton findings on the dispersal of gas-clouds by atmospheric turbulence?

Scorer: No; the cloud of gas does not remain at a fixed height above the ground—it moves with the vertical motion of the air as well as with the horizontal motion, and that is the difference.

Rainey (partly communicated): This is a point which Dick Scorer and I have often argued about since Frank Ludlam first raised it (*Q. Jl R. met. Soc.*, **85** : 171–173); and I agree that the kind of concentrating effect of thermals which they suggest may well fit some swarms some of the time. Thus in swarms travelling with the speed of the wind (as did a number of the larger ones we have studied), one can certainly envisage the miniature convergent wind-systems of each of a series of successive thermals arising within the swarm (with the additional metabolic and solar heat received via the locusts perhaps significant at times [39]) each making some temporary contribution to the continued cohesion of such a swarm. But what I find more difficult to envisage is any effective contribution from such thermals to the maintenance of cohesion of a swarm travelling (like most of those we have studied in detail) at a ground-speed only 10 to 50 per cent of the speed of the wind in which it is flying (like the first swarm in Fig. 5.1), so that not only thermal up-currents but also the corresponding down-currents must be continually travelling, with the mean wind, along such swarms from trailing edge to leading edge, and accordingly subjecting the swarm not only to a temporary concentrating effect in the convergence below each up-current but also to a temporary dispersing effect in the divergence below the descending air. I do not myself know of any kind of wind-system with the sort of range of mobility relative to the general wind which would seem necessary if it is to contribute significantly to the continued cohesion of such slower-moving swarms.

Dr R.J.Wootton: You speak of swarms travelling at speeds less than the speed of the wind. Does this mean in effect that the wind is blowing through the swarm from back to front?

Rainey: Yes indeed, because of the proportion of temporarily settled locusts to be found beneath these flying swarms. This was a point which forced us to reconsider the original idea of putting down a spray-line, for the locusts to fly through, in front of a swarm, where it would in fact move away from the swarm; the spray-line had instead to be put down through the rear part of the swarm. This is an example of how critical observations on insect behaviour can be an essential element in research and development work on control methods.

Gunn: Dr Rainey has touched on a sore point; we had made an elaborate scheme for locusts to fly through a spray-sheet, and the spray-sheet would have moved away from them.

REFERENCES

[1]   ALMA, P.J. (1970). A study of the activity and behaviour of the Winter Moth *Operophtera brumata* (L.) (Lep., Hydriomenidae). *Entomologists mon. Mag.*, **105** (1969): 258–265.

[2]   ASAHINA, S. & TURUOKA, Y. (1970). Records of the insects visited a weather-ship located at the Ocean Weather Station 'Tango' on the Pacific. V. Insects captured during 1968. *Kontyû*, **38** : 318–330. (In Japanese with English summary.)

[3]   BHALOTRA, Y.P.R. (1963). Meteorology of Sudan. *Sudan Met. Service Memoir, no.* 6 : 113 pp.

[4]   BOWDEN, J. & GIBBS, D.G. (1973). Light-trap and suction-trap catches of insects in the northern Gezira, Sudan, in the season of southward movement of the Inter-Tropical Front. *Bull ent. Res.*, **62** : 571–596.

[5]   BROWN, E.S., BETTS, E. & RAINEY, R.C. (1969). Seasonal changes in distribution of the African armyworm, *Spodoptera exempta* (Wlk.) (Lep., Noctuidae), with special reference to eastern Africa. *Bull. ent. Res.*, **58** : 661–728.

[6]   BROWNING, K.A., HARROLD, T.W. & STARR, J.R. (1970). Richardson-number limited shear zones in the free atmosphere. *Q. Jl R. met. Soc.*, **96** : 40–49; discussion **97** : 257–258.

[7]   BURNETT, G.F. (1951). Field observations on the behaviour of the Red Locust (*Nomadacris septemfasciata* Serville) in the solitary phase. *Anti-Locust Bull.*, no. 8 : 36 pp.

[8]   BURTT, E.T. & CATTON, W.T. (1956). Electrical responses to visual stimulation in the optic lobes of the locust and certain other insects. *J. Physiol.*, **133** : 68–88.

[9]   CHAPMAN, R.F. (1959). Observations on the flight activity of the Red Locust, *Nomadacris septemfasciata* (Serville). *Behaviour*, **14** : 300–334.

[10]   CHAPMAN, R.F. (1971). *The insects: structure and function*. 2nd ed., English Universities Press London: xii + 819 pp.

[11]   COCHEMÉ, J. (1966). Assessments of divergence in relation to the Desert Locust. *Tech. Notes Wld met. Org.*, no. 69: 23–41.

[12]   DOWDESWELL, W.H., FISHER, R.A. & FORD, E.B. (1949). The quantitative study of populations in the Lepidoptera : 2. *Maniola jurtina* L. *Heredity, Lond.*, **3** : 67–84.

[13]   EASTWOOD, R.F. (1974). Personal communication.

[14]   FLOHN, H. (1965). Studies on the meteorology of tropical Africa. *Bonn. met. Abh.*, 1965 (5) : 57 pp.

[15]   GOODMAN, L.J. (1959). Reflex responses in flying insects. *Anim. Behav.*, **7** : 113–114.

[16]   GUNN, D.L., GRAHAM, J.F., JAQUES, E.C., PERRY, F.C., SEYMOUR, W.G., TELFORD, T.M., WARD, J., WRIGHT, E.N. & YEO, D. (1948). Aircraft spraying against the Desert Locust in Kenya, 1945. *Anti-Locust Bull.*, no. 4 : 121 pp.

[17]   GUNN, D.L., PERRY, F.C., SEYMOUR, W.G., TELFORD, T.M., WRIGHT, E.N. & YEO, D. (1948). Behaviour of the Desert Locust in relation to aircraft spraying. *Anti-Locust Bull.*, no. 3 : 70 pp.

[18] HAGGIS, M.J. (1971). Light-trap catches of *Spodoptera exempta* (Walk.) in relation to wind direction. *E. Afr. agric. For. J.*, **37** : 100–108.

[19] HASKELL, P.T. (1957). The influence of flight noise on behaviour in the Desert Locust, *Schistocerca gregaria* (Forsk.). *J. Insect Physiol.*, **1** : 52–75.

[20] HASKELL, P.T. (1966). Flight behaviour. *3rd Symp. R. ent. Soc. Lond.*, London 1966 : 29–45.

[21] JACKSON, C.H.N. (1941). The economy of a tsetse population. *Bull. ent. Res.*, **32** : 53–55.

[22] JOHNSON, C.G. (1969). *Migration and dispersal of insects by flight.* Methuen, London : 763 pp.

[23] JOYCE, R.J.V. (1968). Possible developments in the use of aircraft and associated equipment. *Chemy Ind.*, 27 January 1968 : 117–120.

[24] KENNEDY, J.S. (1939). The behaviour of the Desert Locust (*Schistocerca gregaria* (Forsk.)) (Orthopt.) in an outbreak centre. *Trans. R. ent. Soc. Lond.*, **89** : 385–542.

[25] KENNEDY, J.S. (1951). The migration of the Desert Locust (*Schistocerca gregaria* Forsk.). I: The behaviour of swarms. II: A theory of long-range migrations. *Phil. Trans. R. Soc. Ser. B.*, **235** : 163–290.

[26] KENNEDY, J.S., AINSWORTH, M. & TOMS, B.A. (1948). Laboratory studies on the spraying of locusts at rest and in flight. *Anti-Locust Bull.*, no. 2: 64 pp.

[27] KISIMOTO, R. (1971). Long distance migrations of plant hoppers, *Sogatella furcifera* and *Nilaparvata lugens. Proc. Symp. Rice Insects, Tokyo* 1971 : 201–216.

[28] LAWSON, T. (1971). Personal communication.

[29] MACCUAIG, R.D. & WATTS, W.S. (1963). Laboratory studies to determine the effectiveness of DDVP sprays for control of locusts. *J. econ. Ent.*, **56** : 850–858.

[30] MACCUAIG, R.D. & YEATES, M.N.D.B. (1972). Theoretical studies on the efficiency of insecticidal sprays for the control of flying locust swarms. *Anti-Locust Bull.*, no. 49 : 34 pp.

[31] McDONALD, D.J. (1955). The East African airspray campaign January–May 1955. Unpublished T/S : 20 pp.

[32] MALLAMAIRE, A. & ROY, J. (1959). La lutte contre le Criquet Pélerin (*Schistocerca gregaria* Forsk.) en Afrique Occidentale Française. *Bull. Prot. Vég., Dakar*, **1958** : 1–113.

[33] PEDGLEY, D.E. (1966). The Red Sea Convergence Zone. Part I: The horizontal pattern of winds. Part II: Vertical structure. *Weather, Lond.*, **21** : 350–358; 394–406.

[34] RAINEY, R.C. (1947). Observations on the structure of convection currents: meteorological aspects of some South African gliding flights. *Q. Jl R. met. Soc.*, **73** : 437–452.

[35] RAINEY, R.C. (1948). Coastal fronts in South Africa. *Q. Jl R. met. Soc.*, **74** : 199–200.

[36] RAINEY, R.C. (1951). Weather and the movements of locust swarms: a new hypothesis. *Nature, Lond.*, **168** : 1057–1060.

[37] RAINEY, R.C. (1955). Observations of Desert Locust swarms by radar. *Nature, Lond.*, **175** : 77.

[38] RAINEY, R.C. (1958). The use of insecticides against the Desert Locust. *J. Sci. Fd Agric., London*, **9** : 677–692.

[39] RAINEY, R.C. (1958). Some observations on flying locusts and atmospheric turbulence in eastern Africa. *Q. Jl R. met. Soc.*, **84** : 334–354. Discussion, *Ibid.*, **85** : 171–173.

[40] RAINEY, R.C. (1960). Applications of theoretical models to the study of flight-behaviour in locusts and birds. *Symp. Soc. exp. Biol. Cambridge*, **14** : 122–139.

[41] RAINEY, R.C. (1962). The mechanisms of Desert Locust swarm movements and the migration of insects. *11th Int. Congr. Entomol. Vienna* 1960, **3** : 47–49.

[42] RAINEY, R.C. (1963). Aircraft reconnaissance and assessment of locust populations. *2nd Int. agric. Aviat. Congr., Grignon* 1962 : 228–233.

[43] RAINEY, R.C. (1963). Meteorology and the migration of Desert Locusts: applications of synoptic meteorology in locust control. *Anti-Locust Mem.*, no. 7 : 115 pp. (Also as *Tech. Notes Wld met. Org.* no. 54.)

[44] RAINEY, R.C. (1965). The origin of insect flight: some implications of recent findings from palaeo-climatology and locust migration (Abstr.). *12th Int. Congr. Ent., London* 1964 : 134.

[45] RAINEY, R.C. (1969). Effects of atmospheric conditions on insect movements. *Q. Jl R. met. Soc.*, **95** : 424–434.

[46] RAINEY, R.C. (1972). Wind and the distribution of the Desert Locust, *Schistocerca gregaria* (Forsk.). *Proc. Int. Conf. Acridology, London* 1970 : 229–237.

[47] RAINEY, R.C. (1973). Airborne pests and the atmospheric environment. *Weather, Lond.*, **28** : 224–239.

[48] RAINEY, R.C. (1974). Biometeorology and insect flight: some aspects of energy exchange. *A. Rev. Ent.*, **19** : 407–439.

[49] RAINEY, R.C. (1974). Flying insects as ULV spray targets. *Br. Crop. Prot. Counc. Monogr.* No. 11 : 20–28.

[50] RAINEY, R.C. & ASHALL, C. (1953). Note on the behaviour of Desert Locusts in a light beam. *Br. J. Anim. Behav.*, **I** : 136–138.

[51] RAINEY, R.C. & ASPLIDEN, C.I.H. (1963). The geographical distribution and movements of Desert Locusts during 1954–55 in relation to the corresponding synoptic meteorology. *Anti-Locust Mem.*, no. 7 (Also as *Tech. Notes Wld met. Org.*, no. 54) : 54–103.

[52] RAINEY, R.C., BROWN, E.S., BETTS, E., HAGGIS, M.J., MOHAMED, A.K.A., ONYANGO-ODIYO, P., WINDSOR, D.E., WINSTANLEY, D. & YEATES, M.N.D.B. (1971). Experimental work with aircraft: exploring wind-fields for flying insects. *Rec. Res. E. Afr. Agric. For. Res. Org.*, **1970** : 106–112.

[53] RAINEY, R.C. & SAYER, H.J. (1953). Some recent developments in the use of aircraft against flying locust swarms. *Nature, Lond.*, **172** : 224–228.

[54] RAINEY, R.C. & WALOFF, Z. (1948). Desert Locust migrations and synoptic meteorology in the Gulf of Aden area. *J. Anim. Ecol.*, **17** : 101–112.

[55] RAINEY, R.C., WALOFF, Z. & BURNETT, G.F. (1957). The behaviour of the Red Locust (*Nomadacris septemfasciata* Serville) in relation to the topography, meteorology and vegetation of the Rukwa Rift Valley, Tanganyika. *Anti-Locust Bull.*, no. 26 : 96 pp.

[56] REGNIER, P.R. (1933). Les invasions d'acridiens au Maroc de 1927 à 1931. *Def. Cult., Rabat*, no. 3 : 139 pp.

[57] SAWYER, K.F. (1950). Aerial curtain spraying for locust control: a theoretical treatment of some of the factors involved. *Bull. ent. Res.*, **41** : 439–457.

[58] SAYER, H.J. (1956). A photographic method for the study of insect migrations. *Nature, Lond.*, **177** : 226.

[59] SAYER, H.J. (1962). The Desert Locust and tropical convergence. *Nature, Lond.*, **194** : 330–336.

[60] SAYER, H.J. (1965). The determination of flight performance of insects and birds and the associated wind structure of the atmosphere. *Anim. Behav.*, **13** : 337–341.

[61] SCHAEFER, G.W. (1970). Radar studies on locust, moth and butterfly migration in the Sahara. (Abstr.). *Proc. R. ent. Soc. Lond.* (C), **34** : 33. Discussion, *Ibid.* : 39–40.

[62] SCHEEPERS, C.C. & GUNN, D.L. (1958). Enumerating populations of adults of the Red Locust, *Nomadacris septemfasciata* (Serville), in its outbreak areas in east and central Africa. *Bull. ent. Res.*, **49** : 273–285.

[63] SIDDORN, J W. & BROWN, E.S. (1971). A Robinson light-trap modified for segregating samples at predetermined time intervals. *J. appl. Ecol.*, **8** : 69–75.

[64] SIMPSON, J.E. (1964). Sea-breeze fronts in Hampshire. *Weather, Lond.*, **19** : 208–220.

[65] SIMPSON, J.E. (1967). Aerial and radar observations on some sea-breeze fronts. *Weather, Lond.*, **22** : 306–316, 325 327.

[66] SIMPSON, J.E. (1967). Swifts in sea-breeze fronts. *British Birds*, **60** : 225–239

[67] SKELLAM, J.G. & RAINEY, R.C.T. (1973). Personal communications.

[68] TAYLOR, L.R. (1960). The distribution of insects at low levels in the air. *J. Anim. Ecol.*, **29** : 45–63.

[69] TAYLOR, L.R. (1962). The absolute efficiency of insect suction traps. *Ann. appl. Biol.*, **50** : 405–421.

[70] WALLINGTON, C.E. (1961). *Meteorology for glider pilots.* Murray, London: 284 pp.

[71] WALOFF, Z. (1958). The behaviour of locusts in migrating swarms. *10th Int. Congr. Ent.*, *Montreal 1956*, **2** : 567–569.

[72] WALOFF, Z. (1963). Field studies on solitary and *transiens* Desert Locusts in the Red Sea area. *Anti-Locust Bull.*, no. 40 : 93 pp.

[73] WALOFF, Z. (1966). The upsurges and recessions of the Desert Locust plague: an historical survey. *Anti-Locust Mem.*, no. 8 : 111 pp.

[74] WALOFF, Z. (1972). Orientation of flying locusts (*Schistocerca gregaria* Forsk.) in migrating swarms. *Bull. ent. Res.*, **62** : 1–72.

[75] WALOFF, Z. & RAINEY, R.C. (1951). Field studies on factors affecting the displacements of Desert Locust swarms in eastern Africa. *Anti-Locust Bull.*, no. 9 : 1–50.

[76] WIGGLESWORTH, V.B. (1963). *In* Discussion on 'The origin of flight in insects'. *Proc. R. ent. Soc. Lond.* (C), **28** : 23–32.

[77] WILLIAMS, C.B. (1930) *The migration of butterflies.* Oliver & Boyd, Edinburgh: 473 pp.

6 · Forecasting infestations of tropical migrant pests: the Desert Locust and the African Armyworm

ELIZABETH BETTS

Centre for Overseas Pest Research, London

Evidence has been accumulating during the present century that insects of many species from a wide range of orders regularly migrate distances of hundreds and sometimes thousands of kilometres, and that such migrations are wind-borne [27, 28, 31, 39]. To the individual farmer, grazier or forester it is the number of insects present at one time in one particular place which matters; and, while there are undoubtedly fluctuations in the total numbers of insects of a migrant species, it is immigration (including concentration) and emigration which dominate the characteristic contrast, at any one place, between long periods of total absence of these species and sudden devastating infestations.

Two such pests which characteristically appear suddenly in dense concentrations are Desert Locusts, *Schistocerca gregaria* (Forsk.) (Orthoptera Acrididae), and African armyworms, the larvae of *Spodoptera exempta* (Wlk.) (Lepidoptera Noctuidae). The scale of locust invasion has been described frequently, in terms ranging from the poetry of Joel [30],

> 'for a horde has overrun my land
> mighty and past counting',

to radar observations of flying locusts covering some 1,400 km² at densities of about one locust per 10 m³ up to a height of 1,500 m within 50 km of Delhi on 28th July, 1962 [43], with still more swarms present over a further 600 km across India and Pakistan. Such Desert Locust infestations attack a wide range of crops, as well as pastures and trees, within an area of some 30 million km² of sixty different countries in western, northern and eastern Africa and south-western Asia. Somewhat similarly, African armyworms have repeatedly been observed at densities of 100 larvae per m², and occasionally up to 3,000 per m², covering areas ranging from a few square metres to tens of square kilometres. A total area of some 15 million km² is subject from time to time to such infestations, which cause serious damage to cereal crops and pastures, most importantly in the eastern half of Africa from Ethiopia to South Africa; the species is also widely distributed throughout Africa south of the Sahara, and farther east as far as Hawaii [20].

For both these species, effective control requires warnings of when and where such large and dense populations are likely to occur. For both, it has been found possible to forecast in general terms the probable and possible occurrence of populations large enough to be economically serious, immediately or potentially. For Desert Locusts, warnings are needed both of the arrival of immigrant adult populations and of the onset of breeding, since this species is a pest in both the adult and the nymphal stages and since control is carried out against both these. For armyworms, on the other hand, warnings are needed only of the presence of larvae, as it is only these which damage crops and pastures, and there is no control of eggs, pupae or, at present, of moths. In both species services providing effective warnings of infestation have been evolved from studies of population displacements. These have been begun by utilizing the degree of seasonal regularity shown by population movements and breeding, and have later been extended by the use of the association of these movements with wind-systems.

The present paper outlines the methods which have been found effective in forecasting infestations of the Desert Locust over the past thirty years and the African armyworm during the past four years; discusses the varying forecasting problems these species have presented and the progress made so far to resolve them; and provides examples of the light this work has thrown on the ecology and population dynamics of these two species.

Desert Locust forecasting

Systematic and centralized collection and mapping of data on the changing distributions of the Desert Locust were begun in 1929 by the late Sir Boris Uvarov and Miss Z. Waloff. By 1943 sufficient was known of the seasonal changes in distribution [47] for regular locust situation reports, including forecasts, to be issued monthly by what later became the Anti-Locust Research Centre (now Centre for Overseas Pest Research), primarily to aid the planning of control operations by the wartime Middle East Anti-Locust Unit [46]. From 1958 to 1973 this service was supported by the countries concerned through the Food and Agriculture Organization of the United Nations, and in 1961 it was extended with the support of the UN Special Fund to enable current weather data to be used to assist in the interpretation and more detailed forecasting of the Desert Locust situation [25]. In July 1973 responsibility for locust forecasting was transferred to five regional organizations which, under the co-ordination of FAO, collectively cover the area subject to invasion by the Desert Locust [26].

Methods

Locust forecasts are formulated by assessing the current locust situation in the fullest possible detail (including the way in which it has developed as well as what developments are currently in progress), and then envisaging what developments might arise from this situation, mainly as a result of locust redistribution and often over substantial distances, but taking account also of any evidence of probable results of breeding and, more recently, of control operations. These assessments are based on reports of swarms (of adult locusts), bands of hoppers (nymphs), and scattered locusts, observed by the staff of national and regional locust control and plant protection departments, together with reports from local administrative officers, ships and various unofficial sources, amounting to thousands of sightings in months of heavy locust infestation. Even these represent only intermittent contacts with the locust populations present, even when these are large and still more so when they are low, for locusts even in large numbers can escape reporting for months at a time during displacements of many hundreds of kilometres [39, 47], particularly in the less inhabited parts of the Desert Locust invasion area. Continuity in the assessment of past and present locust distributions has consequently been found essential. This is particularly true during recession periods, when locust numbers are generally low but swarms and bands are occasionally reported within the populations migrating between and reproducing in seasonally complementary breeding areas.

Analysis of data to provide the necessary continuous interpretation of the changing spatial distributions and fluctuations must begin with visualization, that is with mapping of all near-contemporary records, usually monthly: on small scales (e.g. 1 : 11,000,000; 1 : 30,000,000) to show the spatial relationships of locusts reported in the whole distribution area of the species; and on medium scales (e.g. 1 : 500,000 to 1 : 4,000,000) to show the available basic details of the distribution and age of the populations present [3, 39]. Records of eggs and nymphs need to be related in time and space to the records of the relevant winged populations of the preceding and current generations, by subtracting or adding numbers of days appropriate to the durations of the egg and larval instars and the stages of development reached [9, 45*, 46a*, 48], respectively from or to the dates when eggs or hoppers were recorded. Added to the maps of reports of locust imagos, these inferred dates of egg-laying and estimated dates of potential emergence make possible an integrated assessment of the overall situation, enabling the forecaster to distinguish more clearly those of the populations previously reported which have apparently completed their migrations and breeding from those which are still likely to reappear and breed, as well as indicating where and when adults of the next generation can be expected to appear.

Such interpretations of the displacements and breeding that have led to the current locust situation are based on knowledge of the history of previous locust developments [47] and understanding of the way these have been governed by the weather [39, 40], and these in turn form the basis of the forecasts. Historical locust data are available in various degrees of analysis and synthesis, most effectively presented on maps for rapid appreciation and comparison. At the Centre for Overseas Pest Research, these include the detailed monthly maps described above; a wall-mounted collage of small-scale maps, providing a month-by-month summary of 30 years of Desert Locust history over the whole distribution area, representing a quick guide to the fuller data on potential analogues; maps such as Fig. 6.1 for each month [3] and each quarter [31], showing the seasonal probabilities of occurrence (and reporting) of swarms and of breeding in all areas; an atlas of small-scale maps summarizing the history of successive generations of locusts over periods totalling some 14 years; a series of comprehensive retrospective analyses of all available data over limited periods, e.g. May 1954–May 1955 [39], providing results such as those illustrated in Fig. 6.2 and Fig. 5.5; and tabulations of the dates when swarms have first reached various countries at certain seasons of the year.

Case studies of locust and weather situations [e.g. 39, 40, 41] illustrate the ecological principles governing locust migrations and breeding, and provide analogues to current situations. Evidence of concentration of insects in wind-convergence zones has already been presented by Rainey (p. 83); and Desert Locust migrations typically follow changes in the positions of these convergence zones, on scales ranging from a few kilometres to a few thousand kilometres [39, 47]. If changes of wind-field have occurred whilst the locusts have been 'earth-bound' either because it has been too cold for them to fly (below about 20°C), or during the juvenile stages (for some $1\frac{1}{2}$ to 4 months), particularly striking displacements occur when the locusts become airborne again, into a different wind-system, and are carried away towards convergence zones which by then may sometimes

* Unrealistic durations derived theoretically [45, 46] could be avoided by using observed values.

9

be thousands of kilometres distant. Such a mechanism appears very largely to account for the frequent but by no means universal association of long-distance migration with post-teneral flight in this species (and perhaps also in others), an association which Johnson [31] regards as especially significant in insect migration.

Weather developments likely to affect the current distribution of locusts can most readily be appreciated on synoptic charts mapping the weather observations [50], which are taken simultaneously all over the world, transmitted by radio and teleprinter in the course of standard meteorological forecasting practice, and can be made available anywhere within a few hours of the time of observation. In addition to pressure and frontal analyses of the surface observations, techniques and levels of upper-air analysis that have been found useful include streamline analysis [36, 37] of the 600 m pilot-balloon observations to show wind-fields roughly appropriate to the heights of flying locusts [4, 18, 19, 32]; contour analyses of the 850 mb pressure surface (approximately 1500 m above sea level), with corresponding temperatures, to show the major synoptic features and wind-fields [37, 39]; and similar analyses of the 500 mb surface (approximately 5500 m a.s.l.) and cloud observations, developed by D.E.Pedgley at COPR to facilitate recognition of rain producing situations.

Forecasting can begin when the probable scale, distribution and stage of development of current locust populations have been established. The first point is to note when, where and approximately on what scale new adult populations will be appearing and becoming ready to fly (about one week after fledging) using the methods indicated earlier. Thereafter longer-term forecasts, for about two months and sometimes further ahead, depend chiefly on seasonal probabilities derived from historical evidence as illustrated by Figs 6.1 and 6.2. Such material provides a guide to impending or continuing migrations of the winged locusts already present and of any others appearing during the forecast period; the areas where they might remain for a few months; and when and where they might breed. The use of meteorological data allows more precise shorter-term forecasts, for example when weather systems of types which are known to have been associated with migrations in previous years are recognized on the synoptic charts in or approaching areas where locusts are known or inferred to be present [39, Chap. 4], and/or when rainfall to the extent necessary for locust breeding (about 20 mm [34]) is recorded. Such warnings have been despatched by cable to countries threatened (Table 6.2).

EXAMPLES

Major swarm redistribution, September to November 1968

Between June and September 1968 (Fig. 6.3) breeding within the Inter-Tropical Convergence Zone (ITCZ) was particularly important in the Sudan and adjoining districts of north-western Ethiopia, and many young swarms were produced. On the basis of the swarm movements which had been established at this season in earlier years, as already illustrated for example in Fig. 6.2, the major populations were considered likely to fly from Sudan towards west, north-east and/or south-east, and from northern Somali peninsula south-westwards, breeding en route. These movements were accordingly forecast, and duly occurred, although the locusts did not get as far as Kenya and Tanzania, for which invasion had been indicated as a possibility.

Concentrating now on the most extensive of these long-distance displacements, towards the west (Fig. 6.4), the first of the intermittent contacts with this migrating

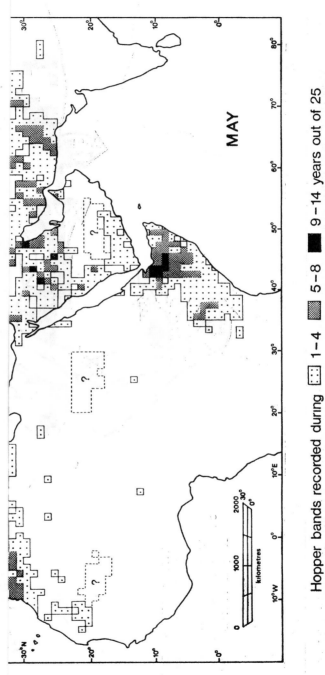

Hopper bands recorded during ⬚ 1–4 ▨ 5–8 ■ 9–14 years out of 25

MAY

Fig. 6.1. **Seasonal breeding areas of the Desert Locust.** Number of years during 1939–63 in which bands of nymphs were reported in each degree-square in each month; queries indicate areas from which reports are largely or wholly lacking.

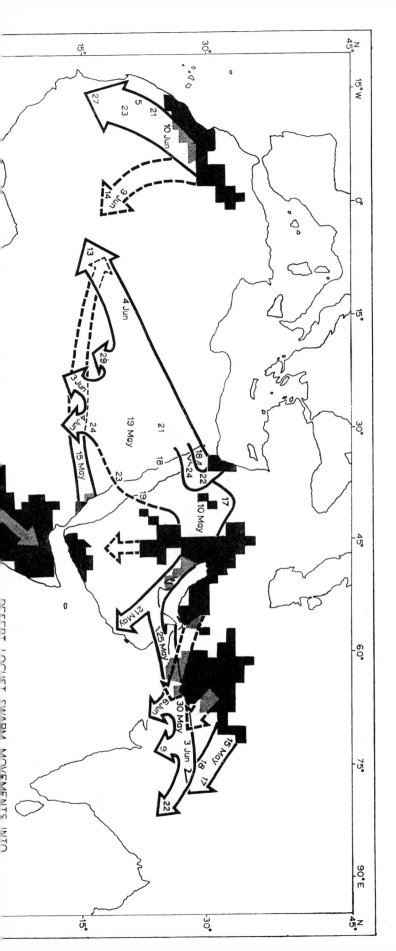

DESERT LOCUST SWARM MOVEMENTS INTO

population was an observation in Chad on 2nd October that alerted the West African locust control organization* to the possibility of further westward movement in their area (six days before the main wave of swarms in fact reached Niger), and a few days later enabled the Desert Locust Information Service in London to infer, because of

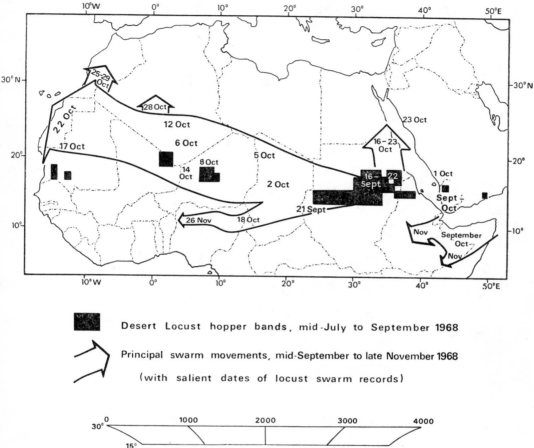

Desert Locust hopper bands, mid-July to September 1968

Principal swarm movements, mid-September to late November 1968

(with salient dates of locust swarm records)

FIG. 6.3. **Major redistributions of Desert Locust swarms;** late September to November 1968. Country names in Fig. 6.4. See Fig. 8.43 for radar picture of swarm in NW Niger on 15 October.

south-easterly winds, that swarms were moving north-westwards to Algeria (as indeed they were) and to warn Morocco and Spanish Sahara, some 3000 km distant, 15 days before they were invaded.

It is clear in retrospect, taking into account data available only later, that this generally westward migration occurred as a series of zig-zags, like those shown on a smaller scale by the tracks of individual swarms moving across Somalia in the ITCZ in September 1953, as in Fig. 5.7. Thus in mid-September 1968, a temporary northward oscillation of the ITCZ in Sudan appears to have caused a northward displacement of swarms, with many reported in the north along the Nile valley during 16th–22nd September at the

* Organisation Commune de Lutte Antiacridienne et de Lutte Antiaviare, representing and serving Mauritania, Senegal, Mali, Niger, Chad, Cameroun, Ivory Coast, Upper Volta and Dahomey.

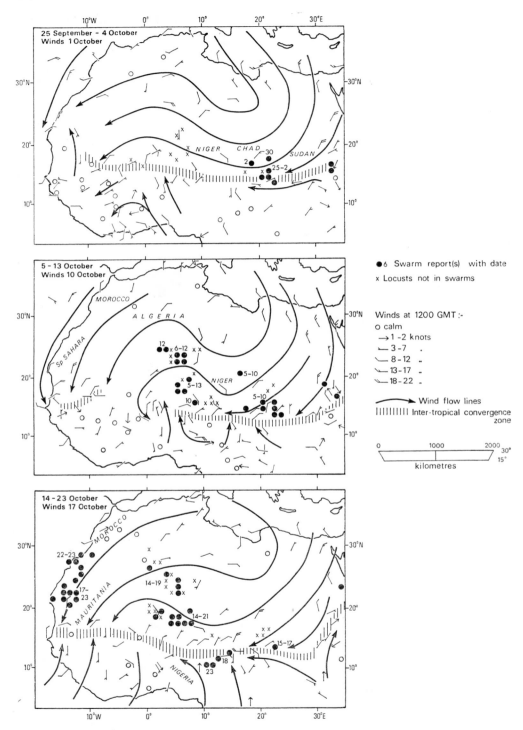

FIG. 6.4. **Wind-flow and Desert Locust migration:** Sudan to western Africa, 1968. Varying winds that transported locusts during this period included successively east-north-easterlies, south-easterlies and north-easterlies; see pp. 116–119.

same time as a disappearance of reported swarms from the inhabited areas in western Sudan, perhaps across the deserts to the north; later, north-easterlies brought swarms into Chad, sighted from 21st September onwards (though the date of the swarm sighting that was first to reach the locust control headquarters was not until 2nd October). Earlier recognition of the northward movement and of its significance could have allowed Chad up to five days' warning of imminent invasion. Next, swarms moved with south-easterlies into north-western Chad on 5th October and southern Algeria from 6th October and farther north on 10th–11th October. Later swarms apparently crossed the sand deserts of the western Sahara with east-north-easterlies and appeared again, in Mauritania, from 17th October. Subsequently there was a divide: some swarms continued moving to the south-west in Mauritania, still with north-easterlies; others came under the influence of south-westerly winds associated with a depression moving from the Atlantic towards the Mediterranean, and were carried into Morocco. Some swarms were still being produced in Sudan until early October and these apparently fed into an east-north-east wind and moved west-south-west, one of them through northern Nigeria as in 1956, but without being forecast on the 1968 occasion. Arrangements for more rapid transmission of data on breeding, especially hatching dates (plentifully recorded in Sudan), could have allowed a forecast that swarms were still being produced in early October and liable to migrate out in this direction.

Problems of assessment from incomplete data

A main factor in the quality of any forecast is the reliability of assessment of the current situation on which the forecast is based. This assessment has to take account of varying degrees of completeness or absence of data, including recognition of areas where locusts may frequently 'go to ground', and allowing for delay in receipt of reports. Such difficulties can occur at all levels of infestation but are greatest when populations are low, as is illustrated in the following three forecasting situations concerning comparable locust displacements but widely differing numbers of locusts (Fig. 6.5). In the heaviest of these seasons, 1961, swarms estimated to total some 500–650 km^2 had been produced in the Somali peninsula by January [33], many had moved towards the Ethiopian highlands, and by April oviposition had begun in eastern Ethiopia and north-western Somali Republic, with other swarms recorded farther east. In the assessment of the locust situation in early May, the most important unknown factor to be taken account of was the extent of the reservoir of swarms thought to be present in the Ethiopian highlands, where swarms have frequently gone to ground in the past and from which most of the reports for March–April 1961 were not available until later. A large population had entered the area, and since there was no evidence of any subsequent immigration or emigration, it was inferred that a substantial population remained unaccounted for. In previous seasons south-westerly winds, extending across the Somali peninsula as the ITCZ moves seasonally northwards, have frequently brought such swarms eastwards, to oviposit in the Ogaden and Somali Republic, and in 1955 the onset of such a displacement of swarms had been noted to have begun when the south-wester became established at Belet Uen. The appearance of south-westerlies at this or neighbouring stations was accordingly looked for when daily weather observations began to be used to aid locust forecasting, and was successfully used in 1961 to provide warning of an extension of the locust breeding area some 400 km eastwards [39].

Assessment of the extent of the swarms available to appear and oviposit in the same area at the same season of the next year was complicated by mutiple migrations, associated with inevitably incomplete data. Thus swarms produced in southern and central Somali peninsula early in 1962, probably on a smaller scale than the previous year, had divided, some (perhaps totalling 200 km²) going into East Africa [23], others to the Ethiopian highlands, where they appear to have been augmented in late January to

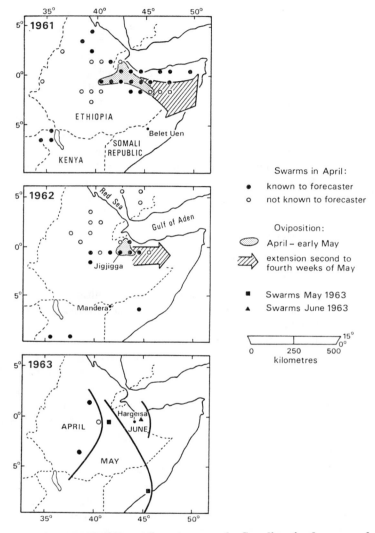

FIG. 6.5. **May–June spread of Desert Locusts across the Somali peninsula:** years of declining infestations and decreasing information.

early February by swarms derived from breeding in the coastal areas around the Gulf of Aden and southern Red Sea [23], and then depleted in mid-March by swarms emigrating northwards across the Red Sea towards Saudi Arabia [38]; breeding began in eastern Ethiopia in late March and north-western Somali Republic in April. The proportions of locusts involved in these various activities was unknown. To these problems were added interruptions in radio transmissions of meteorological data, which delayed comprehensive evidence of the onset of the south-west monsoon across the Somali peninsula until 9th May, although the very few observations for 6th May included south-westerlies in north-eastern Kenya (Mandera) and eastern Ethiopia (Jigjigga), and it is clear in retrospect that it was from this day that the ovipositing swarms began to shift some 250 km farther east (although much of the laying had already occurred [23]). In these circumstances the warning of such a shift in the breeding area, issued on 9th May, was both timid and too late.

As locust swarms became comparatively rare during the following year the problems of assessing the situation became correspondingly more acute, and centred on the reliability of the few reports received. These included a few records of swarms, some confirmed, others not, in seasonally appropriate areas of eastern Africa in every month from June 1962 until January 1963 and again in April, as well as a few records of hopper bands. Two of the three Ethiopian reports for April were available to the forecaster, but both messages were initially corrupt, possibly due to radio interference: most importantly the fact that authentic specimens of *S. gregaria* had been removed from the car radiator of the casual observer who had driven through one of the swarms was omitted from the first message, so that this critical evidence of identification was delayed. Both reports had initially to be regarded as 'unconfirmed'; nevertheless it was still inferred that some locusts remained and might possibly reappear in Ogaden and north-western Somali Republic as the south-west monsoon set in. Weather data were again incomplete on the critical days but included a south-wester at Hargeisa on 4th May, two days before a swarm appeared east of the April records and closer to and thus more accessible to the Hargeisa control base. Warning of an eastward movement was however not issued until fuller evidence of the advance of the south-west monsoon was received on 10th May. Subsequently there were two further reports of swarms farther east in May and June, suggesting that a seasonal eastward shift had indeed occurred.

Thus, consideration of the uncertainties often facing the forecaster draws attention to the need in assessing the locust situation to take account of *all* available observations, including reports still awaiting confirmation, and to maintain confidence in the assessments made, despite paucity of data, as well as to be continuously aware of the locust precedents.

VERIFICATION: ASSESSMENT OF FORECASTS

Subsequent verification of forecasts enables the forecaster to pin-point the causes of forecasting errors, particularly inadequacies in the interpretation of the available data, as well as absence of critical data or delay in transmission of reports. Precise elucidation of these difficulties indicates the action needed, in the form of research as well as of administrative measures. The recent division of forecasting responsibilities (p. 114) may well provide opportunities for forecasters to arrange for locust data to be transmitted more effectively, rapidly and completely, particularly where responsibility for forecasting now resides with the organizations directly responsible for locust control.

The second purpose of forecast verification is to indicate to recipients and subscribers the reliability of the service provided. Tables 6.1 & 6.2 summarize the results of Desert Locust forecasting over a six-year period, which included invasions of 33 countries, often several times [24]. 85 per cent of predictions of high probability of swarms or hopper bands in a country materialized, as did 64 per cent of all 1280 predictions made; on only 15 occasions did a country fail to receive advance warning of impending arrival of swarms or of impending breeding. Forecasting during a period when swarms were widespread was more confident than when swarms were sparse, but was in effect no less reliable under recession conditions. The above discussion of some of the events of 1961–1963 in eastern Africa suggests ways in which this confidence could be improved.

TABLE 6.1. **Verification of monthly Desert Locust forecasts** during a 6-year period of varying levels of infestation [24]

	Predictions of probability			Total
	high	intermediate	low	
	% correct (number)	% correct (number)	% correct (number)	% correct (number)
Swarm activity widespread (summaries issued August 1960 to March 1961)	85 (110)	45 (73)	12 (26)	62 (209)
Infestations heavy in east only (April 1961 to July 1963)	88 (300)	62 (167)	34 (116)	70 (583)
Recession (August 1963 to October 1965)	74 (104)	48 (142)	42 (242)	51 (488)
Total (August 1960 to October 1965)	85 (514)	54 (382)	40 (384)	64 (1280)

Each 'prediction' was a forecast that locusts would be present and/or breeding in a country, included in the summaries published in 63 successive months by the FAO Desert Locust Information Service of the Anti-Locust Research Centre (DLIS).

High probability	Intermediate probability	Low probability
predictions expressed by:	predictions introduced by	predictions expressed by:
'will' 'expected' 'probable' 'likely' 'imminent'	'may'	'possible' 'possibly' 'the possibility cannot be excluded' 'perhaps' 'must be sought' 'looked for' 'vigilance needed' 'need to be alert'

On 15 occasions locusts were recorded in individual countries without prior warning.

TABLE 6.2. **Verification of cabled Desert Locust warnings**
DLIS: March 1961 to December 1965 [24]

Correct	23
Partially correct	22
Developments as expected but had already occurred when cable sent	9
Incorrect	9
Information insufficient for verification	10
Total	73

ARMYWORM FORECASTING

Armyworm forecasting in East Africa was similarly developed from a study of the migrations of the species. Following heavy and unexpected infestations there during 1961, full-time research on armyworm ecology was begun at the East African Agriculture and Forestry Research Organization (EAAFRO), led by the late E.S.Brown of the Commonwealth Institute of Entomology. The migration controversy in the earlier literature [14] was resolved by Brown and Swaine [17], who concluded, from evidence of changing distributions of moths and larvae, and observations of the emigration of newly-emerged moths from a pupation site, that *S. exempta* does migrate. At this stage research was expanded, in co-operation with the Anti-Locust Research Centre (now Centre for Overseas Pest Research) and with the encouragement and assistance of the Desert Locust Control Organization for Eastern Africa and particularly Mr R.J.V.Joyce, then Director of the latter organization, to take account of the experience acquired during the development of locust forecasting (p. 114–122). Adaptations of these methods included both more intensive studies of specific armyworm situations and a more extensive approach to the seasonal and other changes of distribution of *S. exempta* in the whole of Africa south of the Sahara and elsewhere. Dominating effects of wind-systems on moth migrations were established; records of moths and larvae were integrated into assessments of overall situations; and seasonal exchanges of population between East Africa and Ethiopia were elucidated [15], although probable migrations between East Africa and areas to the south and west are still little understood [7, 8].

As a result, it became practicable to issue weekly forecasts of infestations of armyworm larvae in East Africa during the season November 1969 to June 1970 as an experiment [12, 13], and from the beginning of the following season as a regular service, now run by EAAFRO with the support of the agricultural departments of Kenya, Tanzania and Uganda [16, 35].

METHODS

The experimental forecasts were based on assessments of the armyworm developments expected in the light of all available data on the recent distribution of the species. Essential for determining the latter were nightly records of moth catches at Robinson-type light-traps (125 W MV and 160 W mixed tungsten and MV bulbs) at some 19 sites spread across East Africa [15], which were telegraphed to EAAFRO each week and there presented visually on a wall graph showing each night's catch at every trap. These moth catches were complemented by estimates of the dates of potential moth emergence at recorded infestations of larvae, also presented visually on both the wall graph and on maps, which thus indicated the distribution in space and time of the major source areas of moths. Inferred dates of oviposition at known infestations assisted in the interpretation of recent events. Finally, examination of the East African Meteorological Department's synoptic and rainfall charts indicated areas of probable wind-convergence and (on the hypothesis that moths move down-wind) of possible moth concentration.

Having decided where large moth populations were probably present, it was next necessary to consider whether egg-laying had occurred or was occurring at densities sufficient to cause outbreaks of larvae. Current catches were compared with those made at the same traps in previous years on about the same date, noting whether these earlier catches had been followed by infestations, and taking into account catch-reductions associated with full moon.

When forecasting began in November 1969, armyworms had been sparse in eastern Africa for three seasons and in central and southern Africa during one season, so that to begin with there was no reason to expect the 1000 km^2 of infestations (Fig. 6.6) which appeared progressively across East Africa between December 1969 and June 1970. In mid-December moths began to be taken at Ilonga trap, the first there for over three months, and catches rose rapidly (within five nights) to 324 moths in a night. Since:
(1) these were clearly immigrant and not locally produced moths,
(2) in five preceding seasons catches of more than 100 at this trap had been followed by infestations of larvae in the surrounding Kilosa district [10, 11], and
(3) infestations had occurred in Kilosa and/or the neighbouring Morogoro district in six out of eight preceding Decembers (though not in the last two),
a high probability warning was clearly indicated and duly issued, ahead of the first recorded infestations of the season, which were some 85 km from the trap (Fig. 6.9).

At the beginning of the next season, in November 1970, the first important moth catches were at Mombasa. Because:
(1) there had been infestations in Ethiopia in May–June and Somali Republic in June,
(2) there is evidence of a seasonal exchange of populations between Ethiopia/Somali Republic and East Africa in some years [15, pp. 685–687],
(3) The first moths in East Africa have often appeared at the leading edge of advancing north-easterlies, and have been followed by infestations [15, pp. 696–703],
and though (unlike the previous example) infestations of larvae in this area and at this season were unprecedented, Brown and Odiyo issued a successful warning of infestations which subsequently appeared in the Kenya coastal area [16, 35].

Turning now to problems of mid-season forecasting [11], *S. exempta* moths were caught in at least one and often many of the light-traps in East Africa, on every night from December 1969 to June 1970 inclusive, indicating a continued presence of moth populations throughout the season. Marked changes of catch level, however, at traps in different areas, and the evidence provided by changes in the distribution of the infestations of larvae reported during this period (Fig. 6.6), indicated large-scale immigration into the region at least until January, and emigration from April onwards; there were also substantial redistributions of populations within East Africa, mainly from Tanzania towards Kenya, with a minor overflow into Uganda. By early February 1970 heavy infestations of armyworms had become widespread in northern Tanzania and south-western Kenya (Fig. 6.7). Head-capsule measurements on samples of larvae established the instar and enabled the age of these infestations to be inferred, and thus it could be estimated that moths probably emerged from some of these infestations by mid-February and from the majority during the period late February to early March. Within a few hours of emergence they are likely to have emigrated from their breeding areas [15, 17], launching themselves into the wind-fields of east Africa, probably flying throughout the night [6] and possibly on most nights for about a week [15].

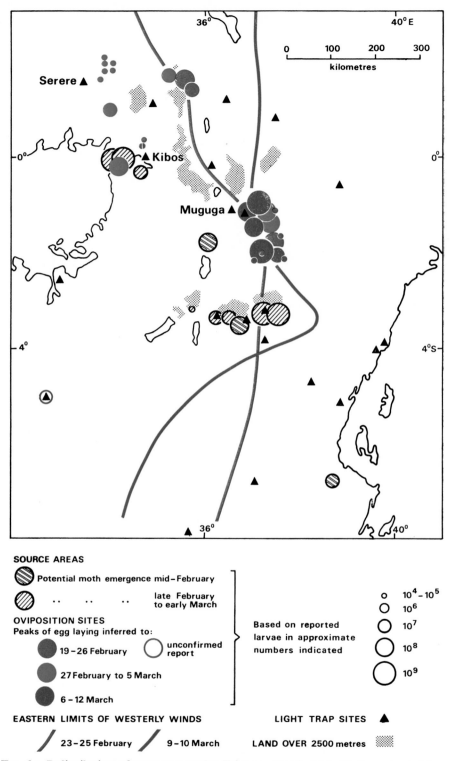

SOURCE AREAS

◐ Potential moth emergence mid–February

◑   ″   ″   ″   late February to early March

OVIPOSITION SITES
Peaks of egg laying inferred to:

● 19–26 February   ○ unconfirmed report

◐ 27 February to 5 March

● 6–12 March

Based on reported larvae in approximate numbers indicated

○ 10^4–10^5
○ 10^6
○ 10^7
○ 10^8
○ 10^9

EASTERN LIMITS OF WESTERLY WINDS

╱ 23–25 February   ╱ 9–10 March

LIGHT TRAP SITES ▲

LAND OVER 2500 metres ▒

FIG. 6.7. **Redistributions of armyworm moths**: February–March, 1970. Moths emigrated from the source areas with predominantly south-easterly winds, interrupted by two major incursions of westerlies of which the eastern limits are shown; most of the subsequent breeding occurred near these limits.

[*facing page* 125]

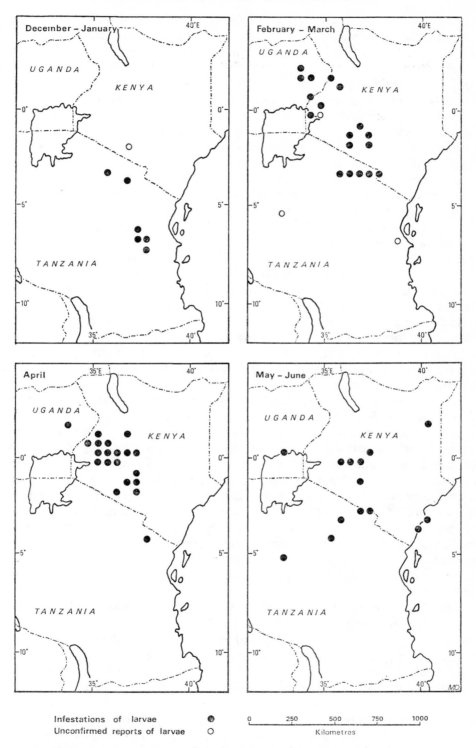

Infestations of larvae ●
Unconfirmed reports of larvae ○

0   250   500   750   1000
Kilometres

FIG. 6.6. **Infestations of armyworm larvae in East Africa in 1969–70.**

At this season of the year the ITCZ begins to return northwards, and predominantly south-easterly winds progressively replace predominantly north-easterly winds across much of East Africa. In addition, the Congo Air Boundary sometimes moves eastwards, allowing temporary incursions of westerly winds to extend across the Kenya highlands (p. 96); such incursions were particularly common during the 1969–1970 armyworm season [29]. Superimposed upon these synoptic-scale wind-systems are the diurnal circulations of alternating land- and sea- or lake-breezes, and of anabatic (daytime up-slope) and katabatic (night-time down-slope) winds. Convergence associated with all these systems is liable to produce storms which provide more intense though less persistent zones of convergence of their own, particularly as down-draught winds spread out ahead of storms (Fig. 8.35, p. 182 and Fig. 5.23, p. 107).

Winds of all these types appear to have affected the displacements of the moths concerned, which emerged, in the main, into southerly and south-easterly winds, and accordingly showed a general tendency to drift towards the north and north-west. Two major incursions of westerly winds, on 23rd–25th February and 9th–10th March 1970 (Fig. 5.16) reached the eastern limits shown in Fig. 6.7, and large-scale infestations of larvae subsequently occurred near these limits. The dates of peak egg-laying inferred for most of the major infestations in Kenya fell at about the time the two incursions reached their eastern-most limits, and no laying was reported beyond these.

Light-trap catches of moths at three traps near areas of larval infestation (Fig. 6.7) during this period are shown in Fig. 6.8. In interpreting these catches, two factors must be taken into account, though neither can yet be assessed quantitatively. First, catches vary from trap to trap because of site differences and differences of operating procedure; data from different trap-sites must accordingly be assessed by taking account of the history of trapping at each individual site, in these cases available for seven previous years at the time when the forecasting experiment began. The Muguga trap had a history of high catches, Serere of moderate catches, and Kibos of much lower catches, partly because it had been run during evening hours only, and less frequently. Accordingly, in the period under study, the peak catches at Kibos and Serere were considered more important than the catches of comparable numbers at Muguga during 4th–7th March. Secondly, catches are less at full moon than at new moon [15]. Thus the catches at Muguga on 23rd and 24th February, the third and fourth nights after full moon, can be interpreted as indicating a large population in the vicinity, but the zero catches at Kibos and Serere three nights after full moon cannot by themselves be taken as evidence of low populations. Conversely, the falling catches shortly before new moon at Kibos and shortly afterwards at Serere can be interpreted as likely to indicate real reductions in local populations.

Allowing for these qualifications, substantial changes in moth distribution are revealed by the light-trap data presented in Fig. 6.8. Catches at Kibos in western Kenya increased during what was both the estimated period of emergence from the main infestations in Nyanza, some 75 km west of the trap (although the beginning of this increase in airborne moth populations may have been missed by the trap because of the effect of moonlight), and also a time when new oviposition was occurring, in a small area some 12 km south of the main emergence area. Catches decreased at the end of the estimated period of peak emergence in Nyanza. Moths also arrived, probably from this source-area, in a new breeding area some 200 km farther north-west, in Teso district of east-central Uganda, as revealed by the rapid rise in catch at Serere. This was at the time of inferred egg-laying at a number of points within 45 to 100 km of that trap; by 9th–10th March,

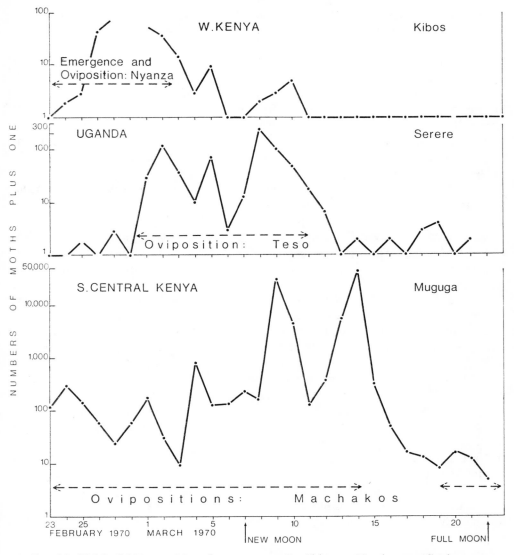

FIG. 6.8. **Nightly light-trap catches of armyworm moths;** February–March, 1970. Catches are dated to the evening of trapping.

at about the time egg-laying was ceasing in these areas, catches at Serere dropped. In south-central Kenya, in the area to which moths emerging from the heavy infestations in northern Tanzania and neighbouring areas of Kenya appear to have been migrating, the Muguga trap indicated that moths were present on every night during the period shown, consistent with the inferred dates of egg-laying in Machakos and neighbouring districts. Particularly abundant populations were indicated both by the catches on 23rd and 24th February, which were unusually high for nights so close to full moon, and the unpredecentedly high catches on 9th and 14th March. The evidence of these light-trap catches at Muguga was supported by catches in a low-level (1·5 m) suction-trap, which amounted to 200 and 70 moths respectively on the nights of 9th and 14th March, the highest catches in 5¼ years' operation of this trap and indicating an exceptionally high aerial density of moths locally (pp. 96–98); likewise on 24th February 8 moths had been taken in this suction-trap, a number exceeded on only 1 per cent of all the nights the trap operated.

These two main periods of high catch at Muguga coincided with the two incursions of westerly winds across the area, and on such nights, including that of 9th–10th March (Fig. 5.17), catches often increased in the actual hour when westerly winds set in [29]. Further afield, the arrival of moths at Serere (and also at Kawanda near Kampala) appears to have been delayed until the westerlies retreated in late February, and the second incursion of westerlies appears to have caused high catches at Serere as the convergence zone moved across, followed by an abrupt disappearance of moths from traps in Uganda and western Kenya.

Such were the main features of a situation in which moths emerging from known infestations appear to have bred mainly some 150–250 km distant from their source-areas, though in one instance perhaps only 12 km distant.

How much of this was the forecaster able to interpret at the time and how complete were the warnings? The areas and periods of moth input were predicted, and information on the main weather features was available. The significance of the second westerly incursion was picked up, but not that of the earlier one, partly because it occurred in the full-moon blind spot of the light-trap data and partly because of difficulties with communications. It was accordingly possible to give a strong warning of the heavy infestations in south-central Kenya and to warn of the other infestations in Kenya and Uganda, but no warning was given of the infestations in West Pokot district, the most northerly in Kenya, lying 75 km from the nearest trap, Kitale, where the catches had risen but only to four moths. Furthermore infestations were unprecedented in this particular district in March, although they had been recorded in previous years both in February and April, a point overlooked by the forecaster. Unfulfilled warnings of moderate and low probability had been issued for a number of other districts, including the previously infested areas of northern Tanzania on the strength of high moth catches in this area and the fact that breeding has often continued here until May. In fact, all the populations of mid-March apparently emigrated from this area without breeding. The question of distinguishing, in an area where high catches undoubtedly represent locally emerging moths, between circumstances in which emigration will become complete and others in which further breeding will occur in the same general area, is one on which further work is required.

VERIFICATION AND DISCUSSION

Verification of armyworm forecasts during the first experimental season in 1969–70 (Fig. 6.9) showed most infestations were forecast to the right district and the right week, and similar results continued through subsequent seasons [35, 35a].

The armyworm forecasts issued so far have, in the main, attempted predictions only one generation ahead, as this was considered likely to be both sufficiently reliable and far enough ahead to be useful. However, helpful longer-term inferences about future moth migrations can occasionally be drawn, on the basis of the recent distribution of larger populations and knowledge of the more usual seasonal changes in distribution. Thus populations in East Africa (Kenya, Tanzania and Uganda) have frequently migrated towards Ethiopia as the ITCZ has moved seasonally northwards [15]. Accordingly, in March 1970, following heavier infestations in East Africa during the previous three

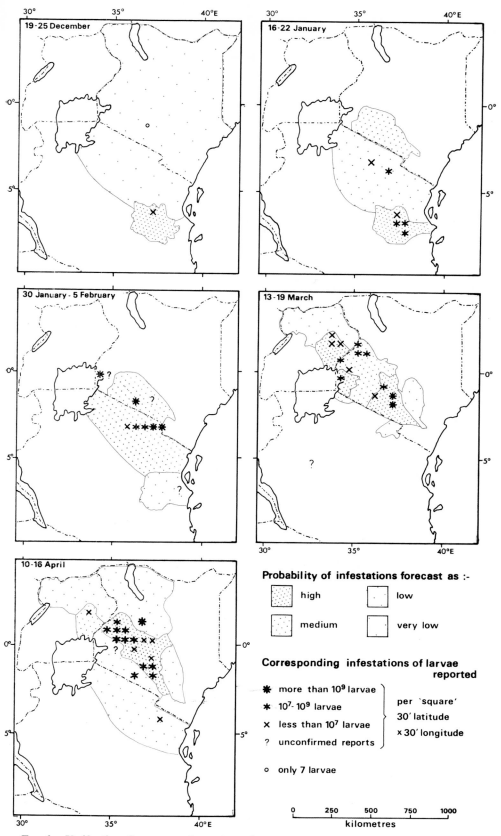

FIG. 6.9. **Verification of armyworm forecasts:** 1969–70.

[*facing page* 128]

months than at any time during the previous $3\frac{1}{2}$ years, a general warning that moths on a scale sufficient to cause serious infestations of larvae might begin to invade Ethiopia, from mid-March onwards, was sent to the central Ethiopian agricultural station concerned. Here it was used to warn other Ethiopian stations, before the occurrence of infestations of larvae in May 1970, the first to be recorded in Ethiopia in two and a half years. The same general warning was also sent to the Desert Locust Control Organization for Eastern Africa which subsequently participated in control operations against these armyworm infestations both in Kenya [21] and later in the season in Ethiopia [22].

Similar forecasting systems would be technically practicable for other areas subject to *S. exempta* infestations, provided records of moths and larvae were available for a minimum of some four or five seasons of reasonably widespread infestations. In fact, a successful warning of imminent armyworm infestations in Rhodesia, following a high catch of moths in a light-trap at Salisbury, had already been issued by Whellan as far back as December 1955 [49]. A network of light-traps has more recently been established in South Africa and neighbouring countries with possible armyworm forecasting as one of its aims (44), and some warnings have recently been issued in Botswana [1, 2].

CONCLUSIONS

These few examples show how forecasts of infestations of migrant pests need to be based on a continuous assessment of population distribution, including both movement and breeding, interpreted with knowledge of the species' history and the factors, notably current weather systems, governing such developments. These interpretations are usually based on analogues from the history of the species, but the mechanisms of these migrations now appear to be understood sufficiently for successful forecasts to have been made even in situations without recorded precedent.

Forecasts have not normally been made more than one generation ahead, involving a longer time-scale for Desert Locust than for African armyworm, because the life-cycle of the former lasts longer than that of the latter. Successful longer-range forecasts have however sometimes been made, as illustrated by the 1970 armyworm invasion of Ethiopia. Again, warning of the first appearance of Desert Locust swarms in Kenya in late 1950, after an absence of over three years, had been given by P.R. Stephenson some seventeen months ahead [39], on the basis of a likely succession of locust migrations of the kind then already on record.

Verification of forecasts has been found essential for two purposes: to show up forecasting errors and omissions as a pre-requisite for forecast improvement; and to guide recipients as to the reliability of the warning service provided. Detailed examination of what went wrong and why is the most useful approach for the former purpose. Two methods, tabulated percentages (Table 6.1) and maps (Fig. 6.9) to show which forecasts materialized, provide some guidance in the latter, although neither is an entirely satisfactory form of presentation. For practical purposes it has been found more profitable to concentrate on learning how to improve forecasts, than to disentangle the three closely inter-related variables in forecasting success; the probabilities of an infestation occurring, of it being forecast, and of it being reported. Successful pest forecasting depends upon a continuously developing interpretation of the fluctuating fortunes of the pest, so that the forecaster, working always and inevitably with data less complete than would be wished, may acquire, retain and increase skill at picking out the significant pointers to current and future developments.

Clearly these approaches can be valuable to the problems of forecasting other species of migrant pests.

These two forecast systems evolved from studies of insect migration, and their operational needs have provided a vast feed-back of information on the behaviour and population dynamics of the species concerned, over extensive areas and long periods. This has provided a very much better picture for example of long-range population displacements than is available for species for which the observations depend on fewer persons, working with less co-ordination, in more restricted areas and often less continuously. Thus it has been possible to keep track of the history and development of successive generations of most of the major populations of the Desert Locust throughout the last thirty years—even when populations were low. This appears also to be true of the African armyworm for much of the year within East Africa, but study of this species has been a more recent venture, and we are still ignorant of where the major populations are from August to October in most years.

ACKNOWLEDGEMENTS

My greatest debts are to the countless people who have suffered the depredations of locusts and armyworms and the very many who have been concerned with controlling these pests, for they have provided the need and stimulus for pest forecasting as well as the basic data used. I wish to record my thanks to the late Sir Boris Uvarov, K.C.M.G., F.R.S., and to Miss Z.Waloff, O.B.E., whose Desert Locust forecasting team I was privileged to join; to Dr R.C.Rainey who metamorphosed locust forecasting by introducing the use of daily weather data and who has greatly encouraged my contributions to pest forecasting, including inviting this paper; and to many colleagues in the former Desert Locust Information Service. The development of the application of meteorology to locust forecasting was financially supported by more than forty countries liable to locust invasion and by the UN Special Fund (now Development Program) co-operating through the Food and Agriculture Organization, its Desert Locust Control Committee and its Locust Specialist, Mr Gurdas Singh, to all of whom my grateful thanks are due.

For the opportunity to investigate armyworm migrations and develop armyworm forecasting I wish to acknowledge the ready support of Dr O.Starnes and Dr B.Majisu, past and present Directors of the East African Agriculture and Forestry Research Organization; the Chief Research Officers and other staff of the ministries and departments of agriculture in Kenya, Uganda and Tanzania; the painstaking co-operation of light-trap operators and the staff of the Armyworm Division at EAAFRO; and particularly the support of the former Head of that Division, the late Mr E.S.Brown, whose field investigations provided the basis of this work.

I am grateful to the Director of the Centre for Overseas Pest Research for permission to contribute to this symposium.

DISCUSSION

Mr P.Onyango-Odiyo (speaking at the invitation of the Chairman, **Dr Gunn**): I inherited the work of Miss Betts' initial programme on armyworm forecasting in East Africa and although there have been problems, particularly with light and irregular outbreaks in subsequent seasons, the forecasts have been sufficiently successful for sustained support from the governments of Kenya, Tanzania and Uganda.

Professor J.W.S.Pringle: May I make a small criticism, perhaps a suggestion? Data presented on the success or otherwise of a forecasting service is difficult to appreciate in the form you gave it to us [an earlier version of Fig. 6.9]. Would it be possible to get some sort of percentage figure for what you reckon to be your success and failure rates?

Betts: On account of time I omitted a slide [Table 6.1] that gave just such data on locust forecasting. It shows, for example, that during a six-year period 85 per cent of a total of 514 predictions of high probability of occurrence of gregarious infestations in individual countries actually materialized.

Pringle: But that is not enough.

Betts: Doing this on a numerical basis raises considerable difficulties in deciding just what such percentages really mean. In practice, for providing guidance for those who are directing control operations, the forecaster needs to learn all the time where a forecast went wrong, and this I think is one of the chief values of verification.

Gunn: When an area is forecast to be invaded, and only part of it is actually reached, is this 100 per cent failure or 100 per cent success—how can you measure it?

Pringle: But it seems to me a forecasting service ought to put effort into devising a method of testing whether it is a good service or not.

Betts: I have tried a number of different ways of doing this for armyworm forecasts, but in the end had concluded that it is a graphical presentation such as that of Fig. 6.9 which can be appreciated most readily.

Onyango-Odiyo (partly communicated): Field verification of the armyworm forecasts may appear less complete than might be wished, but it should be explained that although the EAAFRO forecasting centre at Muguga issues the forecasts, it is the staff of the three separate national governments who are primarily responsible for investigating and reporting outbreaks, and in the absence of special follow-up programmes, involving field scouts trained for verification, the armyworm forecaster can only assume that what has been reported by agricultural personnel, farmers and the public represents what was in the field. We have to depend on what was reported and to assume that if nothing was reported, then nothing occurred.

Gunn: This is always a difficulty about such a service; you may ask for 'nil' reports, but when nobody bothers you do not know for certain whether nothing has really happened.

Onyango-Odiyo: The basic object of the scheme is to save the immediate crops, and, with appreciation of increasing successes in the course of the four years of forecasting so far, expressed by ministers from Kenya, Uganda and Tanzania, we are encouraged to try to do even better. The EAAFRO Armyworm Division has also fostered awareness of the problem through the publication of an illustrated booklet, in Swahili and in English, as well as through the weekly press reports which are often broadcast over the radio and sometimes by television.

Rainey: On this vital point of verification: it is certainly a complicated subject, but, from the point of view of the man concerned for example with directing the locust control organization in a particular country, one can narrow down his requirements to wanting to know, crudely speaking, whether swarms are going to come or not—whether or not he will need more resources. And what he would like to know in addition is, when he is told that the arrival of swarms is 'likely', does this mean a 70 per cent or a 50 per cent or a 30 per cent chance of swarms arriving? Now the degree of verification which we undertook for the FAO locust warning service, in Tables 6.1 and 6.2, did in fact provide just that kind of evidence to those responsible for control operations, enabling them to put a figure on what was the probability of swarms arriving when they were forecast with any of the three grades of probability quoted.

[Communicated] This kind of verification also provides a convenient rough indication of the degree of usefulness of the service to the recipient countries, and in my view may well have been a factor in the sustained support for this service provided by FAO and all countries concerned throughout the period that verification of this kind was regularly undertaken and reported. But where the real complications begin is in any attempt to calculate a rigorous 'figure of forecasting merit', which must not only take account of the proportion of forecast events which actually occur (as well as of events which occur without having been forecast), but also, for each separate class of event, must weight this proportion of successes inversely both by the long-term probability of occurrence of the event and by the amount and value of directly relevant current information available to the forecaster. We have therefore come to regard detailed examination of the manner in which the actual development of the situation has differed from that envisaged in the forecast, as more useful than attempting to evaluate any such quantitative figure of merit.

Professor R.S.Scorer: I should like to support Miss Betts' point, that the learning process is the most important. What she and her colleagues are in effect doing is participating in the control of these pests, and if you simply take out the forecasting element and try to assess their success by that element alone, you really are not measuring the value of the effort.

Betts: What you need to learn from your verifications are the organizational matters which may need attention (like arrangements for payment for telegrams from light-trap operators!), as well as the occasions on which you have failed to recognize or use available analogues, or for which data from earlier seasons providing potentially relevant analogues were available but insufficiently studied or assimilated. While in principle I agree with Professor Pringle, in practice, if there's only a small team available you have to concentrate on the main issue of continuously trying to improve your understanding of what is actually happening.

REFERENCES

[1] ANONYMOUS (1973). Fears of armyworm expressed. *Agrinews Agriviews*, **4** (3) : 15.
[2] ANONYMOUS (1973). Armyworm outbreak. *Agrinews Agriviews*, **4** (4) : 10.
[3] ANTI-LOCUST RESEARCH CENTRE (1966). *The locust handbook*. Loose-leaf, London.
[4] ASPLIDEN, C.I.H. (1963). The work of the WMO technical assistance mission for Desert Locust control. In Bargman, D.J. (ed.) *Tropical meteorology in Africa (Proceedings of the Symposium jointly sponsored by the World Meteorological Organization and the Munitalp Foundation, Nairobi, December 1959)* : 420–426. Munitalp Foundation, Nairobi.
[5] BETTS, E. (1965). Locust mapping techniques. *Tech. Notes Wld met. Org.*, no. 69 : 20–22.

[6] BETTS, E. & BROWN, E.S. (1972). Behaviour of moths in the field. *Rec. Res. E. Afr. Agric. For. Res. Org.,* **1971** : 203–204.

[7] BETTS, E. & HAGGIS, M.J. (1970). Seasonal changes in distribution of *S. exempta. Rec. Res. E. Afr. Agric. For. Res. Org.,* **1969** : 109–110.

[8] BETTS, E., HAGGIS, M.J. & ODIYO, P. (1969). Seasonal distribution of *S. exempta* throughout Africa. *Rec. Res. E. Afr. Agric. For. Res. Org.,* **1968** : 122–123.

[9] BETTS, E. & LAZLO, J. (1961). Geographical variation in the rate of development of the Desert Locust. In *Anti-Locust Research Centre, Annual report for* 1960, paper no. ACALR. 18/61. (Unpublished.)

[10] BETTS, E. & ODIYO, P. (1968). A preliminary note on the use of light-trap data for forecasting outbreaks. *Rec. Res. E. Afr. Agric. For. Res. Org.,* **1967** : 111–112.

[11] BETTS, E., ODIYO, P. & RAINEY, R.C. (1969). Development of forecasting methods. *Rec. Res. E. Afr. Agric. For. Res. Org.,* **1968** : 123–124.

[12] BETTS, E., RAINEY, R.C., BROWN, E.S., MOHAMED, A.K.A. & ODIYO, P. (1970). Inauguration of an experimental armyworm forecasting service in East Africa. *Rec. Res. E. Afr. Agric. For. Res. Org.,* **1969** : 110–112.

[13] BETTS, E., RAINEY, R.C., BROWN, E.S., MOHAMED, A.K.A. & ODIYO, P. (1971). Armyworm forecasting service. *Rec. Res. E. Afr. Agric. For. Res. Org.,* **1970** : 103–105.

[14] BROWN, E.S. (1962). *The African armyworm* Spodoptera exempta *(Walker) (Lepidoptera, Noctuidae): a review of the literature.* Commonwealth Institute of Entomology, London: 57 pp.

[15] BROWN, E.S., BETTS, E. & RAINEY, R.C. (1969). Seasonal changes in distribution of the African armyworm, with special reference to eastern Africa. *Bull. ent. Res.,* **58** : 661–728.

[16] BROWN, E.S., ODIYO, P., BETTS, E., SØNDERGAARD, K.M.M. & ONYANGO, J. (1972). Armyworm forecasting service, 1970/71 season. *Rec. Res. E. Afr. Agric. For. Res. Org.,* **1971** : 199–200.

[17] BROWN, E.S. & SWAINE, G. (1966). New evidence on the migration of moths of the African armyworm. *Bull. ent. Res.,* **56** : 671–684.

[18] COCHEMÉ, J. (1965). Assessments of divergence in relation to the Desert Locust. *Tech. Notes Wld met. Org.,* no. 69 : 23–41.

[19] COCHEMÉ, J. (1966). *Wind opportunities for locust transport and concentration in India and West Pakistan, June 20th–30th, 1961: a case study of an invasion of the summer breeding area of the Desert Locust.* FAO Progress Report no. UNSF/DL/RFS/6, 59 pp.

[20] COMMONWEALTH INSTITUTE OF ENTOMOLOGY (1972). *Spodoptera exempta* (Wlk.). *Distrib. Maps Insect Pests,* no. 53 (Revised).

[21] DESERT LOCUST CONTROL ORGANIZATION FOR EASTERN AFRICA, 1971. Eighth Annual report of the Director (1st July, 1969 to 30th June, 1970). Asmara, Ethiopia: 63 pp.

[22] DESERT LOCUST CONTROL ORGANIZATION FOR EASTERN AFRICA, 1972. Ninth Annual report of the Director (1st July 1970 to 30th June, 1971). Asmara, Ethiopia: 65 pp.

[23] EAST AFRICAN COMMON SERVICES ORGANIZATION. (1963). *Final report of the Desert Locust Survey: 1st June, 1961 to 30th September, 1962.* Nairobi, 19 pp.

[24] FAO (1966). Report of the fourteenth session of the FAO technical advisory committee on Desert Locust control held in Rome, Italy, 26th May–3rd June, 1966. Meeting report no. PL/1966/M/3: 57 pp.

[25] FAO (1968). *Desert Locust Project: final report.* Report no. FAO/SF: 34/DLC: 142 pp.

[26] FAO (1972). *Sixteenth Session of the FAO Desert Locust Control Committee held in Rome, Italy, 23rd–27th October,* 1972. Report no. AGP: 1972/M/7: 43 pp.

[27] FLETCHER, T.B. (1910). Lepidoptera, exclusive of the Tortricidae and Tineidae, with some remarks on their distribution and means of dispersal amongst the islands of the Indian Ocean. *Trans. Linn. Soc. Lond.,* **13** : 263–323.

[28] FLETCHER, T.B. (1925). Migration as a factor of pest outbreaks. *Bull. ent. Res.,* **16** : 177–181.

[29] HAGGIS, M.J. (1971). Light-trap catches of *Spodoptera exempta* (Walk.) in relation to wind direction. *E. Afr. agric. For. J.,* **37** : 100–108.

[30] JOEL. 1, v. 6. *The New English Bible.* O.U.P.; Cambridge U.P., 1970.

[31] JOHNSON, C.G. (1969). *Migration and dispersal of insects by flight.* Methuen, London: 763 pp.

[32] JOHNSON, D.H. (1965). The meteorological implications of downwind movement. *Tech. Notes Wld met. Org.,* no. 69 : 6–19.

[33] JOYCE, R.J.V. *et al.* 1962. *Report of the Desert Locust Survey, 1st June, 1953–31st May, 1961.* East African Common Services Organization, Nairobi : 112 pp.

[34] MAGOR, J.I. (1962). *Rainfall as a factor in the geographical distribution of the Desert Locust breeding areas, with particular reference to the summer breeding area of India and Pakistan.* Ph.D. thesis, Edinburgh.

[35] ODIYO, P.O. (1972). Reliability of the first full-scale [armyworm] forecasting service. *Rec. Res. E. Afr. Agric. For. Res. Org.,* **1971** : 200–202.

[35a] ODIYO, P.O. (1974). Reliability of the forecasting service. *Rec. Res. E. Afr. Agric. For. Res. Org.,* **1972** : 133–137.

[36] PALMER, C.E., WISE, C.W., STEMPSON, L.J. & DUNCAN, G.H. (1955). *The practical aspect of tropical meteorology. Air Force Surveys in Geophysics,* no. 76. Air Force Cambridge Research Center, Bedford, Mass. : 195 pp.

[37] PETTERSEN, S. (1956). *Weather analysis and forecasting.* McGraw-Hill, New York. 2 vols : 428 & 266 pp.

[38] RAINEY, R.C. (1962). Weather and swarm movements in eastern Ethiopia and neighbouring countries in mid-March 1962. Appendix I in *Advisory visits to the United Arab Republic, Sudan, Ethiopia, Lebanon and Iran,* 1962. FAO Report no. UNSF/DL/RFS/1 : 23 pp.

[39] RAINEY, R.C. (1963). Meteorology and the migration of Desert Locusts: applications of synoptic meteorology in locust control. *Tech. Notes Wld met. Org.,* no. 54 : 115 pp. (Also as *Anti-Locust Mem.,* no. 7.)

[40] RAINEY, R.C. (1969). Effects of atmospheric conditions on insect movement. *Q. Jl R. met. S.,* **95** : 424–434.

[41] RAINEY, R.C. (1972). Airborne pests and the atmospheric environment. *Weather, Lond.,* **28** : 224–239.

[42] RAINEY, R.C. & ASPLIDEN, C.I.H. (1963). *in* Rainey, R.C. 1963. *op. cit.,* 54–103.

[43] RAMANA MURTY, Bh.V., ROY, A.K., BISWAS, K.R. & KHEMANI, L.T. (1964). Observations on flying locusts by radar. *J. scient. ind. Res.,* New Delhi, **23** : 289–296.

[44] ROOME, R.E. (1974). The establishment of a light-trap grid in southern Africa. *J. ent. Soc. sth Afr.,* **37** : 63–66.

[45] SYMMONS, P.M., GREEN, S.M., ROBERTSON, R.A. & WARDHAUGH, K.G. (1973). *Incubation and hopper development periods of the Desert Locust* (156 Sheets and Guides). Centre for Overseas Pest Research, London.

[46] UVAROV, B.P. (1951). Locust research and control 1929–1950. *Colon. Res. Publ.,* no. 10 : 67 pp.

[46a] SYMMONS, P.M., GREEN, S.M., ROBERTSON, R.A. & WARDHAUGH, K.G. (1974). The production of distribution maps of the incubation and hopper development periods of the Desert Locust *Schistocerca gregaria* (Forsk.) (Orthoptera Acrididae). *Bull. ent. Res.,* **64** : 443–451.

[47] WALOFF, Z. (1966). The upsurges and recessions of the Desert Locust plague: an historical survey. *Anti-Locust Mem.,* no. 8 : 111 pp.

[48] WARDHAUGH, K., ASHOUR, Y., IBRAHIM, A.O., KHAN, A.M. & BASSENBOL, M. (1969). Experiments on the incubation and hopper development periods of the Desert Locust (*Schistocerca gregaria* Forskål) in Saudi Arabia. *Anti-Locust Bull.,* no. 45 : 38 pp.

[49] WHELLAN, J.A. (1958). Report of the Chief Entomologist for the year ending 30th September, 1956. *Rhodesia agric. J.,* **55** : 302–313.

[50] WORLD METEOROLOGICAL ORGANIZATION (continuously updated). *Vol. A. Stations. Vol. C Transmissions.* WMO/OMM, no. 9TP4.

7 · Insect flight in relation to problems of pest control

R.J.V. JOYCE

*Ciba-Geigy Agricultural Aviation Research Unit,
c/o Cranfield Institute of Technology*

Currently-accepted principles of plant protection were expressed sixty years ago by Kurdyumov [51] who considered that the appearance of pests in large numbers was a result of the conditions of agricultural production in which man provided a food base, a microclimate and protection for pests. The same view was expressed by Uvarov [87] and this is the view which Clark *et al.* [13] consider it customary to accept. Many insects are indeed pests because they multiply in a favourable environment created by crop husbandry, but many others are pests because of the densities immediately resulting from the scale of their immigration into crops. Conversely the decline and eventual disappearance of populations—the 'fifth phase' of Polyakov (see below)— cannot be attributed solely to reduced survival rate or be understood without considering effects of emigration [88].

Present world-wide crop protection practice still emphasizes increase in numbers as the prime cause of pest outbreaks and takes little account of the contribution to population dynamics of immigration and emigration. It is the purpose of this paper to examine the impact of recent work on insect flight and migration on the presently accepted principles of crop protection and to suggest that this must prescribe far reaching changes in existing practice.

THE BASIS OF PRESENT CROP PROTECTION PRACTICE

Polyakov [60–62] defines five phases constituting the cycle of population dynamics of a pest species. These are:
1. period of depression—only reservations inhabited;
2. colonization of places outside reservations;
3. mass breeding accompanied by further colonization;
4. rapid increase in density of the population in inhabited territories (habitats);
5. decline in breeding rate and reduction in survival rate leading to extinction of the population in temporarily infested habitats,
and these phases may presumably be followed by pests originating in the various ways described by Clark *et al.* [13].

Recent discussions on the main causes for increase in the economic significance of pests continue to emphasize the destruction by agriculture of 'historically formed biogeocœnoses' which had their own mechanisms for regulating insect numbers [4, 60–62]. Such destruction may result in more insects but fewer species [87] being available to colonize crops, and, once within them, finding conditions favourable for rapid multiplication. Johnson [36] however points out that the contribution of migration to the dynamics of insect populations 'has been much neglected'. The National Academy of Sciences [55], for example, still regards migration as a homeostatic response that 'in the case of crop species, adjusts the pest population in advance to the capacity of the crop habitat'. Pest outbreaks are thus still envisaged as the process through which an insect species exploits its environment by multiplication within it to the maximum numbers the environment can sustain, before the species dies out in or leaves that environment.

Whilst these principles of plant protection might have been expected to lead to the development of logical systems of integrated management of insect populations—indeed in the 1930's particular significance appears to have been attached to farming techniques and Uvarov [87] stressed the limitations of control applied only within a crop—nevertheless emphasis shifted towards developing methods for direct destruction of injurious species within the crop and here chemicals provided the most powerful and versatile weapon. Since World War II the immense expansion in the numbers of chemicals available for insect control has vastly exceeded the acquisition of knowledge concerning the causes of pest outbreaks, so that a purely empirical approach to crop protection has come to be accepted by most economic entomologists—that is to apply a regulator (normally a chemical) and to measure the response (ultimately yield).

Present crop protection practice thus relies heavily on defining the 'economic threshold'—the population level that will cause sufficient damage to make control economic—and then to treat the crop chemically so as to maintain it as an environment lethal to the pest. The economic threshold varies in time and place, is sensitive to weather, agricultural practice, market conditions and individual choice. It also may have to anticipate short-term population trends. Thus it may be imposed at phase 2, 3 or 4 of the Polyakov cycle. If, however, control measures are applied at any one time to only part of the population, its effects may be submerged by later redistribution of numbers derived from other parts of the population treated at different times, as well as by immigration by new populations. It follows that the concept of 'economic threshold' is tenable only under conditions where it can be applied to the whole population—defined as 'a number of organisms of the same species forming a more or less frequently interconnected group, separated more or less clearly from other groups of the species' [78].

A single population, however, is rarely confined to a field or to plots of a size suitable for agronomic experiments by which alone accurate measurements can be obtained of yield response to treatments, although this may happen amongst those insects whose flight muscles are autolyzed after migration or during maturation, such as *Dysdercus intermedius* (C.G.Johnson, personal communication) and several species of aphids [30–32]. The first requirement of the agronomist in the interpretation of field trials, namely that all plots should be independent, is met only exceptionally in experiments in insect control.

Evidence of the dangers of interpreting yield response from sprayed and unsprayed plots without considering the flight activity of the insect populations concerned was provided in the 1950's by small plot experiments [39] and field trials [48] in the Sudan, and later in Uganda [70]. The Sudan experiments showed that the numbers of the jassid

Empoasca lybica de Berg and *Caliothrips fumipennis* Bag. on unsprayed cotton, as well as the yield response to the control of these pests, increased with the size of the untreated relative to the treated plots. In Uganda, Reed [70] emphasized the important effect on parasitoids and predators, which were mobile, of the establishment of a neighbouring lethal environment by spraying which depleted populations of these insects in unsprayed cotton, and he concluded further that in many spraying trials unsprayed controls serve no useful purpose. An economist has stated 'agricultural pest control is still handled as though pests were immobile . . . ' [26].

A result of the conventional approach is that insect control in agriculture is usually accepted without question as being an individual responsibility to be applied (or withheld) by a farmer at his discretion. But insect mobility means that the application of pesticides has effects beyond the fields or the plots to which they have been applied. This is the basis of the conflict between the environmentalists and the pesticide users. It is not the weapon (pesticides), however, which is at fault; the use of some kind of insecticide in regulating insect numbers seems at present inescapable, but the trouble derives from the way in which the weapon is wielded. A farmer seeks increased yields and measures the value of his efforts in these terms; he has no means by which he can evaluate the wider environmental effect of his crop protection practice and this indeed is outside his competence. He is impelled by economics to adopt the cheapest method of pest control irrespective of wider consequences, and this is usually by the use of pesticides. As agricultural production is intensified and extended so also is the significance of insect pests as a factor affecting yield. The corresponding increased use of pesticides increases the dissatisfaction felt throughout the world concerning present methods of crop protection, and the time is ripe for an examination of the basic principles on which current practice is founded.

PEST OUTBREAKS AND ADAPTIVE DISPERSAL

The implications of the concepts of adaptive dispersal developed by Johnson [33–36]— and regarded by him as synonymous with migration—have scarcely been assimilated by applied entomologists engaged in crop protection. The homeostatic response concept has tended to be assumed except when it is obviously inapplicable, and insect pests have been envisaged as increasing in density mainly by completing a number of generations within the crop. Insects which disperse adaptively, however, engage in the flight activity of the species at a particular time in their life history, typically, according to Johnson, as young adults prior to sexual maturation. Such flight not only causes population redistribution within a field but may well result in large-scale displacements of whole populations. Insects flying in the "boundary layer", in the sense of L.R. Taylor (i.e. the air near the earth's surface "in which air movement is less than the insects' air-speed, or within which the insects' sensory mechanisms and behaviour permit active orientation to the ground," [81]) are unlikely to disperse widely so that their populations undergo limited redistribution. On the other hand, the dispersal of insects whose flight-habit takes them out of this boundary layer will inevitably be subject to the same factors (particularly wind-systems) which have been found to dominate the distribution of the Desert Locust, *Schistocerca gregaria* Forsk. [63, 66 *et seq.*] and the African armyworm *Spodoptera exempta* (Wlk.) [10]. Johnson [35] lists fifteen examples from six orders of insects which spiral upwards after becoming airborne. Greenbank, in unpublished observations on the flight behaviour of spruce budworm moths, *Choristoneura fumiferana* Clem., has directed particular attention to that of the females which take off abruptly during the evening and climb almost vertically until above the boundary layer, in contrast with the males seen to remain buzzing around the crowns of the trees. The recent radar observations on this species by Schaefer (p. 174 *et seq.*) quantify the rate and extent of climb under varying conditions, up to flight ceilings which may be several hundreds of metres above ground level. The spruce budworm is amongst the few insects with which attempts have been made to quantify the effects of migration on its population dynamics: during 1949–1955 moth movements at the time of egg-laying were found to result in population changes ranging from a loss of more than 90 per cent by emigration to an increase of 1700 per cent by immigration [21], effects on numbers considerably larger than those of other factors considered in model life-tables [54]. Johnson [36] quotes many examples of mass influx of airborne insects associated with meteorological conditions; a notable example is *Empoasca fabae* Harris, which arrives in Illinois particularly in association with cold fronts [27]. Mass invasions of such species can constitute a damaging outbreak without further breeding, as with locusts and grasshoppers. They provide moreover, opportunities for high reproductive rates, and their importance in the further development of outbreaks of such pests as *C. fumiferana* may well have been under-rated.

Dispersal is to 'scatter . . . in different directions' (OED), but can be a misleading term [49] for adaptive wind-borne transport; a Gaussian probability distribution from a source area is an inadequate model for predicting the movements of airborne insects out of contact with the ground. Locust swarms have been described as exhibiting a 'protean variability of structure' reflecting the turbulence of the air in which they flew [65]. It has been estimated [43] that between June and August 50 per cent of the entire population of Desert Locusts in eastern Africa in some years could be found each afternoon in a belt some 36 km wide on each side of the 09° 42′ N parallel between Hargeisa and Borama in the Somali Republic. This population became concentrated into swarms during the afternoon but became scattered by turbulence and wind during the mornings, and it was suggested that the structure and activity of the Inter-Tropical Front over northern Somalia at this time of the year provided the mechanism for this daily concentration of locust populations into swarms and the subsequent daily disintegration [72]. More recently [67–69], Doppler navigation equipment in a Pilatus Turbo-Porter aircraft has been used to investigate, in greater detail than hitherto possible, the relevant structure of a number of such wind discontinuities, including the Inter-Tropical Front (Figs. 5.8 to 5.14) and Red Sea Convergence Zone (Fig. 5.19) in the Sudan, the African Rift Convergence Zone in Kenya (Fig. 5.18) and the Sea-Breeze Front in England (Fig. 5.20). Wind convergences recorded were at times capable of concentrating any airborne insects constrained (e.g. by air temperatures) to remain within a few kilometres of the ground at a rate giving a sixfold increase in area-density in each hour, and much higher values of convergence are now known to occur (e.g. [11]). Insect trapping using a specially designed collecting net on the aircraft [80] confirmed changes in insect density associated with wind discontinuities (p. 90 *et seq.*). Schaefer records ground radar observations of concentrations of airborne insects associated with cold outflows from storm centres (Fig. 8.35), temperature inversions (Fig. 8.46) and frontal systems (Figs. 8.36 & 39) suggesting increases up to a hundred-fold in volume density.

The air cannot be considered as a homogeneous featureless medium in which insects become randomly dispersed. Wide departures from the exponential decrease of insect density with height demonstrated by Johnson [36] have been shown by Schaefer's radar observations in the Sudan Gezira, where even in the absence of wind convergence evening profiles regularly showed insect densities which were relatively constant up to a ceiling of 1000 m or so (Fig. 8.28), while substantial increases in density with height occurred at the ITF (Fig. 8.41 etc). An increase in insect density with height was also recorded by suction-traps at three levels on a 15 m tower (Fig. 5.12), associated with an increase in numbers caught as the ITF passed over the tower site [7]. More extended trapping observations were subsequently made in the Gezira by Russell-Smith (unpublished): traps, mounted on the same tower, were operated continuously from 3rd–27th October 1971, and catches separated every two hours. High catches (i.e. with more than half the total catch for the day caught within a single two-hour period) in association with the passage of the ITF were shown independently by all the seven taxa which constituted the bulk of the catch and were separately recorded, namely whitefly (largely or wholly *Bemisia tabaci* Genn.), aphids (mainly *Aphis gossypii* Glov. from cotton and *Longiunguis sacchari* (Zehnt.) from *Sorghum*), thrips, mainly *Caliothrips* and *Frankliniella* (recorded separately), all flying predominantly by day; and Cicadellidae/Cixiidae/Delphacidae, Staphylinidae, and other Coleoptera, all flying predominantly by night. Eight separate passages of the ITF over the trap site were each associated with high catches of one to three of these taxa; all save possibly three of these occasions showed some increase in insect density with height; and two of the eight occasions were at times of day well removed from that of any regular diurnal peak of flight activity. Eight further notable suction-trap catches were noted as associated with other wind-shifts well to the south of the ITF, most of them being recorded also on a second anemograph some 30 km away and thus representing wind-features of at least tens of kilometres in extent.

The nature of pest outbreaks cannot be elucidated without considering these causal relationships between redistribution of insect populations and the nature of the wind-fields which transport and concentrate (or disperse) them. In developing the control of the cotton jassid *Empoasca lybica* in the Sudan Gezira, Joyce [42] showed from trapping and survey evidence that outbreaks were more likely to derive from immigration from a distance than from the gradual spread previously postulated [16] from local sources such as gardens. At the time, however, these data were interpreted as showing that once cotton was colonized flight activity was reduced. This was shown by Schultz [73], using improved trapping procedures, to be in error; in fact the numbers of flying adults trapped increased as the multiplication on cotton proceeded. The rapid rate of increase of jassid numbers on cotton in the northern Gezira to damaging levels early in the season [40], and the subsequent rapid decline [14], in contrast with the slow rate of increase in the southern Gezira, are more readily explained by population redistribution, under the influence of wind-systems, than by variations in rate of breeding as postulated earlier [40]. Such transfer of population is not necessarily by 'mass migration' in the traditional sense, but must occur, and must be dominated by wind-fields, whenever the flight activity of the insect removes it from its boundary layer. Thus the southerly winds which cover the Gezira during most of October may be expected to cause an accumulation of jassid populations in the northern Gezira, at a time when young cotton is most damaged by jassid feeding. A subsequent transfer of adult jassids to the southern Gezira apparently takes place after the passage of the ITF in November, with the appearance there of large populations but too late to affect yield [76]; and cotton has been found to respond to insecticide spraying better in the northern than in the southern Gezira [41].

A contrary situation occurs with the cotton whitefly *B. tabaci* which is most serious in the southern Gezira and of greatest economic importance during later stages of growth of the crop [19]. Unlike *Aleyrodes brassicae* in England, found (in unpublished work by El Khidir) to migrate en masse only in autumn, *B. tabaci* in Sudan was recorded in suction-traps throughout the season. During the day the whitefly evidently became dispersed vertically, with aircraft catches indicating a near-logarithmic profile up to 60 m in the late afternoon, but with the subsequent breakdown of convective mixing they became concentrated at lower levels towards sunset (Rainey and Haggis, unpublished). The suction-traps showed that during October adult whitefly were airborne for 20 hours per day, with peak catches in the late afternoon and at sunset; during November catches comparable with the large October ones occurred only around mid-day and sunset; and during December, when the weather was cold, catches were confined to a few hours around mid-day (Russell-Smith, unpublished). Thus during October redistribution of whitefly populations also can be expected on a scale at least as great as the whole Gezira, with populations displacing largely north, but, in November, with the winds then north-easterly, a reverse transfer must occur, resulting in breeding populations which are largely static in December and heaviest in the south. There have been occasions, however, when massive outbreaks have occurred on very young cotton within clearly restricted belts across the Gezira, as in 1951 [38]. Such an outbreak could have arisen only from a mass arrival of adults, producing a damaging population in a single generation. Thus the incidence of *B. tabaci* on cotton in the Gezira may be considered as reflecting aerial transportation rather than effects of particularly favourable micro-climates as have been inferred in the past [e.g. 20], and consequently economic thresholds of infestation have limited relevance to the problem.

Reporting on the Rothamsted Insect Survey, in which airborne populations over Great Britain are sampled by a network of suction- and light-traps, Taylor and French [84] state 'species are already showing profound differences in the extent of their re-distribution, and the way annual trends in population density interact with distribution makes it clear why attempts to interpret population change at a single site are usually futile. Mobility is an essential, highly specific component of the population dynamics of winged insects'. The monitoring of the density of airborne insects in this survey is now proving a more reliable method of determining the date of arrival of aphids on cereals than any practicable field sampling technique, and is to be used by ADAS to advise farmers with regard to chemical control [1, 83].

IMPLICATIONS OF ADAPTIVE DISPERSAL FOR CROP PROTECTION

The transport and concentration of airborne insects in wind-fields, found to be of regular occurrence in Schaefer's radar observations, diminishes the importance to be attached to numerical increase as a prime cause of damaging pest attacks.

It seems very unlikely that swarms of locusts, for example *Schistocerca gregaria*, did not exist before man practised agriculture, or that numbers in their distribution area have increased through man's activities. Intermittently locusts are still a threatening pest, not necessarily because of numerical increase, but because the expanding areas under crops increase the insects' opportunities for crop damage. Similarly, with the grasshopper *Aiolopus simulatrix* Wlk. which breeds in short grass areas of Sudan where rainfall is inadequate for crop production and which subsequently invades cropped areas [37, 79], there is no evidence that cultivation has increased numbers, indeed the reverse is probable [cf. 87]. The species attained pest status in Sudan when increased areas of cultivation provided opportunities for damage. The situation resembles that of the weaver bird *Quelea quelea aethiopica* whose swarming habits are adaptive and whose pest status is incidental [91]. Similarly, outbreaks of the spruce budworm *Choristoneura fumiferana* in Canada (where some knowledge extends back to 1770) constitute a regular part of the cycle of forest regeneration [54], and it has been suggested by C.A.Miller that it is man's effort to protect forests from such naturally occurring outbreaks [53] that has in New Brunswick transformed an epidemic to an endemic situation. Again in Sudan, outbreaks of the *dura andat, Agnoscelis versicolor* F. appear to be unrelated to cropping activity: the adults invading and feeding on ripening *Sorghum* cause crop loss, so that, as with grasshoppers, the major problem is what determines their transport to crops rather than their overall numbers (which appear to be affected mainly by rainfall rather than cropping practice). Yet again, the Middle East Sunn pest *Eurygaster integriceps* Put. migrates from mountain hibernation sites to wheat fields in spring [8, 9]; the success of breeding determines the numbers available for hibernation, but the location of damage and subsequent crop loss are determined by aerial transport and concentration, though the extent may depend on breeding rates which are affected by wheat variety [62].

It appears that numbers of insects adequate to constitute, or cause, an outbreak can derive from concentration by wind-fields and not depend on absolute population increase. Insects which disperse adaptively outside the boundary layer can increase in numbers within a crop only if that crop extends over distances comparable with those covered by the adults prior to completion of oviposition, now known to be of the order of tens or hundreds of kilometres in the case of many pest species. The concept of an economic threshold has no value in the control for example of stemborers (such as *Tryporyza* spp.), or gall midges (such as *Pachydiplosis* spp.), or bollworms (such as *Heliothis* spp.) whose larvae perpetrate much or all of their damage within hours of hatching from the egg. Only control applied on a scale comparable with that of adult displacement could reduce the numbers available for the next generation, and when this scale extends over hundreds of kilometres, as in the Sudan Gezira in relation to jassids and whitefly, pest outbreaks can arise from adult concentrations occurring during flight. The demarcation of areas receiving such concentrations, and the application in them of a regulator such as an insecticide, could be expected to effect maximum influence on the pest population and thus maximum crop protection.

Consideration of adaptive dispersal leads to the conclusion that crop protection to be effective must be applied on a scale determined by the insects' flight activity rather than by artificial constraints imposed by field or farm size, and at a time dictated by knowledge of the factors which have determined the arrival of numbers of economic importance rather than by arbitrarily determined economic thresholds, which assume rates of multiplication to be known and populations to be static.

INSECT CONTROL STRATEGY IN RELATION TO INSECT FLIGHT

The procedure by which data are obtained on the spatial and temporal extent of a pest population has been connoted 'Synoptic Survey' [45]. Such survey defines the crop protection problem and the control strategy most likely to be effective and economic. Strategy, which is essentially a matter of scale, dictates tactics of application in which methods are selected or developed, tailored to the scale. Such methods resulting in control synchronized over the whole distribution of the threatening population will involve any method by which contact between the insect and the chosen regulator can be achieved with the maximum economy and minimum unwanted effect. The control of agricultural pests is not simply a matter of crop-spraying designed to maintain crops as an environment lethal to pest species, but involves the development of methods by which pest outbreak may be anticipated and curtailed.

DESERT LOCUST

The strategic needs for the control of that notoriously mobile pest, the Desert Locust, have been reviewed [44] and attention drawn to situations arising from time to time when the bulk of the total population of the species may be concentrated in very limited areas. The large numbers of insects, their cohesion into discrete swarms, and their mobility dictate the use of aerial application of concentrated insecticide as the only practical means by which the required number of toxic doses can be delivered to the insects in the time available. Such insecticide can be most efficiently applied to the flying locust [64].

Such a strategy of control of an agricultural pest is far removed from conventional crop-spraying, which would clearly be futile for crop protection against such an insect. The Desert Locust is a conspicuous insect and engages in adaptive dispersal during most of its life as an adult, lasting up to six months. It is now clear, however, that the wind systems responsible for the Desert Locust's spectacular displacements and concentrations are exploited by other insects, many of them pests, the resultant population redistributions being determined by the flight habit of the species, including the time spent during adult life in adaptive dispersal.

PLAGUE GRASSHOPPER IN SUDAN

In attempts to develop methods of control of *Aiolopus simulatrix* in the great *Sorghum* growing areas of eastern Sudan various workers, particularly Joyce, have in the past [37] sought the ecological conditions in that area which, favouring successful breeding, cause outbreaks. It now appears that successful breeding may be of smaller importance in causing outbreaks than the occurrence of the type of wind-system which is able to concentrate airborne adults of widely separated origins, and deposit them in the vicinity of crop areas. It is known that this species breeds along a 5000 km belt extending to the west coast of Africa, and the recent radar and aircraft observations in Sudan (p. 181) indicate that the adults are likely to cover distances of the order of 100 km per night, with volume densities at times comparable with those of locust swarms, and thus to provide potential airborne targets worthy of further study for aerial spraying. Effective crop protection is considered more likely to be achieved by attacking such airborne concentrations than by attempting control efforts throughout the range of distribution of the potentially dangerous population.

SPRUCE BUDWORM

Similar opportunities appear to exist to attack airborne concentrations of *C. fumiferana* over forests in eastern Canada (p. 104). In this case, however, females have already laid a proportion of their eggs before their first flight [54], and the significance of dispersal in relation to outbreaks has yet to be elucidated. Nevertheless, knowledge of the areas which have received a large influx of adults would permit a more precise definition of the size of the area in which damaging larval populations are to be expected and thus the scale of the required larval control operation.

RICE STEMBORERS IN JAVA

The importance of the scale on which control operations are conducted is illustrated by experience in the control of *Tryporyza incertulas* Wlk. on rice in Java. This insect displays all the characteristics of classical pest outbreak. As with stemborers generally [6], the crop cannot be considered in isolation from other graminaceous hosts, though in Java the succession of rice crops probably provides the 'reservations' of Polyakov [29]. Moth flight has previously been described as of limited duration [22, 58] and numbers increase from breeding in the crop, the insect in Java completing three generations each in about 30 days [77]. It is to be expected therefore that application of a suitable regulator at the correct threshold level of infestation will suppress further breeding and prevent crop loss; however, to achieve this will depend on the scale on which the regulator is applied and this in turn on the extent of moth flight. Damage by *T. incertulas* is perpetrated by the first instar larvae, which may enter the stem within 30 minutes of hatching [58]. Destruction of these larvae in a plot of rice after they have destroyed the tiller provides no crop protection in itself, and can be of value in reducing the numbers available for the next generation only if that plot is not recolonized from outside. Singh and Sutyoso [75] established kerosene light-traps in 1000 m² observation plots in rice fields in Java; half the plots were within areas which were sprayed by aircraft so that large areas were treated synchronously, and half were in areas where farmers were responsible for their own crop protection, which was therefore unsynchronized, and similar quantities of insecticide were used in both areas. Three regions, West, Central and East Java, were sampled, ten pairs of observation plots being established in each region. The mean weekly light-trap captures show (Fig. 7.1) that where the application of the control was not synchronized in this way, the incidence of *T. incertulas* showed much more marked peaks. These were moderated to the greatest extent in West Java, where almost continuous paddy over 100,000 ha enabled up to 15,000 ha/day to be treated synchronously [47], by ultra-low-volume spraying with electronic track-guidance, and least in Central Java where a smaller proportion of the total paddy areas was suitable for synchronous aerial spraying

TABLE 7.1. **Control of stemborers and rice yields in Java** [*after* 75]. Mean paddy yield in kg/ha; December 1969– May 1970

Region	Control	
	Synchronized	Unsynchronized
WEST JAVA		
Indramaju*	2933	1335
Bekasi-Krawang	3023	2186
CENTRAL JAVA		
Demak*	3217	2279
EAST JAVA		
Bodjonegoro	2691	2088

* Yield differences between areas of synchronized and unsynchronized control significant at the 5 per cent level.

in this manner. Correspondingly the yield from areas treated synchronously reflected better control (Table 7.1), two of the four regions showing the large yield-increases needed to reach the 5 per cent level of significance under these conditions.

In order to determine the size of area which had to be treated synchronously to secure most effective control, Singh sprayed by aircraft areas of 5000, 1000 and 500 ha, with results summarized in Table 7.2; the 5000-ha areas of synchronous control showed reductions in moth-catch and increases in yield (relative to the areas treated by individual farmers) which both reached the 5 per cent level of significance.

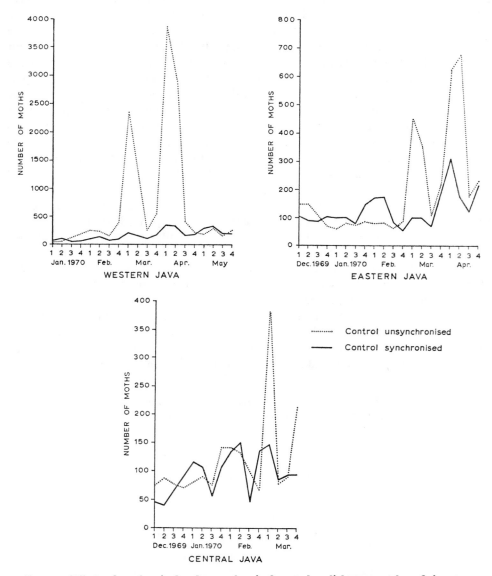

FIG. 7.1. **Effects of synchronized and unsynchronized control on light-trap catches of rice stem-borer moths.** Synchronous control: ULV aircraft spraying. Unsynchronized control: ground-spraying by individual farmers. Number of moths is mean *Tryporyza* count per night per trap, from weekly catches at ten traps [from 74].

Evidently the adult female of *T. incertulas* is sufficiently mobile to show this degree of escape from unsynchronized control. None of the rice stemborers (*Tryporyza, Chilo* or *Sesamia* spp.) appear to have been recognized as showing adaptive dispersal, but they have been described as strong fliers [57], and air-speeds up to 2·15 m/s have been recorded for females of *C. suppressalis* [28]. These species are caught at light-traps in the early evening and towards dawn [57, 74], but their whereabouts between these times is not known. An unpublished study of the wind-systems of Java available for transporting and concentrating airborne insects, by Wales-Smith, has shown that nocturnal convergence may occur off and parallel with the north Java coast, marked by lines of ragged cumulus clouds; and these belts may move inland with any increase in the on-shore wind (for example during the NW monsoon in December–March) and at dawn. Further, gust-fronts can be produced in the evening by katabatic winds from the mountains, and could collect and concentrate airborne insects (p. 183).

TABLE 7.2. **Control of rice stemborers in Java in relation to extent of area sprayed synchronously** (from Singh S.R., unpublished)

Method of treatment	Size of area sprayed synchronously (ha)	Moth catch at light-traps (%)	Paddy yield (%)
Aerial (ULV)	5000	29·1*	194·8*
Aerial (ULV)	1000	63·0	140·1
Aerial (ULV)	500	60·0	141·0
By individual farmers		100·0†	100·0†

* Significantly different from †, at 5% level.

How far *Tryporyza* in Java may in fact be transported and concentrated by these wind-systems is unknown; the observations of Singh *et al.* [75] suggest flight on a scale such that the moths would have difficulty in avoiding them, and experience in the control of this pest emphasizes the need for synoptic survey to determine the spatial and temporal extent of the parent moths, followed by synchronous control on the appropriate scale.

COTTON PESTS IN THE SUDAN GEZIRA

Similar problems of synoptic survey to determine the scale on which operations must be synchronized to secure effective crop protection have been studied in relation to the control of cotton pests in the Sudan Gezira, which occupies an area of 25,000 km^2 between the Blue and White Niles (Figs. 5.8–14) in which about 800,000 ha of irrigated crops include annually some 200,000 ha of cotton, grown in precise rotation (Fig. 8.21) by co-operative farming under the common administration of the Sudan Gezira Board. In an early review of insect pests of cotton, F. Crowther [85] recorded thrips (*Caliothrips* (*Hercothrips*) spp.), and jassids (*Empoasca lybica*) as the most important in the 1940's. From 1946–1965 both pests were controlled by a single application of DDT, with marked yield responses [15, 40, 76]. From 1965 the bollworm *Heliothis armigera* Hb., which had been recorded for many years on cotton throughout the Sudan [3] and which from time to time caused important loss of immature fruit, chiefly amongst early sown cotton [38], began to assume increasing importance following a severe outbreak in Managil (then a new extension) in the 1962/63 season, and by 1965 became considered the major pest throughout the Gezira [25]. *Heliothis* therefore replaced jassids as the primary control target and the multiple applications of DDT (four to five sprays per season) needed to maintain larval numbers at a low level were found by experiment to produce an economic yield response [25], in contrast with failure during the 1950's to establish a yield benefit from even two sprays [39]. Concurrently with this expanded spraying schedule, *Bemisia tabaci* and *Aphis gossypii* increased in importance as late season pests [42], to which were primarily attributed [19] serious 'stickiness' in cotton lint, a condition which severely affected saleability.

An association between *Heliothis* oviposition and the flowering of many host plants [59], including cotton, is well established elsewhere. It has been held by many workers (e.g. Way, personal communication) that this change in the status of *Heliothis* is due to the introduction of groundnuts into the Gezira rotation in place of the later-planted *Dolichos* beans. Groundnuts, flowering in September and October, bridge the host-plant gap between *Sorghum* which flowers from mid-August to mid-September, and cotton which flowers from mid-October to February. It has therefore been suggested that man's activity has, in the classical way, provided an opportunity for *Heliothis* to attack cotton in increased numbers.

Detailed studies of the incidence of *Heliothis* eggs and larvae on Gezira cotton were undertaken during the four seasons 1970/71 to 1973/74, in conjunction with spray-trials, as part of a co-ordinated research project involving the use of aircraft and radar. The layout of the Gezira [85] is particularly well suited for this type of investigation since the standard cotton 'Number' is of regular shape (1350 m × 280 m), and divided by water channels into nine tenancies (*howasha*) of 150 × 280 m, and each tenancy in turn into fourteen plots (*angia*) of 150 × 20 m, within and between which the variability of incidence of infestation may be conveniently recorded. The variance of the mean number of *Heliothis* eggs per plant increased with the mean in a characteristic way [81] and the aggregation recorded was such that, even when infestations reached an average of nearly one egg per three plants, 85 per cent of the plants remained uninfested, and about half the eggs recorded were on plants infested with only one egg. A typical example of the distribution of the incidence of *Heliothis* eggs within and between fields is given in Table 7.3 where counts were taken during three days on 900 plants in 9 Numbers over an

TABLE 7.3. Analysis of variance of incidence of *Heliothis* eggs on irrigated cotton (log n + 1 : eggs per 100 plants). Sudan Gezira—Kumor block: 12th–14th October, 1971

Sources of variance; between	Degrees of freedom	Sums of squares	Mean square	F ratio
Dates	2	3·77	1·88	37·6***
Fields (Numbers)	2	0·29	0·15	3·00*
Positions of tenancy (*howasha*) in field	1	0·03	0·03	0·60
Positions of plot (*angia*) in tenancy	4	0·52	0·13	2·60*
Error (+ 1st, 2nd & 3rd order interaction)	890	41·45	0·05	

*** = significant at 0·001.   * = significant at 0·05.

area of about 40 km². Although small significant differences in infestation could be detected between the fields (probably due to spray treatments), and the edge *angias* tended to lighter infestations than the others, by far the greatest source of variance was that between dates; and this has remained strikingly true throughout these studies. Thus the whole of the 1971 experimental area, which comprised Kumor and Radma Blocks (Fig. 5.13) and extended over about 250 km² containing 4000 ha of cotton, was divided for the purpose of insect assessment into a 5 × 5 Latin Square design. Each of these 25 units occupied about 10 km² and contained 4 or 5 cotton Numbers, with the rows of the Latin Square representing roughly the East-West and columns the North-South axes. The distribution of egg-laying over the experimental area during September–November is shown diagrammatically in Fig. 7.2. There was an effect of spraying on subsequent egg infestation; and Kumor Block, with earlier-sown and more vigorous cotton, on better land with more nitrogenous fertilizer, tended to higher and earlier levels of infestation than Radma Block. But, out of 14 counting occasions between 18th September and 4th November, only 4 showed any significant differences in egg-distribution along the N-S or E-W axes, and these differences did not coincide with boundaries either of sprayed areas or of the administrative Blocks. The distribution of egg-laying was in fact dominated by a common temporal pattern of infestation over the whole of the experimental area (Fig. 7.2). Examination of the corresponding Gezira Board routine insect survey data on *Heliothis* egg-counts, by M.J.Haggis [24], has suggested that this pattern was common to a much bigger area, with egg-laying peaks occurring on the same dates, and with the heavy laying of 5th–7th October, for example, experienced over an area which extended beyond Radma and Kumor for some 70 km from east to west. Data collected in 1970/71, 1972/73 and 1973/74 provided similar patterns, found in further unpublished work by M.J.Haggis and N.Russell-Smith to extend over areas varying from hundred to thousands of square kilometres, usually involving three to four egg-laying cycles lasting for about 10 days and separated by 7–14 days; the progress of egg-laying within an individual area during a single cycle usually followed that found [2] to be characteristic of an individual female. This is consistent with the occurrence of a single population in each such area and cycle.

The life cycle of *Heliothis* in the Gezira during September and October occupies 29 days, from egg to gravid female [2], so that the moths responsible for all except perhaps the last of these successive egg-laying cycles could not have been generated within the cotton fields, though they could have come from other, neighbouring crops within the experimental area. The first egg-laying cycle on cotton occurred in late September, before the formation of fruiting buds, the second in early October when bud formation was beginning, and only after mid-October were flowers available. In each season oviposition was trivial after early November when flowering was becoming widespread. Thus, in the Sudan Gezira oviposition on cotton bore no relation to the flowering cycle, nor was there any evidence of increase in numbers by breeding in the cotton crop. Surveys on *Sorghum* gave peak numbers of *Heliothis* larvae during the second half of September and trivial numbers in October, as the grain hardened. Whilst numerous young larvae were recorded during regular surveys of groundnuts particularly during October, the absence of subsequent large larvae casts doubts on the ability of *Heliothis* to complete its development on this crop in the Gezira during all seasons (Russell-Smith, personal communication).

A fertilized female can lay about 500 eggs in seven days of active oviposition [2], so that a population of 600 moths/ha (sex ratio 1 : 1) is enough to account for the heaviest

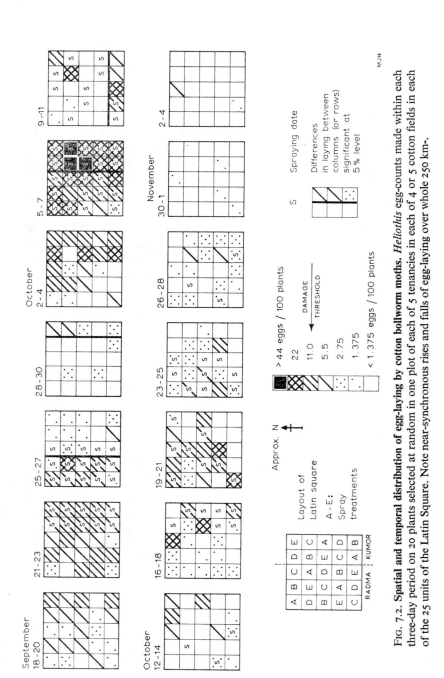

FIG. 7.2. **Spatial and temporal distribution of egg-laying by cotton bollworm moths.** *Heliothis* egg-counts made within each three-day period on 20 plants selected at random in one plot of each of 5 tenancies in each of 4 or 5 cotton fields in each of the 25 units of the Latin Square. Note near-synchronous rises and falls of egg-laying over whole 250 km.

egg-laying recorded in 1971/72, and egg infestations regarded as of economic importance (7 per 100 plants) could be provided by 100 gravid females per hectare.

Besides *Sorghum* in the Gezira which (according to data collected during regular sampling in 1972) could generate 10,000 moths per ha, innumerable other sources of *Heliothis* exist, especially outside the Gezira in the tall grass/*Acacia* plains; here it breeds on the dominant grass *Sorghum purpureo-sericeum* which extends over hundreds of thousands of hectares. Elsewhere there is circumstantial evidence that *H. armigera* can be wind-transported over considerable distances; for example moths of this species caught in England have been back-tracked to North Africa [18]. The related species *H. zea* Bod. was collected in Georgia in black-light traps, especially designed so that insects caught had to fly at least as high as the traps, which were located at fifteen different levels between 8 and 340 m [12]; in both seasons of observation *H. zea* was the most numerous insect species at the highest trapping level. In East Africa, records from ten of the light-traps (Fig. 6.7) established by E.S.Brown [10] included some *Heliothis* at most traps throughout the year, but there were marked seasonal changes, with peak catches coinciding with the rainy seasons along with peak catches (p. 96) of the well known migrant *Spodoptera exempta* (Russell-Smith, unpublished). The highest catches were at Tengeru, Tanzania, and Muguga, Kenya, where monthly totals reached 5000 and 45,000 in April and May respectively; at Muguga between 75–85 per cent of the females caught were unfertilized, and most of these had undeveloped eggs. In the Sudan Gezira and in striking and so far unexplained contrast, less than 50 *Heliothis* moths were captured in eight similar light-traps in 40 days [7] during October and November 1970—the time of maximum incidence of the species on Gezira crops; and comparably small numbers were caught there by Russell-Smith during the following season. In both series of observations most of the females caught had already oviposited. An anomaly of the *Heliothis* problem in the Gezira is indeed why so few moths oviposit on the cotton crop, considering the large numbers which should be available from local sources.

The moths observed by Schaefer as radar echoes (p. 186) may well have included *Heliothis* (particularly in 1973), and this directs attention to adaptive dispersal of the moths, outside the boundary layer, as an explanation for the uniform pattern of oviposition on cotton over large areas in the Sudan Gezira. The period during which *Heliothis* oviposits in important numbers on cotton coincides with that when the ITF is over the Gezira during October. *Heliothis* moths are still found after the southward departure of the ITF, but are still more rare, and fail to generate larvae in damaging numbers despite the fact that maximum flowering and boll-formation occur during November. This failure to multiply on cotton in the Sudan could be due to continuous depletion of the Gezira moth populations by dispersal, following the retreating ITF. The changed status of *Heliothis* in the Gezira seems to be due to an advance in sowing date rather than to increased numbers of the insect, the median sowing date in 1971, for example, having been some three weeks earlier than during the 1950's, bringing more cotton to a stage at which it can be damaged by *Heliothis* larvae.

Damage by *Heliothis* to cotton in the Sudan Gezira appears then to derive from successive invasions of moths whose origins may be local or distant, and intensity light or heavy in accordance with the wind-systems that have transported or concentrated them. Conventional crop-spraying can provide crop protection only if repeated very frequently, with the accompanying risk of generating new pest problems. Knowledge of the flight activity of this pest would permit a strategy of control to be planned so as to reduce the numbers of the threatening population throughout the range of its current distribution.

CONCLUSIONS

As insect mobility dictates synoptic survey whereby the distribution of a potentially (or actually) dangerous pest population can be monitored over the entire range of its current distribution, so the corollary must be control synchronized on a scale determined by the survey, otherwise re-distribution of a partially controlled population can nullify the effects of control. In the Sudan Gezira, which enjoys a common administration, synoptic survey has been conducted on cotton for 10 years on a Group basis (that is cotton areas of about 20,000 ha), and control, for administrative convenience, on a Block basis (that is cotton areas of about 2000 ha); Figs. 5.10 & 13 illustrate Blocks and Groups. Plans are being studied for integrating synoptic survey to treat the entire Gezira as a unit, and for the development of control tactics which match the scale of the problem in both space and time [47]. Had it been the policy of the Gezira Board to leave the individual tenant to apply insecticides to his 4 ha holding at his own discretion there can be little doubt that not only would the use of pesticides have been less efficient and probably greater, but also that new and intractable pest problems would have been generated, such as have led to the abandonment of chemical control of cotton pests in Peru and elsewhere [56, 57]. Yet it is regarded in Europe, North America and in other countries with otherwise advanced agricultural techniques, as unthinkable that crop protection should be other than an individual responsibility. This attitude presumably derives from the fact that insects are less mobile in temperate climates, and the belief that pest situations develop *in situ*. Even when this is so (as in the case of many aphids), the Rothamsted work has shown that the first requirement for pest outbreak in many crops in England, namely suitable numbers of adults, is met by immigration of the airborne stages, and that the application of control needs to be synchronized through co-ordination of individual effort.

The case for synchronized control is clearest for pests such as locusts and grasshoppers, where the main damage is done by the immigrant adults. Other pests, such as *Tryporyza*, are important only in the larval stage. If the adult is mobile and moves out of the area in which it is bred, larval destruction provides little crop protection, the pest situation having been created by oviposition by sufficient numbers of adults. A large number of important tropical pests, in particular stemborers and bollworms, fall into this category, and examples are likely to be augmented with increasing knowledge of the flight activity of insects. In some instances the incidence of sufficient numbers has been made possible by agricultural (and other) practices which have caused numerical increase; in other cases numerical increase is apparently unnecessary, perhaps even irrelevant. The common factor is aerial transport and concentration. Until understanding of ecosystems is sufficiently advanced to enable insect numbers to be managed by environmental manipulation (and this is a formidable task in species where dispersal may transport populations over thousands of kilometres), methods of killing insects, almost certainly including the use of chemicals, will continue to remain the basis of crop protection. Up to the present, pest destruction has relied too heavily on spraying the actual crop which is to be protected, so as to create and maintain an environment lethal to the insect pest. Such crop-spraying for insect control is grossly inefficient in terms of the proportion of the applied chemical which is collected by the ultimate target, the insect. This is suggested by Scorer (see e.g. p. 151) to be of the order of perhaps 10^{-8} in relation to bollworm control. This figure may be contrasted with the efficiencies obtained in Desert Locust control where the dead locusts may account for up to 6 per cent of the theoretical number of toxic doses applied [5, 44, 64]. Once the control of crop pests ceases to be regarded as a responsibility of individual cultivators, the management of pest populations need no longer be limited to crop spraying but may be constructed around the best practical procedures which may doubtless include chemical control. Recent data on airborne insects direct attention to the possibility of airborne concentrations as spray targets as well as to the adult in the vicinity of the crop in contrast with the crop itself, highly effective methods having been developed for spraying flying locust swarms [e.g. 23, 51, 53, 64, 71] and have both required and provided new understanding of physiology and behaviour of flying locusts [e.g. 88, 89, 91]. The importance of a study of insect flight to pest control is that the first essential requirement for crop protection—namely defining the spatial and temporal extent of the threatening pest population—cannot be reached without this knowledge. Pest control is not simply a matter of spraying the crop, on which attention has so far been concentrated. On the contrary crop protection poses complex problems in, for example, taxonomy, biometrics, physiology, behaviour, ecology, meteorology (both synoptic and micro), aerodynamics and toxicology, in which the applied entomologist makes demands on the knowledge of specialists in such fields. His task is to assemble the knowledge so as to develop a coherent system which will reduce crop loss. His research and development work has both required and at times generated new information in these specialized fields. The study of insect flight in relation to crop protection is no exception.

ACKNOWLEDGEMENTS

The author is grateful to Dr J.Meierhans, Dr H.Aebi and Mr E.Bernet, Directors of Ciba-Geigy Ltd., Basel, Switzerland, whose personal help and encouragement has enabled the research on which the views expressed in the paper are based to be continued during the past five years. He is also grateful to Dr P.T.Haskell, C.M.G., Director, Centre for Overseas Pest Research, for permitting the collaboration of his staff, particularly Dr Rainey and Miss Haggis, in studies of common interest. He is indebted to many colleagues, mentioned in the text, for the use of unpublished data, and to Dr D.L.Gunn for many helpful suggestions, but above all to Dr C.G.Johnson and Dr R.C.Rainey whose work has provided him with inspiration and guidance for many years.

DISCUSSION

Dr D.L.Gunn (Chairman): NAAS before it became ADAS made an attempt at regional control of pests, which involved an enormous amount of negotiation to secure the co-operation of individual farmers, but was very successful as far as it went.

Professor R.S.Scorer: In crop-spraying, efficiency can be measured in such ways as by the proportion of the spray that goes onto the crop, or the fraction of the crop that is covered by the spray. But the way I would like to measure the efficiency is by the fraction of the insecticide picked up by the insect. In typical crop-spraying this is perhaps of the order of one part in 10^8—multiplied or divided by a factor of 100, but still not very large even at the most optimistic. Now I understand from Dr Rainey*

---

* Quoting somewhat rash verbal comment, which was however based on field counts of dead locusts unambiguously equivalent to 5 per cent or more of the maximum kill theoretically possible (from application of same total quantity of insecticide, uniformly at the median lethal individual dosage), and on guessing at the combined effects of the undoubted accompanying wastage of insecticide by the over-dosing of some locusts and the sub-lethal dosing of many others (both effects inevitable from the nature of the spray distribution), and of locust deaths outside the areas of assessment.—Ed.

that there have been occasions on which flying locusts may have picked up as much as 50 per cent of the spray applied to the swarm, which illustrates how much more efficient it can be to spray swarms in flight than to apply insecticides in more usual ways. With a prospect of an improvement in efficiency from say one part in 10^8 to one part in two, the next item on the programme, dealing with radar detection of flying insects, is of particular interest because of course you cannot spray an insect swarm unless you can find it.

Gunn: Before hearing from Dr Schaefer, can we perhaps begin to envisage a sheet of spray, laid horizontally in a zone of localized wind-convergence and falling at the same speed at which the air is rising so that it remains stationary, with the insects going up into it? This would indeed be a concentrated insect target.

Joyce: I think that is an excellent idea—originally envisaged in fact by John Sayer (*Nature*, 1962, **194**: 330–336) in the light of our experience against locust swarms in the Inter-Tropical Front and similar zones of convergence. I should also like strongly to support Professor Scorer's point: the present use of insecticide in current crop-spraying practice is grossly inefficient, and better methods must be devised. A point arising from Dr Schaefer's work which may provide some consolation to Dr Gunn is an unexpected predominance of down-wind orientations among high-flying insects, shown by radar on a number of occasions (p. 164). This brings back into the realm of possibility Dr Gunn's original concept of curtain spraying, because in such cases the insects are flying at ground-speeds greater than the wind-speed, and would therefore fly through one of Dr Gunn's curtains laid ahead of them. This is something we should like to try out.

Gunn: In fairness I should say not 'my' curtain but Porton's.

Professor J.W.S.Pringle: It seems to have been assumed in what was just said that the flying insects maintain a uniform orientation with the effect of translating the swarm over the ground at a velocity different from that of the wind. Is this assumption proven or necessary?

Gunn: Not in a locust swarm, in which the locusts fly randomly in groups in all directions (see e.g. Fig. 5.3), and often settle on the ground for a time, so that the resultant velocity of the swarm as a whole can be considerably less than that of the wind.

Joyce: But this has not been found so in Dr Schaefer's radar observations on other insects—or on scattered night-flying locusts—in which predominantly down-wind orientations have repeatedly given ground-speeds well above the corresponding wind-speed (p. 163 *et seq.*).

REFERENCES

[1]   ANON (1973). Introduction. *Rep. Rothamsted exp. Stn*, **1972**: Pt. 1 : 25–26.

[2]   BALLA, A.N. (1970). American bollworm in the Gezira and Managil. *In* [74] : 281–292.

[3]   BEDFORD, H.W. (1933). Report of the Government Entomologist for the year 1932. *Bull. Wellcome Trop. Res. Lab.*, no. 36 : 38 pp.

[4]   BEIRNE, B.P. (1969). *Pest management*. Leonard Hill, London : 123 pp.

[5]   BENNETT, L.V. & SYMMONS, P.M. (1972). A review of estimates of effectiveness of certain control techniques and insecticides against the Desert Locust. *Anti-Locust Bull.*, no. 50 : 15 pp.

[6]   BOWDEN, J. (1954). The stemborer problem in tropical cereal crops. *Rep. 6th Commonw. ent. Conf.*, London: 104–107.

[7]   BOWDEN, J. & GIBBS, D. (1973). Light and suction trap catches of insects in the northern Gezira, Sudan, in the season of southward movement of the Inter-Tropical Front. *Bull. ent. Res.*, **62** : 571–596.

[8]   BROWN, E.S. (1962). Research on the ecology and biology of *Eurygaster integriceps* Put. (Hemiptera Scutelleridae) in Middle East countries with special reference to the over-wintering period. *Bull. ent. Res.*, **53** : 445–514.

[9]   BROWN, E.S. (1965). Notes on the migration and direction of flight of *Eurygaster* and *Aelia* species (Hemiptera Pentatomoidea) and their possible bearing on invasions of cereal crops. *J. Anim. Ecol.*, **34** : 93–107.

[10] BROWN, E.S., BETTS, E. & RAINEY, R.C. (1969). Seasonal changes in the distribution of the African armyworm *Spodoptera exempta* Wlk. (Lep. Noctuidae) with special reference to East Africa. *Bull. ent. Res.*, **58** : 661–728.

[11] BROWNING, K.A. & HARROLD, T.W. (1970). Air motion and precipitation growth at a cold front. *Q. Jl R. met. Soc.*, **96** : 369–389.

[12] CALLAHAN, P.S., SPARKS, A.N., SNOW, J.W. & COPLAND, W.W. In Sparks, A. (1972) *Heliothis* migration. *Sth. Coop. Ser. Bull.*, no. 169 : 15–17.

[13] CLARK, L.W., GRIER, P.W., HUGHES, R.D. & MORRIS, R.F. (1967). *The ecology of insect populations in theory and practice*. Methuen, London : 232 pp.

[14] COWLAND, J.W. (1932–1934). Final reports on experimental work, 1931–1933. *Rep. Gezira agric. Res. Serv. Anglo-Egypt. Sudan*, **1932–1933**.

[15] COWLAND, J.W. & EDWARDS, C.J. (1949). Control of *Empoasca lybica* de Berg on cotton in the Anglo-Egyptian Sudan. *Bull. ent. Res.*, **40** : 83–96.

[16] COWLAND, J.W. & HANNA, A.D. (1950). The cotton jassid *Empoasca lybica* de Berg during the dead season in the Gezira, Anglo-Egyptian Sudan. *Bull. ent. Res.*, **41** : 355–358.

[17] CROWTHER, F. (1948). A review of experimental work. *In* [85] : 439–592.

[18] FRENCH, R.A. & HURST, G.W. (1969). Moth immigrations in the British Isles in July, 1968. *Entomologist's Gaz.*, **20** : 37–44.

[19] GAMEEL, O. (1970). The effects of whitefly on cotton. *In* [74] : 265–280.

[20] GEORGE, L. & RIPPER, W.E. (1965). *Cotton Pests of Sudan*. Blackwell Scientific Publications, Oxford : xv+345 pp.

[21] GREENBANK, D.O. (1957). The role of climate and dispersal in the initiation of outbreaks of the spruce budworm in New Brunswick. *Can. Ent.*, **97** : 1077–1089.

[22] GRIST, D.H. & LEVER, R.J.A.W. (1969). *Pests of Rice*. Longmans and Green, London : 520 pp.

23]   GUNN, D.L., GRAHAM, J.F., JAQUES, E.C., PERRY, F.C., SEYMOUR, W.G., TELFORD, E.M., WARD, J., WRIGHT, E.N. & YEO, D. (1948). Aircraft spraying against the Desert Locust in Kenya, 1948. *Anti-Locust Bull.*, no. 4 : 121 pp.

[24] HAGGIS, M.J. (1973). The distribution of oviposition by the cotton bollworm *Heliothis armigera* (Hb.) over the Sudan Gezira: a preliminary analysis. (Abstract) *Pest Artic. News Summs* (PANS), **19** : 419–421 [Conference report].

[25] HASSAN, H.M. (1970). Progress in chemical control of pests of cotton in the Gezira. *In* [74] : 232–246.

[26] HEADLEY, J.C. (1972). Economics of agricultural pest control. *A. Rev. Ent.*, **17** : 273–286.

[27] HUFF, F.H. (1963). Relation between leaf-hopper influxes and synoptic weather conditions. *J. appl. Met.*, **2** : 39–43.

[28] IATOMI, K. & SEKIGUTI, S. (1936). Flying velocity of the rice stemborer. *Oyo-Dobuts-Zasshi*, **8** : 55–56 (*In* [58]).

[29] JEPSON, W.F. (1954). *A critical study of the world literature on the lepidopterous stalk borers of tropical graminaceous crops*. Commonwealth Institute of Entomology, London : vi+127 pp.

[30] JOHNSON, B. (1953). Flight muscle autolysis and reproduction in aphids. *Nature, Lond.*, **172** : 813.

[31] JOHNSON, B. (1957). Studies on the degeneration of flight muscles of alate aphids. I. A comparative study of muscle breakdown in relation to reproduction in several species. *J. Insect Physiol.*, **1** : 248–266.

[32] JOHNSON, B. (1959). Studies on the degeneration of flight muscles of alate aphids. II. Histology and control of muscle breakdown. *J. Insect Physiol.*, **3** : 367–377.

[33] JOHNSON, C. G. (1960). A basis for a general system of insect migration by flight. *Nature, Lond.*, **186** : 348–350.

[34] JOHNSON, C.G. (1963). Physiological factors in insect migration by flight. *Nature, Lond.*, **198** : 423–427.

[35] JOHNSON, C.G. (1966). A functional system of adaptive dispersal by flight. *A. Rev. Ent.*, **11** : 233–260.

[36] JOHNSON, C.G. (1969). *Migration and dispersal of insects by flight*. Methuen, London, xxii+763 pp.

[37] JOYCE, R.J.V. (1952). The ecology of grasshoppers in the east central Sudan. *Anti-Locust Bull.*, **11** : 103 pp.

[38] JOYCE, R.J.V. (1951–56). Annual Reports of the Entomological Section. *Rep. Res. Div. Minist. Agric. Sudan*, 1948–1954.

[39] JOYCE, R.J.V. (1956). Insect mobility and design of field experiments. *Nature, Lond.*, **177** : 282–283.

[40] JOYCE, R.J.V. (1959). The yield response of cotton in the Sudan Gezira to DDT spraying. *Bull. ent. Res.*, **50** : 567–594.

[41] JOYCE, R.J.V. (1959). Recent progress in entomological research in the Sudan Gezira. *Emp. Cott. Grow. Rev.*, **56** : 179–186.

[42] JOYCE, R.J.V. (1961). Some factors affecting the number of *Empoasca lybica* de Berg (Homoptera: Cicadellidae) infesting cotton in the Sudan Gezira. *Bull. ent. Res.*, **52** : 191–232.

[43] JOYCE, R.J.V. (1962). *Report of the Desert Locust Survey, 1st June, 1955–31st May, 1961*. East African Common Services Organization. Nairobi, Kenya.

[44] JOYCE, R.J.V. (1965). Logistics and strategy of Desert Locust control. *Tech. Notes Wld met. Org.*, no. 69 : 285–297.

[45] JOYCE, R.J.V. (1972). A contribution by industry to the study of problems involved in the large scale application of pesticides in developing countries. *Meded. Fak. Landb.-Wet. Gent*, **37** : 399–407. (Abstract in 24th *Int. Symp. Crop Prot. Ghent* 1972 : SP1.4).

[46] JOYCE, R.J.V. (1973). Insect mobility and the philosophy of crop protection with reference to the Sudan Gezira. *Pest Artic. News Summs* (PANS), **19** : 62–70.

[47] JOYCE, R.J.V., MARMOL, L.C., LUCKEN, J., BALS, E. & QUANTICK, R. (1970). Large-scale aerial spraying of paddy in Java. CIBA-Bimas project. *Pest Artic. News Summs* (PANS), **16** : 309–326.

[48] JOYCE, R.J.V. & ROBERTS, P. (1959). The determination of size of plot suitable for cotton spraying experiments in the Sudan Gezira. *Ann. appl. Biol.*, **47** : 287–305.

[49] KENNEDY, J.S. (1961). A turning point in the study of insect migration. *Nature, Lond.*, **189** : 785–791.

[50] KENNEDY, J.S., AINSWORTH, M. & TOMS, B.A. (1948). Laboratory studies on the spraying of locusts at rest and in flight. *Anti-Locust Bull.*, no. 2 : 64 pp.

[51] KURDYUMOV, N.V. (1913). The question of the trend of work at entomological stations. *In* [61].

[52] MacCUAIG, R.D. (1958). Spray collecting area of locusts and their susceptibility to insecticides. *Nature, Lond.*, **182** : 578–579.

[53] MILLER, C.A. (1971). The spruce budworm in eastern North America. *Proc. Tall Timbers Conf. on Ecological Animal Control by Habitat Management* : 169–177.

[54] MORRIS, R.F. (1963). The dynamics of epidemic spruce budworm populations. *Mem. ent. Soc. Can*, no. 31 : 332 pp.

[55] NATIONAL ACADEMY OF SCIENCES (1969). *Insect pest management. Principles of plant and animal pest control*. **3**. Washington D.C.

[56] NEWSOM, L.D. (1970). *Basic biological parameters of pest insect populations for use in integrated control of cotton insects*. Working Paper no. AGP. IPC/70/5 for the Third Session of the FAO Panel of Experts on Integrated Pest Control, Rome, 1970

[57] NEWSOM, L.D. (1972). Theory of population management for *Heliothis* spp. in cotton. *Sth coop. Ser. Bull.*, no. 169 : 80–92.

[58] PATHAK, M.D. (1968). Ecology of common insect pests of rice. *A. Rev. Ent.*, **13** : 257–294.

[59] PEARSON, E.O. & MAXWELL-DARLING, R.C. (1958). *The insect pests of cotton in tropical Africa*. Commonwealth Institute of Entolology, London, x+355 pp.

[60] POLYAKOV, I.YA. (1964). *Forecasting the spread of agricultural crops*. (In Russian). Izd. Kolos, Leningrad, 326 pp.

[61] POLYAKOV, I.YA. (1967). Science protects the harvest. (In Russian). *All Union Inst. for sci. & tech. Infor. on Agriculture*, 44–51.

[62] POLYAKOV, I.YA. (1968). Basic premises of a theory of the protection of plants against pests. *Ent. Obozr.*, **47** : 343–362. English trans: *Ent. Rev., Wash.*, **47** : 200–210.

[63] RAINEY, R.C. (1951). Weather and the movements of locust swarms: a new hypothesis. *Nature, Lond.*, **168** : 1057–1068.

[64] RAINEY, R.C. (1958). The use of insecticides against the Desert Locust. *J. Sci. Fd Agric.*, **9** : 677–692.

[65] RAINEY, R.C. (1958). Some observations on flying locusts and atmospheric turbulence in eastern Africa. *Q. Jl R. met. Soc.*, **84** : 334–354.

[66] RAINEY, R.C. (1963). Meteorology and the migration of Desert Locusts. *Tech. Notes Wld met. Org.* no. 54, 125 pp. (Also as *Anti-Locust Mem.*, no. 7.)

[67] RAINEY, R.C. (1972). Flying insects as potential targets: initial feasibility studies with airborne Doppler equipment in East Africa. *Aeronaut. J.*, **76** : 501–506.

[68] RAINEY, R.C. (1973). Airborne pests and the atmospheric environment. *Weather, Lond.*, **28** : 224–239.

[69] RAINEY, R.C. & JOYCE, R.J.V. (1972). The use of airborne Doppler equipment in monitoring wind-fields for airborne insects. *7th Int. Aerospace Instrumn Symp., Cranfield* 1972 : 8.1–8.4.

[70] REED, W. (1972). Uses and abuses of unsprayed controls in spraying trials. *Cott. Grow. Rev.*, **49** : 67–72.

[71] SAWYER, K.F. (1950). Aerial curtain spraying for locust control: a theoretical treatment of some of the factors involved. *Bull. ent. Res.*, **41** : 439–457.

[72] SAYER, H.J. (1962). The Desert Locust and tropical convergence. *Nature, Lond.*, **194** : 330–336.

[73] SCHULTZ, L.R. (1970). The use of yellow sticky traps for observations on the cotton jassid *Empoasca lybica* de Berg in the Sudan Gezira. *In* [74] : 247–264.

[74] SIDDIG, M.A. & HUGHES, L.C. (eds.) (1970). *Cotton growth in the Gezira environment.* (*Gezira Research Station Symposium*, 1969), Agricultural Research Station, Wad Medani, 318 pp.

[75] SINGH, S.R. & SUTYOSO, Y. (1973). Effect of Phosphamidon ultra-low-volume aerial application on rice over a large area in Java. *J. econ. Ent.*, **66** : 1107–1109.

[76] SNOW, O.W. & TAYLOR, J. (1952). The large scale control of the cotton jassid in the Gezira and White Nile areas of the Sudan. *Bull. ent. Res.*, **43** : 479–502.

[77] SOENARDI, IR. (1967). Insect pests of rice in Indonesia. *Proc. Symp. Int. Rice Res. Inst., Philippines*, 1964 : 675–683, John Hopkins Press, Baltimore.

[78] SOLOMON, M.E. (1949). The natural control of animal populations. *J. Anim. Ecol.*, **18** : 1–35.

[79] SPENCER, S. (1973). *Grasshopper survey and control in the Sudan Democratic Republic*, 1972. ODA Tech. Assistance Project Report, London.

[80] SPILLMAN, J.J. (in preparation). *The design of an insect trap for use on a light aircraft.*

[81] TAYLOR, L.R. (1958). Aphid dispersal and diurnal periodicity. *Proc. Linn. Soc. Lond.*, **169** : 67–73.

[82] TAYLOR, L.R. (1961). Aggregation, variance and the mean. *Nature, Lond.*, **189** : 732–735.

[83] TAYLOR, L.R. (1973). Monitor surveying for migrant insect pests. *Outlk Agric.*, **7** : 109–116.

[84] TAYLOR, L.R. & FRENCH, R.A. (1973). The Rothamsted Insect Survey. *Rep. Rothamsted exp. Stn* 1972, Pt. 1 : 195–201.

[85] TOTHILL, J.D. (ed.) (1948). *Agriculture in the Sudan.* Oxford University Press.

[86] UVAROV, B.P. (1928). *Locusts and grasshoppers.* Imperial Bureau of Entomology, London, xiii+ 352 pp.

[87] UVAROV, B.P. (1964). Problems of insect ecology in developing countries. *J. appl. Ecol.*, **1** : 159–168.

[88] WALOFF, N. (1968). A comparison of factors affecting different insect species on the same host. *Symp. R. ent. Soc. Lond.* no. 3 : 76–87.

[89] WALOFF, Z. (1972). Orientation of flying locusts, *Schistocerca gregaria* (Forsk.) in migrating swarms. *Bull. ent. Res.*, **62** : 1–72.

[90] WALOFF, Z. (1972) Observations on the airspeeds of freely flying locusts. *Anim. Behav.*, **20** : 367–372.

[91] WARD, P. (1971). The migration patterns of *Quelea quelea* in Africa. *Ibis*, **113** : 275–297.

[92] WEIS-FOGH, T. (1956). Biology and physics of locust flight. *Phil. Trans. R. Soc.* (B), **239** : 459–510.

8 · Radar observations of insect flight

G.W.SCHAEFER

Ecological Physics Group, Loughborough University of Technology*

INTRODUCTION

This paper reviews the initial findings from the introduction of radar techniques to the study of insect flight, in field expeditions to the southern Sahara, Australia, the Sudan Gezira (twice) and Canada. Modest radars have been found to make visible a series of marvellous phenomena, over distances up to 70 km, revealing the precision with which great masses of insect migrants become airborne after dark, regularly assume a common orientation related to wind-direction or compass bearing, climb actively for hundreds of metres to reach altitudes of apparently optimum temperatures and winds, continue to fly for periods of hours, and finally land after travelling tens or hundreds of kilometres, repeating this behaviour night after night. The implications of this impressive and precise mobility for population dynamics and for the rational management of pest species are as yet largely unrecognized.

Radar has also revealed the equally impressive effects of convergent wind-systems of a variety of types upon the concentration and transportation of airborne insects. These mechanisms have produced concentrations of swarm extent and density, and of sufficiently frequent occurrence to encourage the belief that they are biologically significant to the reproduction of many species, and that they must be considered and possibly exploited in pest management. Conversely, insect echoes have revealed new features of the detailed structure of these wind-systems.

As early as 1949 it was shown that individual insects could produce radar echoes [6], but it was not until 1966 [13] that the interpretation of many of the clear-air 'angels' on large radars in terms of insects, rather than atmospheric irregularities, became generally accepted [2]. However, the indirect methods involved do not provide satisfying identification of the class of target, not to mention species, and depend upon scarce and unsystematized knowledge of the radar-reflecting properties of insects.

Studies have accordingly been carried out to provide a foundation for radar entomology, including quantitative observations on airborne populations and their flight performance, and the recognition of species and sex (at distances up to 1·5 km) by echo 'signature' analysis as developed by the author for birds [22]. 66,000 radar photographs have been secured, showing 5 million insect echoes and ¼ million ground-tracks; analysis continues.

* Now Cranfield Institute of Technology, Bedfordshire.

Published radar observations of entomological significance record the passage of Desert Locust swarms in India [21], and 'angel' activity over a saltmarsh [7].

Elements of radar entomology

I. Basic methods

EQUIPMENT, PERFORMANCE AND SITING

The chosen criteria for the radar system were
 (i) low cost;
 (ii) compactness and high reliability, for operation in remote areas;
 (iii) a wavelength (λ) not exceeding three times body-length of target species, for best detectability;
 (iv) a high resolving-power, suited to insect aerial densities; and
 (v) a useful range of detection.

The first three criteria are readily met by commercial X-band marine radar sets ($\lambda = 3\cdot2$ cm), and GEC-Marconi Escort 651 radars were used. For high resolving-power a short pulse length was used and the standard marine aerial was replaced by a circular parabolic reflector to give a narrow pencil beam. A suitable resolution was achieved by using a $1\cdot5$ m diameter reflector giving a $1\cdot45°$ half-power beam width; on the first expedition, to the Sahara, a $0\cdot9$ m dish was used, giving a $2\cdot4°$ beam. For individual echoes closer than 3 km, a pulse length of $0\cdot1$ μs was employed, giving a useful resolution of 15 m; for insect concentrations at long ranges, 1 μs pulses were used. The peak pulse power was 20 kW; the pulse repetition frequency, 2000 Hz and 1000 Hz (long-range). The dipole radiator was normally set for horizontal polarization; vertical polarization was used in Australia. The noise factor was $12\cdot5$ dB.

The steerable antennae were mounted on a Land Rover in the Sahara, on heavy-duty trailers in Australia and the Sudan, and on a prime mover in Canada, as illustrated in Fig. 8.1. A typical interior layout is illustrated in Fig. 8.2A. The basic components of the radar system comprize the 9 in. radar display unit—the plan-position-indicator (PPI)—and transmitter/receiver, illustrated in Fig. 8.2B; 35 mm or 16 mm cameras for photographing the radar screen; an oscilloscope, doubling as an A-scope for presenting echo range; azimuth and elevation indicators; and an electronic range-gate to select echo signatures for recording on a frequency-modulated tape-recorder. The cost for a complete mobile station, inclusive of a 2 kW generator and a $1\cdot5$ kW air-conditioner with generator, was about £6000, using government-surplus vehicles.

The performance of the system can be expressed as the maximum ranges for PPI detection of individual aphids, budworm moths, bollworm moths, grasshoppers and locusts, which were found to be respectively $0\cdot2$, $1\cdot0$, $2\cdot1$, $2\cdot1$ and $3\cdot1$ km; wing-beat frequencies were measured up to one-half of these ranges. Insect concentrations have been observed at distances up to 70 km and rain-storms up to the maximum range of the display, 100 km.

The siting of the radar is critical. The beam side-lobes (off-centre radiations) are weak, but may be reflected by large ground targets to produce strong echoes (ground clutter) which compete on the radar screen with the weak echoes from insect targets. By careful siting most ground clutter has been confined within 400 m for elevation angles above 2°.

FIG. 8.1. **Radar sets in the field.**

A, **Sahara**—In Abangharit, Niger Republic; September–October, 1968;

B, **Sudan Gezira**—Radma airstrip; tower with suction-traps in background; October–November, 1971;

C, **Canada**—Chipman, New Brunswick; July 1973.

[*facing page* 158]

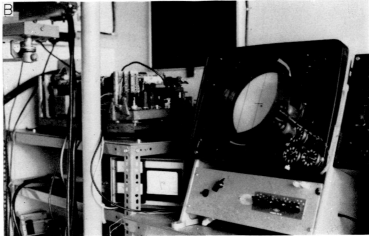

Fig. 8.2. **Radar equipment 1973.**

A, Vehicle interior;

B, Basic radar unit; compact marine radar, with 9″ plan-position-indicator in foreground and transmitter/receiver in background.

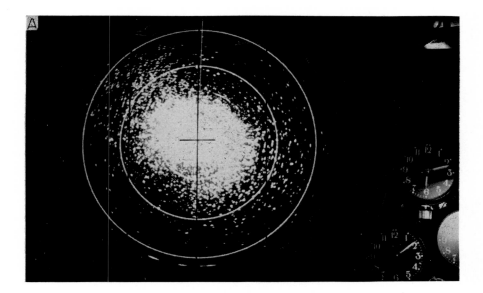

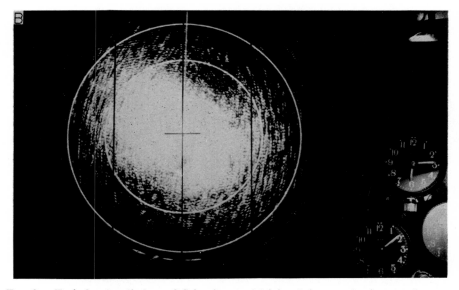

FIG. 8.3. **Typical radar displays of flying insects.** Mainly *Aiolopus* and other grasshoppers, shortly after evening take-off, at Radma, Sudan; range-rings 450 metres (500 yards) apart in range and 145 m in altitude; elevation 18°; 16th October, 1971.

A, 1812 (GMT + 2); single-exposure.

B, 1814; quadruple exposure.

The system has thus been able to detect all insects of weight exceeding about 15 mg, flying within a hemispherical shell of inner radius 400 m and of outer radius given by the maximum range of detection of each species, and above an altitude of 15–20 m over the crop or forest canopy. The volume of this hemisphere is about 2 km³ for individual bud-worm moths and 10 km³ for larger insects. It has been possible to observe down to 10 m over some crops (Fig. 8.23A) and to 50 cm over a smooth surface for aphids and moths.

RADAR DETECTION AND DISPLAY

The operation of radar in the surveillance mode is illustrated in Fig. 8.4, and further details are given in the Appendix. The pencil beam sweeps out a thin conical volume, while a scan line rotates synchronously on the PPI. Short pulses of radio energy are

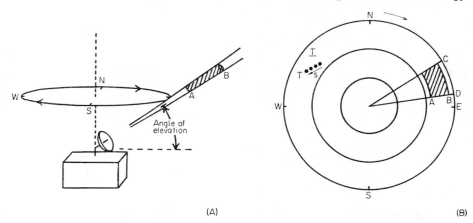

(A)   (B)

FIG. 8.4. **Surveillance by radar.**
   A, Pencil beam rotating about a vertical axis and steerable in elevation;
   B, Radar display on screen of plan-position-indicator, showing range-rings, specification of a standard area, and multiple interception of a target showing speed s and direction T relative to ground.

fired at the speed of light along the beam. A target in the beam reflects some of the pulse energy, and the echo is detected and presented as a spot of light on the radar screen at the instantaneous antenna azimuth and the appropriate echo range. Each spot represents a pulse volume determined by the beam width and the pulse length; a typical resolution cell at a range of 1 km extends 15 m in range and 25 m in elevation and azimuth, with a volume of approximately 10,000 cubic metres.

   A PPI photographic record, single-frame and time-lapse, was obtained by exposing for the 3-sec period of rotation of the antenna. For the study of individual insects (not previously undertaken by PPI photography), the $1\frac{1}{2}$ km display range was generally used; a sample photograph is shown in Fig. 8.3A. False echoes from radar noise were avoided by setting the PPI detection threshold to 10 dB above noise power.

ESTIMATION OF INSECT NUMBERS AND DENSITIES

Absolute density, d, is the number of insects, N, divided by the volume of air space, V m³, in which they appear. The simplest method of measurement is illustrated in Fig. 8.4B. A standard PPI area is chosen, lying between ranges A and B and azimuths C and D. B is chosen to be somewhat less than the maximum range of detection of the target species when flying in side-view. The interval from 1000 to 1500 yd was suitable for all the main species studied except spruce budworm, for which 600 to 800 yd was used. CD was selected to make the area approximately square. The standard sampling volume, V, was obtained by multiplying the standard area by the average detection beam width between A and B (Fig. 8.4A). This average was obtained from Fig. 8.49 for the species concerned and varied between 15 and 35 m, giving standard sampling volumes of 6 million cubic metres for most pest species, and of 0·4 million m³ for spruce budworm moths.

At low density, N was obtained by counting the number of echoes in the standard area, requiring only a few seconds. At high density, with several insects in a resolution volume (10,000 m³), the screen becomes white; an additional radar calibration is then necessary to measure the average volume reflectivity and thence N from the known insect radar cross-sections (RCS; see Appendix). In these studies, N refers to all detectable insects—those with body-length exceeding 8 mm.

For obtaining insect density/height profiles, $d(h)$, the standard density measurement was made at a series of elevation angles, so chosen as to produce overlapping height intervals. The absolute area-density, n (the number per hectare flying above approximately 15 m and averaged over several km²) and the average flight altitude, $\bar{h}$, were calculated by integrating $d(h)$. The flight ceiling was found from the maximum range of echoes at high elevation angles. Variations with time were obtained by repeated elevation scans, at intervals as short as six minutes.

MEASUREMENT OF INSECT GROUND-SPEED AND TRACK-DIRECTION; ESTIMATION OF AIR-SPEED AND ORIENTATION

Each passing insect is intercepted repeatedly by the rotating beam, giving an echo which steps across the PPI screen for 50–500 m, as illustrated in Fig. 8.4B, and is readily recorded by multiple-exposure photography as in Fig. 8.3B. The insect ground-speed, s, is given by the distance between the centres of the first and last echo divided by the time-interval. The track-direction, T, is evident; the ambiguity as to which is the early end is generally resolved by a faint 'footprint' persisting on the screen from the previous rotation. The insect ground-speeds so found have generally exceeded the corresponding wind-speed, by some 2–6 m/s, because of the orientation of the insects having a good component down-wind on most occasions. Insect altitude is calculated from range and elevation angle. The PPI presents on a flat surface (Fig. 8.4B) what is actually happening on a cone (Fig. 8.4A), so that ground-speeds and directions are significantly distorted at elevations above 18°; appropriate corrections have been applied.

'Angel' ground-speeds have also been measured [1] by using a vertical beam, but the analysis is complicated, the sampling rate is minute, directions remain unknown, and 'signatures' were not obtainable for target identification.

The movement of a flying insect relative to the air (its orientation or heading H and air-speed v) can be obtained by subtracting vectorially the velocity (direction D and speed w) of the wind in which it is flying from its direction and speed of movement relative to the ground (track and ground-speed) as measured by the radar:

$$\underline{C}(v, H) = \underline{T}(s, T) - \underline{W}(w, D).$$

A convenient graphical method for this subtraction is illustrated in Fig. 8.5, for an occasion with very light winds. The 26 track-and-ground-speed points were obtained from a quadruple exposure similar to Fig. 8.3B, and are therefore averaged over 9 sec. The errors indicated by the length of the arms of each cross are typical of this method. The time of Fig. 8.5 was just prior to the peak take-off; the insects were grasshoppers, mainly *Aiolopus simulatrix*, and a lesser number of Noctuid moths. Air-speeds and orientations (headings) may be read off directly in the air-stream reference frame. For example, the insect tracking relative to the ground towards 132° at 5·9 m/s (*T*) is heading towards 120° at 5·5 m/s relative to the air (*C*). Most of the air-speeds are between 4 and 6 m/s, probably relating to insects shortly after take-off, when air-speeds can be about 1 m/s higher than later at night; the orientations range from east to west, with the majority heading towards 205 ± 25°, averaging 15° to the right of down-wind. Figure 8.6 presents the headings obtained on an occasion of stronger wind, from a similar analysis

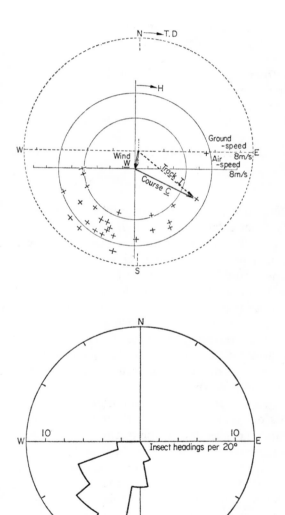

FIG. 8.5. **Radar estimation of orientation and air-speed of individual flying insects.** Use of triangle of velocities for vectorial subtraction of appropriate wind-velocity from measured velocity of radar echo relative to the ground; mainly *Aiolopus*.

Radma, Sudan 1812 13th October, 1971; 40–80 m above ground.

Each cross represents the track-direction and ground-speed of a single insect echo measured from the radar display and plotted from the centre of the dashed circle. The strength and direction of the wind (measured by pilot-balloon as nearly as possible at the same time, place and height) are plotted from the same point, and the third side of the triangle indicates the orientation or heading (course) and the air-speed of this individual insect. The continuous circles, centred at the point of the wind-arrow, show that the air-speeds of most of the insects were between 4 and 6 m/s.

FIG. 8.6. **Orientations of insects flying 300–400 m above the ground.** Mainly *Aiolopus;* from Fig. 8.3B; wind N 7 m/s; average heading 215° (35° from down-wind; standard deviation 20°), computed as in Fig. 8.5.

of the data of Fig. 8.3B. When the insects show a predominance of orientations in one sector (as they often do), there is an independent and powerful method for determining orientations instantaneously from the PPI display, as described in the next section. The air-speeds of scattered night-flying Desert Locusts averaged 3·5–4 m/s in cruising flight; spruce budworm moths gave air-speeds of 2·5 ± 1 m/s with some as low as 1 m/s by the middle of the night.

ELEMENTS OF RADAR ENTOMOLOGY

II. CHARACTERISTICS OF RADAR ECHOES FROM INSECTS

ECHO AMPLITUDE

The Appendix contains comprehensive measurements of the properties of radar echoes from insects, and compares them with a theory of dielectric spheres and spheroids. It is found that:
  (i) radar waves are reflected primarily from biological materials with high free-water content, which are nearly as reflective as metal;
  (ii) reflections from the wings are minute (Fig. 8.54A);
  (iii) appreciable echoes are obtained only from species with body-lengths exceeding one-third of the radar wavelength; and
  (iv) insects observed side-view have echoes 10–1000 times stronger than those seen end-view.

ORIENTATION PATTERN

This last feature (iv) of the insect echo provides a simple and powerful method for the instantaneous recognition of uniformity of orientation among flying insects, and for reading off this predominant heading directly from the PPI pattern. Echoes from insects all flying with the same orientation should produce a roughly elliptical PPI pattern whose axis of symmetry coincides with the head-tail direction, as in Fig. 8.11.

Examples of this pattern may be seen in Figs. 8.3, 8.7–8.10 and 8.39, referring to moths, butterflies, grasshoppers and locusts; they are more obvious with higher densities, the 'dumb-bell' being particularly clear in Fig. 8.7A; the 1350 m PPI range-ring corresponds closely to the circle in Fig. 8.11 representing the range of a sphere of radar cross-section 1 cm². The ambiguity as to which direction is the head and which the tail can be readily resolved by observing the direction of motion, as seen visually on successive rotations or from the 'footprints' in a multiple-exposure photograph.

The interpretation of the pattern in Fig. 8.3A, for example, would be that most insects are heading towards 220° along the axis of symmetry, showing themselves mainly in the NW and SE quadrants where they are seen side-view. If the orientation were completely uniform, Fig. 8.11 indicates that there should be no echoes beyond about 700 m in the head (040°) or tail (220°) direction; the few visible echoes in these two directions would be interpreted as originating from a minority of insects (about 10 per cent) heading approximately at right angles to these two directions. The proof of this interpretation is obtained from Fig. 8.6, the analysis of the ground-tracks lying between 900 and 1350 m in Fig. 8.3B, taken 2 minutes after Fig. 8.3A; see also Fig. 8.53. The average heading of 215° agrees well with the axis of the pattern; all the insect aspects lay within about 45° of side-view as required by theory (Appendix), which also explains the variation of the

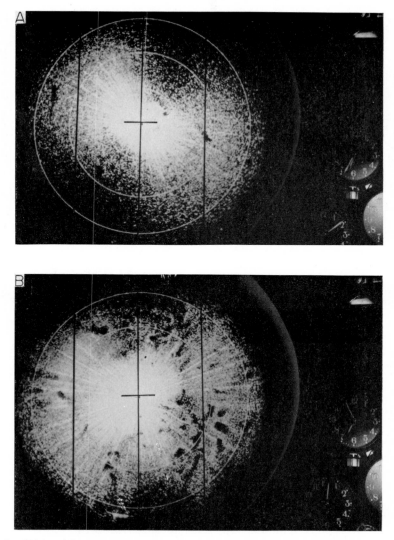

FIG. 8.7. **Flight of Sudan grasshoppers at high density.** Mainly *Aiolopus*, later on same evening as Fig. 8.3, at Radma; single-exposures with range-rings 450 m apart; 16th October, 1971.

A, 1834 at 18° elevation; well-marked 'dumb-bell' display showing that flying grasshoppers were very uniformly orientated, still towards 215° and their overwintering quarters.

B, 1856 at $1\frac{1}{2}$° elevation, with gain reduced by 10 dB because of high density; orientation towards S, down-wind; on this occasion *Aiolopus simulatrix* was taken by aircraft at 1200 m and, with *Catantops axillaris*, at 450 m, at density of 1 grasshopper per 1000 m³, in close agreement with prior radar estimate (p. 173).

Micro-scale atmospheric turbulence was frequently revealed by the insect tracers. Here the photographs show sharply-defined persistent empty 'black holes' of two types; the rounder form moved approximately with the air-stream and insects, the elongated form moved against the insects and was often stationary relative to the ground. These eddies occurred up to 1500 m. Remarkably, an overall uniformity of orientation persisted.

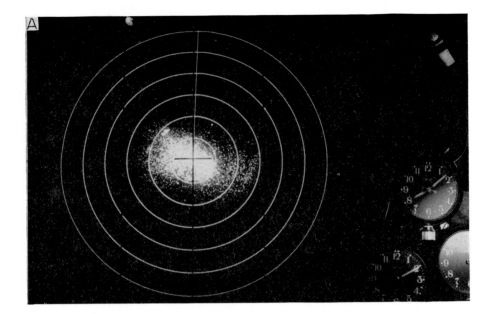

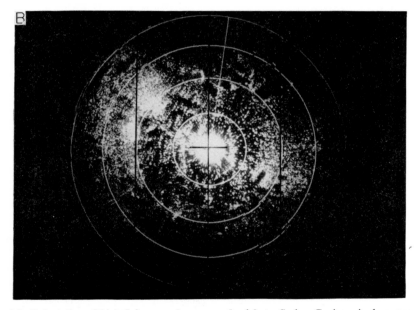

FIG. 8.8. **Orientation of high-flying grasshoppers and crickets.** Sudan Gezira; single-exposures.

A, Radma; 2008 13th October, 1971; range-rings 900 m (1000 yd) apart and elevation 18°.

Mainly *Aiolopus*, up to 1000 m above ground, orientated towards SSW over belt at least 5 km wide, in NW cross-wind of 2 m/s. Turbo-Porter research aircraft also shown, at distance of 1·6 km to NW.

B, Kumor; 1909 4th October, 1973; range-rings 450 m apart and elevation 8°.

Mainly *Aiolopus* and *Gryllus*, up to 260 m; orientation again towards SSW, now directly down-wind; further details in Fig. 8.48. Note persistence of orientation despite turbulence and internal gravity waves. The elongated 'black hole' eddy in SE moved slowly NE, up-wind, with insects concentrating ahead of it.

[*facing Fig.* 8.9]

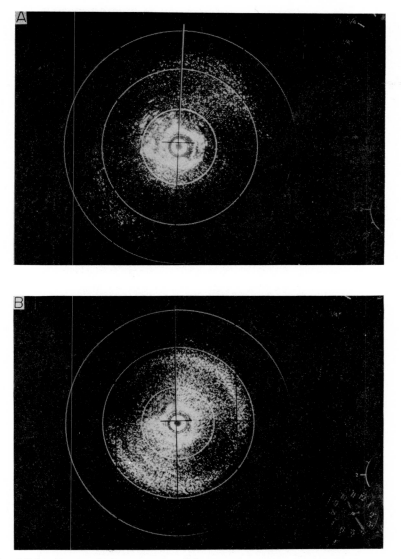

FIG. 8.9. **Flight of spruce budworm moths.** Chipman, New Brunswick; single-exposures with range-rings 450 m apart.

A, 2137 (GMT-3) 6th July, 1973 at 8° elevation; beginning of layer seen at 900–1100 m range; edges of forest clearing also shown, at 100–200 m. This crepuscular flight, in a wind from W (Fig. 8.27), was orientated predominantly towards SE, i.e. 45° from down-wind. Note also plume of moths at 400–700 m to NNE, taking off from a 20 hectare patch of highly infested forest and rising E-wards to join the layer.

B, 2219 10th July, 1973 at 18° elevation; well-marked layer at height of 250±15 m, shown by partial annulus of echoes, at frontal interface (see Fig. 5.21, p. 103 for same occasion); moths orientated down-wind (towards SE), typical of budworm flight behaviour after dark.

[*facing page Fig*. 8.8]

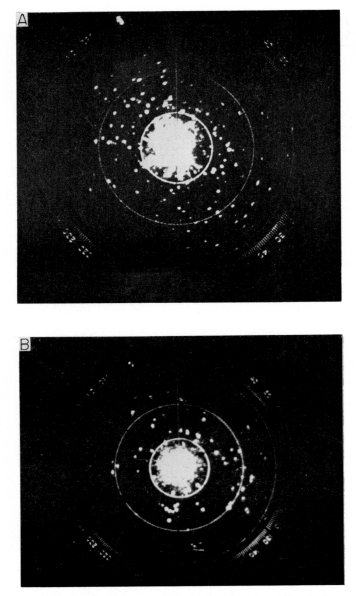

FIG. 8.10. **Migrating insects in Sahara.** In Abangharit, Niger Republic; single-exposures with range-rings 450 m apart.

A, Night-flying Desert Locusts; 1805 (GMT) 13th October, 1968; elevation 5°. Orientation towards SSW (down-wind); elliptical PPI pattern evident even at density of $10^{-6}/m^3$; ground-speeds in Fig. 8.45.

B, Day-flying butterflies (*Catopsilia florella*); 0846 7th October, 1968; elevation 10°. Orientation towards NE (035°—cross-wind); wind from SE (c.8 m/s) and tracks towards N (350° ± 20°); several echo 'footprints' seen between NW and N; elliptical PPI pattern just visible at this very low density of c. $10^{-7}/m^3$; flight ceiling was 1500 m.

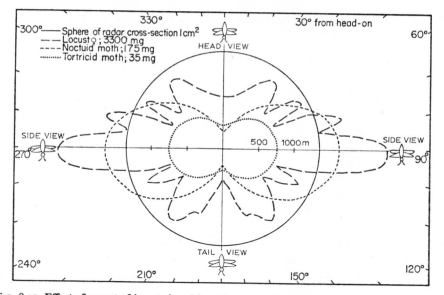

FIG. 8.11. **Effect of aspect of insect viewed by radar on maximum range for detection.** Perfectly-aligned flying insects all heading due south at uniform density would show an elongated or 'dumb-bell' pattern of echoes on the PPI as indicated by these contours; wavelength 3·18 cm and horizontal polarization. This effect provides immediate evidence of any uniformity of orientation of the insects under observation. Randomly orientated insects would produce a circular PPI pattern (see Appendix).

recorded ground-tracks with azimuth which is evident in Fig. 8.3B.

This uniformity of orientation of flying insects, of many taxa, was regularly observed, in moonlight, starlight, on overcast nights, and (with spruce budworm) by daylight also; it was not seen in zones of strong wind-shift nor generally in very light winds.

Scattered night-flying Desert Locusts in the Sahara (e.g. in Fig. 8.10A) were orientated almost exclusively down-wind, with only the occasional individual heading into a very light wind; down-wind orientations were recorded over a very wide range of wind-

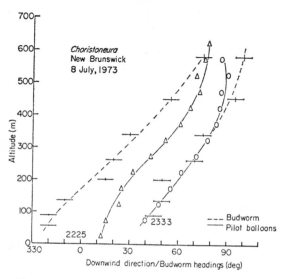

FIG. 8.12. **Orientation of spruce budworm moths: variation with height, time and wind.** Horizontal lines show range of moth orientations as indicated by PPI patterns, varying with height as shown by broken curves; corresponding wind-directions indicated by continuous curves; for wind-speeds see Fig. 8.27.

The winds given by the two pilot-balloon ascents veer with height by 60°; the earlier low-flying moths, just after take-off (as in Fig. 8.24), were initially orientated at about 25° to the down-wind direction, but those that had already climbed to altitude, and all the later moths, showed directly down-wind orientation, which followed the wind-shear.

12

directions, including all four compass quadrants, and in winds so light as to be almost imperceptible at ground level.

Spruce budworm moths showed orientations up to 60° from down-wind (differing from day to day) in daylight, but following the peak take-off after sunset (Fig. 8.24) orientations became consistently down-wind; on one occasion when the wind-direction veered steadily with height the orientations of the moths changed similarly (Fig. 8.12).

Orientations differing significantly from the down-wind direction were at times shown by other species, such as the Pierid butterflies in Fig. 8.10B, heading cross-wind into the Sahara desert! Sudan grasshoppers (Figs. 8.3, 7, 8 & 41), with winter quarters to the south, generally orientated to SSW, in all but strongly-opposed winds. Without this strategy they are not thought likely to achieve their goal in their few nights of flight. Rather similar behaviour was shown by Gezira Noctuid moths (Figs. 8.23 & 39).

Both the mechanism and the ecological significance of such orientation behaviour await further analysis. The long-held view that insects actively orientate only within their 'boundary layer', and that they merely drift down-wind above it, must be abandoned in view of the great biomass of migrants now found to fly above this layer and to show uniformity of orientation.

ECHO SIGNATURES

It has been shown [22] that the echo from a bird contains modulations of amplitude in rhythm with the beating wings; see e.g. Plate 3 in ref. [8]. High-speed photography in the field has shown that bird species can be characterized by a wing-beat pattern, and a library of patterns has been produced to show the use of radar signatures to identify echoes from birds to within three or four species, sometimes uniquely. It was very satisfying to discover from the outset of the Sahara expedition in 1968, and subsequently, that radar echoes from insects also have characteristic signatures at the wavelength used, at least from Desert Locusts through grasshoppers, Sphingids and Noctuids down to small spruce budworm moths.

There have been some earlier comments about frequencies in radar echoes from insects. Several butterflies were automatically tracked by a 3 cm radar during a study of bird echoes [10], and it was concluded that the insect echoes could be distinguished from bird echoes by much smaller fluctuations; none could be seen in the example presented. From further studies by this method it was concluded [3] that insects in general produced no clear fluctuation. Fluctuations observed in echoes from individual insects, thrown out from a high-flying aircraft and tracked by very high-power multiple radars [11], included a number of cases of low-amplitude high-frequency fluctuations, which were explained in terms of the insect gyrating or toppling as it fell. There was also a record of fluctuations at nine per second from a small dragonfly, which were thought to be possible wing-beats, at an ambient temperature of about 8°C, but these fluctuations also are perhaps more likely to have represented body motions resulting from ejection of the insect at high speed and altitude. The present field studies of wing-beat frequencies suggest that a frequency as low as this is unlikely for a small dragonfly (Fig. 8.17), and that wing-beating is indeed unlikely to occur at such temperatures.

Recording and analysis of echo signatures

The simplest method of obtaining a radar signature is to park the radar beam at a suitable angle of elevation and wait for a target to pass through the beam. As it does so the amplitude of the echo increases until the target is in the centre of the beam, and then decreases again, the total duration of an echo being a few seconds, which is in general sufficiently long for the target to give away its identity. More lengthy periods can be obtained by careful manual tracking (or by more expensive auto-tracking radars). As a target begins to appear on the A-scope, the echo is selected by bringing the strobe of the variable rangemarker into coincidence with it. An electronic range-gate triggered by the strobe then reads out the amplitude of the chosen echo, which may be viewed or stored in a variety of ways including magnetic tape-recording, oscilloscope display and/or spectrum-analyzer.

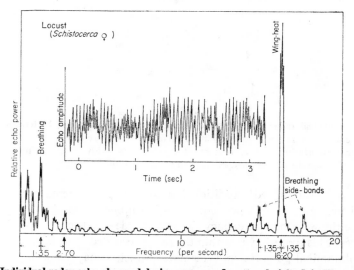

FIG. 8.13. **Individual radar echo observed during passage of scattered night-flying Desert Locusts.** Echo signature shown as recorded in inset, together with spectrum of its constituent frequencies produced by a frequency-analyzer. Frequency of 16·2 per sec is interpreted as that of the wing-beating and frequency of 1·35 per sec as that of respiratory abdominal pumping [29] shown also as first harmonic at 2·7 per sec and as side-bands to wing-beat frequency at 14·85 and 17·55 per sec. Recorded at height of 200 m and distance of 1·2 km. In Abangharit, Niger; 6th October, 1968.

The amplitude modulation of the received echo is termed its signature. Examples are given in the insets of Figs. 8.13 to 8.15. In Fig. 8.13 (inset) the modulations are immediately noticeable, totalling about 50 per cent of the average echo amplitude, with a higher frequency component and a slow modulation of about 1½ per sec. This signal has been passed through a 60 Hz low-pass filter. Although visible modulations like this occur frequently, there are many echoes which show no such obvious modulation, and because of a substantial radar noise component it is necessary to extract the frequencies occurring in a given signature by playing the echo voltage into a spectrum analyzer. Figure 8.13 shows the effect of putting the illustrated echo through a real-time spectrum-analyzer employing the 20 Hz range with a resolution of 0·1 Hz. The most prominent frequency is observed at 16·2 per sec, and evidence follows to attribute this to the wing-beat. Although this frequency is quite visible in the amplitude trace, the degree to which it can be isolated from the background noise is seen in the figure. The peak is as narrow as the resolution of the machine, implying that the wing-beat frequency of this insect did not vary by more than 0·1 per sec in a period of 10 sec or 162 wing-beats—a very accur-

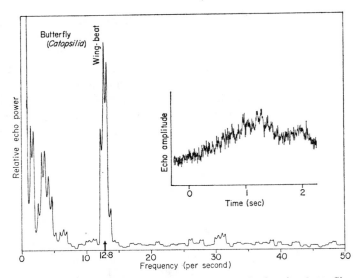

FIG. 8.14. **Individual radar echo observed during passage of migrating butterflies** (*Catopsilia florella*). Echo signature and frequency-spectrum as Fig. 8.13. Frequency of 12·8 interpreted as that of the wing-beating—see p. 167; selected as an echo with wing-beat modulations just visually perceptible. In Abangharit, Niger; 7th October, 1968.

ately tuned oscillator indeed! Additional frequency components are evident, with a fundamental frequency at 1·35 per sec; a first harmonic of this frequency occurs at 2·7 per sec, while the wing-beat frequency shows two side-bands correspondingly displaced by 1·35 per sec. This frequency is within the range recorded in wind-tunnel flight [29] for respiratory abdominal pumping movements, to which it may accordingly be attributed. Scattered night-flying locusts have given values ranging from 1·0 to 1·6 per sec, and

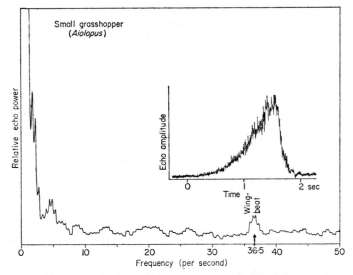

FIG. 8.15. **Individual radar echo observed during passage of night-flying grasshoppers** (mainly *Aiolopus*). Echo signature and frequency-spectrum as above; selected as an individual echo in which frequency-analysis was essential to discriminate from radar noise the frequency of 36·5 per sec, interpreted as that of a small male *Aiolopus simulatrix*. Radma, Sudan; 20th October, 1971.

day-flying locusts in a swarm from 2·0 to 2·5 per sec; such frequencies were shown by about one-third of the echoes studied [23].

An example is given in Fig. 8.15 of an echo signature whose modulations consist predominantly of radar noise with no directly visible discrete frequencies; the value of a spectrum-analyzer lies in separating discrete frequencies from such noise, and in this case a probable wing-beat frequency of 36·5 per sec is seen at approximately twice radar noise level. The species concerned is likely to have been *Aiolopus simulatrix*.

Wing-beat frequencies in the insect echo signature

Evidence that the prominent frequency occurring in an echo signature is that of the wing-beat has been obtained in a number of ways. In the Sahara a thin Desert Locust swarm flew over the radar station at noon on 13th October 1968, when a number of echo signatures of low-flying insects were tape-recorded and a series of high-speed cine photographs were made of individual locusts flying within 50 m of the ground. Figure 8.16 shows that the photographically-observed wing-beat frequencies were identical

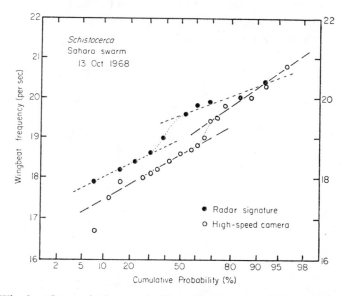

FIG. 8.16. **Wing-beat frequencies in a passing Desert Locust swarm: high-speed photography and radar observations.** The distribution of the wing-beat frequencies recorded photographically is almost identical with that of the frequencies found in the radar echo signatures; both are markedly bimodal, attributed to the difference in size and in turn of wing-beat frequency between the sexes.

In Abangharit, Niger; 1200 13th October, 1968 (see also Fig. 8.17).

with the prominent frequencies in the echo signatures, and that both distributions were markedly bimodal, attributable to sexual dimorphism. Similar techniques were used on the diurnal butterfly *Catopsilia florella* in the Sahara (Fig. 8.14), with equally good confirmation; most of the butterfly echoes had visually perceptible wing-beat modulations, at frequencies between 12 and 14 per sec (Fig. 8.17).

Indirect further evidence on this point has been provided by a wide range of species of grasshoppers and locusts which have been photographed with a high-speed camera and have shown a strong correlation (Fig. 8.17) between the wing-beat frequency (f per sec) and the wing-length (l, in mm), closely fitting the line $f = 400\ l^{-0.78}$. This is close to the $l^{-2/3}$ relationship found by Weis Fogh [28] for Desert Locusts of different sizes. On such grounds it was possible to attribute immediately to medium-sized grasshoppers, at a density of about 1 per 1000 m^3 at a height of 450 m and extending up to 1200 m, a dense display of radar echoes seen in the Sudan on the evening of 16th October 1971 (Fig. 8.7), prior to the direct confirmation of each of these inferences, provided by aircraft catches (p. 91 and Fig. 8.19).

Modulation mechanism

For birds the modulation of the radar echo is most probably mainly due to the oscillation in the size and shape of the pectoral muscles [22].

This is unlikely to be the reason for the modulation of insect echoes, particularly at these wavelengths; the thoracic muscles are far too small to create significant interference changes. Wing movement is unlikely to be the direct cause of modulation, because direct measurements of the radar cross-section (see p. 190–195) of the complete, detached wings of the butterfly *Aglais urticae* (Fig. 8.54A) gave values of only 0·005 cm^2 broadside on and about 0·001 when placed in the plane of polarization, as compared

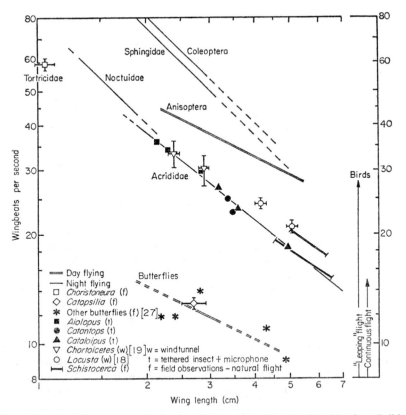

FIG. 8.17. **Wing-beat frequency and wing-length: chart to aid radar identification.** Solid lines show ranges encountered in radar field-work; broken lines extend over size-range of group. 'Other butterflies', from [27], corrected for probable visual under-estimation [9].

with the RCS of 2 cm² given by the complete insect side-view and the head-or-tail RCS of 0·01 cm², and considered in relation to the modulation of approximately 20 per cent found in the field for the butterflies *C. florella* observed side-view (the above wing RCS is approximately 1/1000 of the side-view RCS). Significant wing-beat modulation percentages were obtained when insects were seen head-or-tail view; it is possible that the wings themselves were responsible.

At these wavelengths the larger insects are in the interference scattering region (see Appendix), so that significant distortion of the body could produce appreciable modulations, as may be inferred from Fig. 8.54. The most likely body distortion of elongated insects like grasshoppers or locusts would be the flexing of the longitudinal body axis, as has been observed in wind-tunnel studies. During normal flight there is rapid up-and-down movement of the abdomen, in phase with the movements of the thorax which are largely responsible for wing movements, though the abdominal movements are independently generated by abdominal muscles, at least in the Desert Locust [4], and are not due simply to a mechanical linkage between the abdomen and the thorax. While this is the most likely mechanism, and could easily account for the observed modulation percentages in the Acrididae and Lepidoptera at a wavelength of 3 cm, the degree of modulation is likely to be less at longer wavelengths or for less flexible insects like beetles.

Use of wing-beat frequencies in insect identification

Figure 8.17 has been compiled to aid radar identification of species by relating wing-beat frequency to wing-length for the main groups of insects encountered in the field.

Wing-beat frequencies have been sought in the published literature, and measured in the field either using insects hovering in front of a microphone or by cine-photography of insects in free flight over short distances, including Desert Locusts in a passing swarm (Fig. 8.16), and ascending female spruce budworm moths. Grasshoppers, in particular, and hawk moths give frequencies inversely related to wing-length; beetles and Noctuid moths show frequency ranges fortunately well separated from grasshoppers and locusts.

By correlating the observed echo frequencies with field information on the species present, from surveys, light-traps and aircraft-trapping, it has been possible to separate grasshoppers and locusts from moths and beetles, and to secure quantitative data on airborne populations of grasshoppers and locusts of different wing-lengths and frequently of known species.

Male and female grasshoppers and locusts usually differ significantly in size and hence in wing-beat frequency, making it possible to distinguish the sexes under favourable circumstances (p. 171).

Distinguishing echo signatures of insects from those of birds

Bird signatures are of three main types: continuous wing-beating; intermittent wing-beating—'leaping' flight; and the irregular and intermittent wing-beating of hovering birds such as nightjars [22]. The frequencies of 'leaping' birds extend from approximately 6–30 wing-beats per second, but fortunately this type of pattern has not been encountered in insect signatures, nor has the intermittent hovering type. Bird frequencies with continuous wing-beating extend from approximately 2·8–14 per sec (except for humming-birds), while insect frequencies range from about 8 per sec upwards. On the basis of frequency alone there is therefore some possibility for confusion at frequencies between 8 and 14 per sec. The species involved are moderate to large butterflies, very large moths and the largest Tree Locusts. Being large, these insects have substantial modulation amplitudes like birds of the same frequency, as well as large radar echoes (Fig. 8.50), like the bird species concerned. The latter are the small waders; the best method for differentiation would appear to be air-speed, for these birds fly at approximately 15 m/s compared to approximately 4 m/s for the insects concerned; only rarely would wind-speeds be sufficiently high to obscure this difference.

Echo signatures of Desert Locusts

Approximately 1500 echo signatures were obtained from Desert Locusts in the Sahara in 1968. Wing-beat frequencies of gregarious-phase locusts flying in a low-density swarm at midday in full sunlight are shown in Fig. 8.16. The morphometrics (*E*, elytron length; *F*, femur length) of a sample from this swarm have been applied to the wing-beat frequency formula obtained by Weis-Fogh [28] to provide estimates of the wing-beat frequencies of the males and females, with the results indicated to the left of Fig. 8.18.

In a similar way the wing-beat frequencies of a typical occasion of nocturnal migration of *solitaria/transiens*-phase populations, flying at cruising air-speeds of 3·5–4·0 m/s under a clear sky at 2140 (local time) on 14th October 1968 and at an altitude of approximately 400 m, are shown also in Fig. 8.18, together with the wing-beat frequencies estimated from the elytron and femur lengths of locusts caught in the same period in a light-trap.

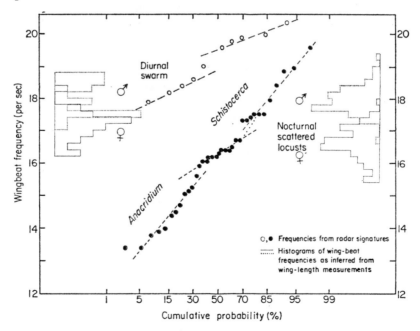

FIG. 8.18. **Inferred wing-beat frequencies of day- and night-flying Desert Locusts.** Radar echo signature frequencies and estimates [28] from wing-lengths measured on samples from same populations; In Abangharit, Niger; 13–14th Oct., 1968. Tree Locusts also present (p. 171).

Sexual dimorphism in wing-length reflected in estimates of wing-beat frequency; radar echo frequencies included similarly bimodal distributions.

These nocturnal wing-beat frequencies are very close to those to be expected from the laboratory experiments of Weis-Fogh [28], which were performed in the absence of solar radiation and at an air temperature of approximately 30°C in subdued light, conditions very similar to those of the cruising nocturnal flight of Fig. 8.18. On the other hand the day-flying locusts showed wing-beat frequencies (by high-speed photography as well as by radar) which were approximately 1·5 beats per sec above those given by the formula, both for females and for males. The radar signatures were obtained from locusts flying at about 125 m with the swarm in rolling flight, i.e. with many locusts temporarily settled, and it is possible that the 8 per cent increase in frequency was associated with a lift force in excess of weight, with the individuals recorded perhaps climbing after being settled on the ground. This 8 per cent increase in frequency is however greater than the maximum increase of 5 per cent recorded at the highest lift in the laboratory experiments. The strong solar radiation in the field is likely to have increased the body temperature above that in the laboratory experiments (possibly by 5° or so), and perhaps resulting in lift beyond the maximum of 140 per cent relative lift attained in the laboratory. Hand-released locusts, recorded by cine-photography, always showed frequencies in excess of 20 wing-beats per second, and are clearly not representative of cruising nor even perhaps of climbing free-flying locusts which have been airborne for at least 30 sec, as in the cine-photography of Fig. 8.16.

The degree to which the sexes of Desert Locust may be distinguished by radar signatures is indicated by both sets of results in Figs. 8.16 and 8.18, where the probability distributions all show changes of slope at about mid-point; frequencies significantly removed from the mid-point are almost certainly either males on the one hand or females on the other. In this way it is fairly certain that the individual locust signature illustrated in Fig. 8.13 is that of a female.

Echo signatures of Tree Locusts

Anacridium melanorhodon was prominant in our Sahara light-trap catches as the largest insect to be caught; elytron length ranged from about 60 to 74 mm, which from Fig. 8.17 would imply frequencies in the range 13·5–15·5 wing-beats per sec. Figure 8.18 shows that frequencies in this range were recorded during the same echo session as the Desert Locust frequencies; however similar frequencies were also obtained from 'leaping' birds during this session, and it is just possible that the frequencies shown lying in the range 13–14 per sec include a few trans-Saharan bird migrants of the wader type, though this is considered unlikely as relatively few bird echoes, with their characteristically higher air-speed, were seen at this particular time; it is most likely that the observed frequencies below 15·5 per sec are from Tree Locusts.

Hand-released Tree Locusts had frequencies in the range 16–19 wing-beats per sec. If these had been inflated by an 'escape factor' of about 25 per cent, as found above for Desert Locusts, the frequencies during normal nocturnal cruising flight of Tree Locusts would be expected around 13–15 wing-beats per sec, in good agreement with Fig. 8.18.

Echo signatures of Sudan grasshoppers

Aiolopus simulatrix was by far the most common insect recorded by all methods during the Sudan observations in October–November 1971; on some occasions as many as

10,000 were caught by the light-trap during the hour of main take-off. It is an important economic pest on *dura* (*Sorghum*) and breeds first in the long-grass savannah south of the Gezira during the early spring rains, then goes through summer and autumn generations in the desert scrub area at the same latitude as the Gezira, some 200–400 km north of the spring breeding range, and returns in autumn to overwinter in the long-grass savannahs [see also p. 142]. Individual insects suspended by the pronotum in front of a microphone in a wind-stream produced the wing-beat frequencies shown in Fig. 8.17. The elytron length of other captured individuals was between 16 and 30 mm, suggesting (p. 168) wing-beat frequencies between 28 and 44 per second.

Of the 1100 echo-signatures recorded during 1971 in the Sudan, some 250 were from several insects simultaneously, some 200 contained no identifiable wing-beat frequencies, and the remaining 641 echoes contained the frequencies shown in Fig. 8.19. The main

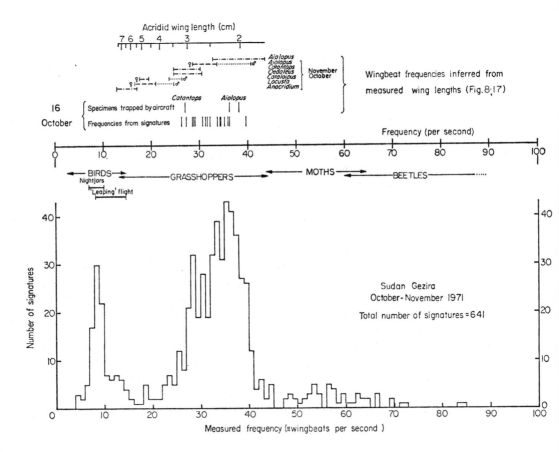

FIG. 8.19. **Echo signatures and wing-beat frequencies in the Sudan Gezira: October–November, 1971.**

Above: wing-beat frequencies estimated from elytron measurements of commoner Acrididae present.

Below: distribution of frequencies recorded in radar echo signatures, with inferred identifications; relative numbers of bird and moth frequencies respectively over- and under-estimated.

Note predominance of frequencies attributable to *Aiolopus simulatrix;* this species was indeed present in very large numbers during this season.

features of this distribution are frequencies attributable to birds, to the left; a larger number of frequencies attributable to grasshoppers, in the centre; and a smaller number of frequencies likely to represent moths and beetles, to the right. The large peak in the bird region, from echoes recorded at heights between 300 and 1000 m and predominantly on moonlit nights, is believed to be due primarily to hovering nightjars, perhaps feeding on passing grasshoppers. Frequencies attributable to birds are limited by approximately 15 per sec; the few between 12 and 15 per sec show the 'leaping' characteristic. All available evidence indicates that the wing-beat frequencies between 15 and 44 per sec were due almost exclusively to locusts, grasshoppers and crickets.

Figure 8.19 suggests that the great majority in this band would represent *Aiolopus simulatrix*. Most of the signatures between 41 and 45 per sec were recorded a month later than the main investigation, on two evenings in November; *Aiolopus* was still

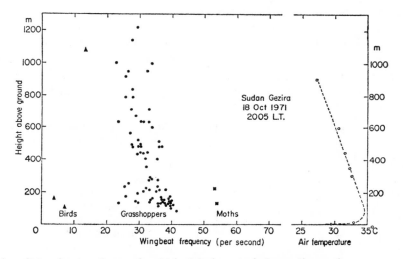

FIG. 8.20. **Echo signature frequencies obtained during a typical recording session.**
Note decrease of average frequency with increasing altitude and decreasing temperature.

dominant but of smaller size, with the elytron about 2 mm shorter, and thus corresponding as indicated in Fig. 8.19 with frequencies up to about 44 per sec. From the frequencies recorded it may also be suggested that *Catantops*, *Oedaleus* and *Acrotylus* together amounted to about 1/6 of the numbers of *Aiolopus*, with *Locusta migratoria*, *Cataloipus* and *Amphiprosopia* amounting to about 1/20 and *Anacridium* to about 1/100 of the *Aiolopus* total.

A little direct evidence on this point was obtained on the only night with sufficiently high density for grasshoppers to be caught in the aircraft net, namely 16 October, at 450 m, when as already mentioned the radar indicated an aerial density of about 10^{-3} grasshoppers per cubic metre and the aircraft net produced a density of $1 \cdot 1 \times 10^{-3}$. The specific individual wing-beat frequencies measured on this occasion are shown in Fig. 8.19; from these data it could be expected that of every 16 grasshoppers there would be 3–5 of the size of *Catantops*. In fact the aircraft captured four *A. simulatrix* and one *Catantops axillaris*.

A prominent group of *A. simulatrix* frequencies was exhibited by the signatures of each recording session in 1971, spanning a period of 15 days with heights up to 1000 m at different times of night. A typical recording session is illustrated in Fig. 8.20. The best represented frequency appears to decrease with altitude, but this effect does not appear to be shown by the larger species with frequencies between 22 and 27 per sec.

OBSERVATIONS ON TAKE-OFF, HEIGHT, DENSITY AND DURATION OF FLIGHT OF INSECT POPULATIONS

EVENING TAKE-OFF

After sunset, on all five expeditions, the radar screen regularly and most impressively changed from entomologically blank to recording larger and larger densities of insect echoes extending to higher and higher angles of elevation. The main species studied—the Desert Locust (*Schistocerca gregaria*), the Australian plague locust (*Chortoicetes terminifera*), the Sudan plague grasshopper (*Aiolopus simulatrix*), the cotton bollworm moth (*Heliothis armigera*) and the spruce budworm moth (*Choristoneura fumiferana*)—were all visually observed taking off, silhouetted against the twilight sky, at the same time as the radar made its first contact at low elevations. However, the radar indicated that peak take-off typically occurred about 15 minutes later than visual observations were possible,

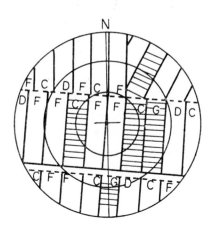

FIG. 8.21. **Map of crops around Sudan Gezira radar site.** C, cotton; D, *dura (Sorghum);* F, fallow; G, groundnuts.

Kumor; October, 1973; with positions of 450, 900 and 1350 m range-rings. Compare permanent echoes in Fig. 8.23; note nearer edges of neighbouring cotton and of *Sorghum* further away; note also Ramadan canal, running to WNW and passing 700 m S of radar. Figure 8.23A, at low elevation, also shows some of smaller irrigation channels.

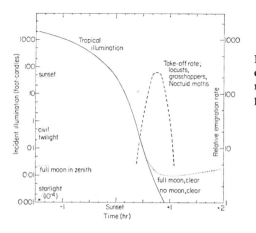

FIG. 8.22. **Light-intensity and insect take-off on clear tropical evenings.** Representative measurements from Sahara, Australia and Sudan expeditions.

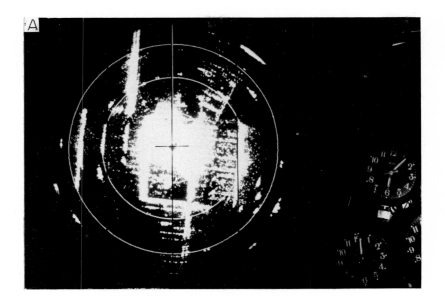

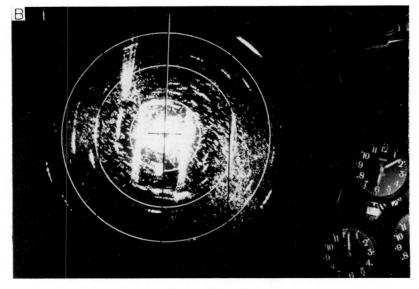

FIG. 8.23. **Beginning of evening take-off in the Sudan Gezira**

Mainly Noctuid moths (*Heliothis armigera*, *Spodoptera littoralis*, etc.).

Kumor; 11th October, 1973; range-rings 450 m apart.

A, 1806; single-exposure at $1\frac{1}{2}°$ elevation.

B, 1808; triple exposure at $3°$ elevation.

First compare Fig. 8.21 and note permanent echoes particularly from nearer edges of the two nearest fields of cotton, and, further away, from the edges of the *Sorghum* fields to NW, W and E; then note insect echoes in B, most of which occurred in a patch the shape of a field which could be traced back to the groundnut field some 500–700 m E of the radar. The patch moved towards SSE, in a NW wind of 6 m/s, with a calculated common orientation towards SSW and an airspeed of 5 m/s, and ascending at about 1 m/s.

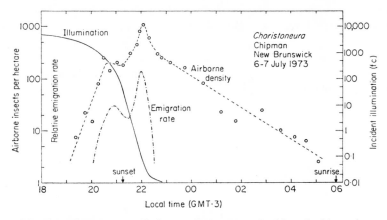

FIG. 8.24. **Light-intensity and take-off of spruce budworm moths.** Note double peak.

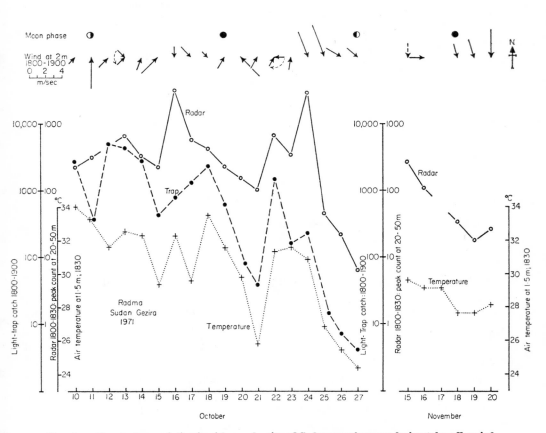

FIG. 8.25. **Day-to-day variation in airborne density of Sudan grasshoppers during take-off period.**
Mainly *Aiolopus simulatrix;* winds at 2 m, averaged over period 1800–1900; main wind-changes
associated with passages of Inter-Tropical Front. Peak insect numbers on radar, on 16th and 24th
October, not reflected in light-trap catches.

and in the case of the spruce budworm direct confirmation on this point was provided by the observations of Dr Greenbank and Mr Russell-Smith using a military night-viewing device.

In the early take-off period radar scans at the lowest elevations often showed patches of insect echoes at high density rising as plumes from particular areas. In the Sudan Gezira, with its standard rectangular fields (Fig. 8.21) and its regular rotation of cotton,

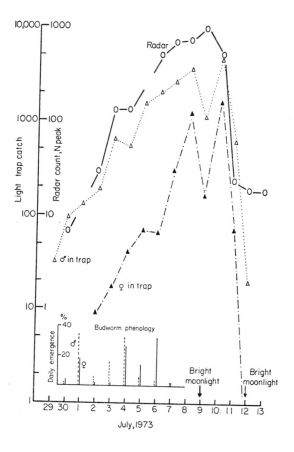

FIG. 8.26. Day-to-day variation in air-borne density of spruce budworm moths during take-off period throughout 1973 flight season. Peak radar count at 30 m above forest canopy during take-off period; nightly moth catch in a light-trap in the canopy; and daily emergence rates of male and female moths from locally sampled pupae. Chipman, New Brunswick.

Peak insect numbers on radar, on 9th July, not reflected by light-trap catches.

fallow, *Sorghum* and groundnuts (or other legume), it was found possible in 1973 to trace such plumes back to individual fields. In particular the Noctuid moths *Heliothis armigera* and *Spodoptera littoralis* (verified by hand-catching during take-off and by daytime ground observations) were seen regularly between 8th and 18th October, as in Fig. 8.23, arising as a radar plume from a field of groundnuts which were in flower and probably the only major source of nectar in the vicinity. The plumes reached their full development some 45 minutes after sunset; they appeared (Fig. 8.23) to show at first the shape of the field from which they rose, and reached 150–200 m above the ground before becoming diffuse. The moths orientated to SSW ($\pm 30°$), as deduced from vectorial subtraction of measured winds, plume geometry and PPI 'dumb-bell', in winds from all quadrants.

The association between take-off and changing light intensity at sundown is illustrated in Figs. 8.22 and 8.24, respectively for tropical and temperate conditions.

Radar evidence on day-to-day variations in the density of airborne insects during the take-off period has been secured over two extended periods, for grasshoppers in the Sudan and for spruce budworm moths in Canada, and is presented in Figs. 8.25 and 8.26, together with corresponding light-trap catches.

The Sudan radar observations in Fig. 8.25 show two mass movements, together involving the majority of all the insects so recorded, on evenings following a change of wind from southerly to northerly, and there were two further, lesser peaks also near the surface position of the Inter-Tropical Front (p. 185). The corresponding light-trap catches do not reflect either of the two major peaks shown by the radar; the light-trap catches were progressively reduced, relative to the numbers seen on radar, during the period following new moon, and appear also to have been reduced in higher winds and at lower temperatures. The light-trap also appeared to provide relatively smaller samples of the larger grasshoppers; the relative numbers of smaller (*Aiolopus*), medium-sized (*Catantops*, *Oedaleus* and *Acrotylus*) and larger grasshoppers (e.g. *Amphiprosopia*, *Locusta*) taken in the light-trap were about 1000 : 5 : 1, compared with the ratio of 20 : 3 : 1 given by the radar signature frequencies over the same period (p. 173).

The spruce budworm data in Fig. 8.26 cover the entire flight season of the species in central New Brunswick, and the airborne moth counts show rapid rises two days after the peaks of emergence from local pupae. The light-trap catches and radar counts are in general agreement except for the two nights of bright moonlight (9th and 12th July), with an 8 to 10-fold reduction in the light-trap catch in each case. Figure 8.27 shows the winds encountered by these moths throughout their season of flight activity.

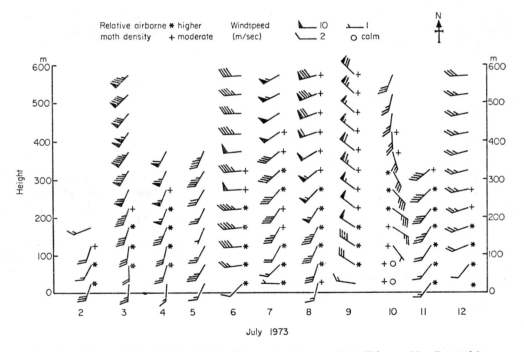

FIG. 8.27. **Upper winds and flying heights of spruce budworm moths at Chipman, New Brunswick, throughout the 1973 flight period.** Daily observations of winds encountered by moths, at about 2200, around the time of peak take-off.

FLYING HEIGHTS AND DENSITIES AFTER THE TAKE-OFF PERIOD

Following take-off, changes in flying height and density during evenings of steady winds are illustrated in Figs. 8.28 and 8.29. Figure 8.28 is representative of numerous observations on locusts and grasshoppers: at 1800, early in the take-off period, many insects were only just airborne, with few at any substantial altitude—this gives a roughly exponential profile of the type which has often been described in the literature, though usually in convective conditions by day. At peak take-off (1821), very many insects were still close to the ground, but the numbers at altitude had increased substantially and the ceiling (level of the topmost insects) had risen. While it has not been possible to track an individual insect for long enough to provide a direct measure of its rate of climb, the rate of rise of the flight ceiling during the take-off period can be measured, and has provided a minimum estimate of the rate of climb of the individuals concerned; this has given a value of about 0·4 metres per second for the rate of climb of *Aiolopus*. Within a quarter of an hour after peak take-off the topmost insects had reached maximum ceiling and the high densities near the ground had extended upwards, giving a distribution similar to that shown for 1915. This type of profile, with density almost constant up to maximum ceiling, generally continued throughout the night. Spruce budworm moths showed a rather different, 'bullet-like' height distribution, illustrated in Fig. 8.29; the ceiling is around 600 m with mean heights of flight of 200–250 m, tending to rise in the course of the night. In the Sudan Gezira, the maximum ceiling for *Aiolopus* was about 1500 m in late September, descending to 900 m by mid-November as temperatures fell, and similar ceiling levels were found for *Chortoicetes* in New South Wales. In Fig. 8.30 it will be seen that the average flying height of the Sudan grasshoppers, like their ceiling, rose steadily to a maximum about an hour after early take-off, but then remained fairly constant until an hour or so before dawn. The mean heights of flight for *Schistocerca*, *Aiolopus* and *Chortoicetes* were respectively about 400, 600 and 500 m during the

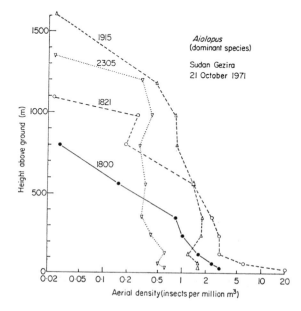

FIG. 8.28. **Typical variation of height/ density profile of Sudan grasshoppers during the evening.** Mainly *Aiolopus simulatrix;* Radma.

Note relatively constant density up to more than 1000 m after peak take-off; similar results found in Sahara and Australia.

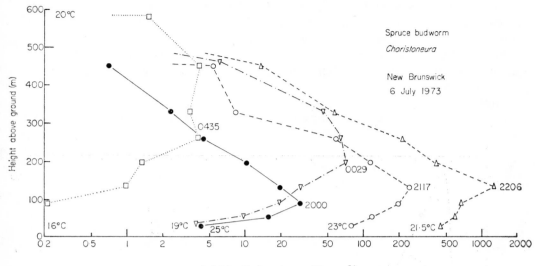

FIG. 8.29. **Typical variation of height/density profile of spruce budworm moths in the course of the night.** Air temperatures at 30 m at times indicated, and at 600 m at 2200.
Level of maximum moth density rises in the course of the night, as lower air cools.

periods studied. The maximum ceiling for night-flying Desert Locusts in the Sahara came down from about 1800 m in mid-September to 1000 m in mid-October. Moths in the Gezira showed ceiling levels and average heights about two-thirds of those attained by *Aiolopus*, at times with high densities within 50 m of the ground as in Fig. 8.41.

Volume-densities of the Sudan grasshoppers at peak take-off averaged about 1 grasshopper per 10,000 m³ at the lowest level, with a range from about one per thousand to one per million cubic metres. Mid-evening densities, showing little change with height, were almost always near to one per 500,000 m³. Desert Locusts in the southern Sahara flew at densities ranging from 3 per 100 thousand to 1 per 100 million m³ by night, and from 3 per 1,000 m³ to 1 per 10,000 m³ in thin swarms by day. Spruce budworm moths

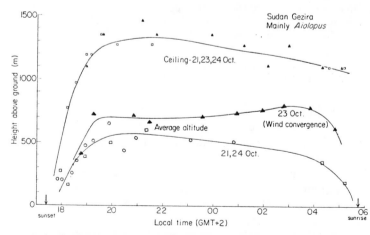

FIG. 8.30. **Variation in height of flight of Sudan grasshoppers in the course of the night.** Radma; similar results in Sahara and Australia.

at their nocturnal peak of flight activity reached densities of about 3 moths per 1000 m³ up to a height of about 200 m. In Australia, after exceptional rains, insects of many species gave total densities intermediate between those recorded for the Sudan grasshoppers and those for the budworm moths. Radar can determine volume-densities down to about one insect per hundred million cubic metres—i.e. of the order of ten insects per cubic kilometre.

Integrating height/density profiles such as Figs. 8.28 and 8.29 with respect to height provides estimates of the area-density of the airborne insects, illustrated by Fig. 8.31.

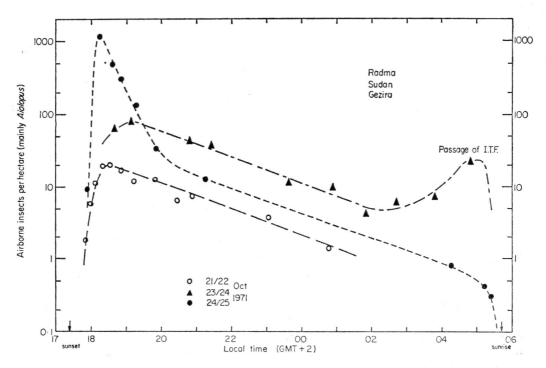

FIG. 8.31. **Variation in area-density of airborne Sudan grasshoppers during the night.** Integrated over all heights; note characteristic exponential decay, and 20-fold increase in numbers during passage of Inter-Tropical Front (cf. Figs. 8.38–8.41).

Scattered night-flying Desert Locusts averaged about one per hectare, reaching about 30 per hectare in wind-shift lines (p. 183); spruce budworm moths flew at about 15,000 per hectare in peak flight periods. The Sudan grasshoppers ranged from 2000 down to 5 per hectare in flight; changes in the course of the night (Fig. 8.31) were characterized by a peak during the first hour after take-off and an exponential decay (with area-densities halved about every 1·6 hours) during the rest of the night until dawn. The duration of individual flights can be estimated to have averaged about 1½ hours during the take-off peak and about 5 hours for the exponential period. Spruce budworm moths, after a double peak of take-off, showed subsequent changes in area-density (Fig. 8.24), basically similar to Fig. 8.31; initial flights averaged about 1½ hours, but the exponential decay was more rapid than with the grasshoppers, with area-densities of the moths halved about every hour, and flight durations averaging about 3 hours over this period.

Combining these estimates of average duration with the radar observations of ground-speed provides estimates of average distances covered per flight. The Sudan grasshoppers with an average ground-speed of about 8 m/s would cover 40–150 km per night; spruce budworm moths, flying for about 1½ hours at a ground-speed of about 10 m/s, would cover about 60 km per night; the Desert Locusts, at ground-speeds around 15 m/s (Fig. 8.45), would cover 100 km in an average flight-time of 2 hours, and the longer-duration fliers could cover 400 km in a night!

From these area-densities and the mass of the corresponding insects, the biomass in flight per hectare can be estimated as up to 150 g spruce budworm moths and 1500 g of *Aiolopus simulatrix* per hectare, substantially more than the corresponding figures for the smaller insects sampled over the same periods by aircraft-trapping. In these areas at least it would appear that most biomass moves at night, without assistance from thermal convection currents [see also p. 272].

OBSERVATIONS ON CONCENTRATION AND TRANSPORTATION OF AIRBORNE INSECTS

CONCENTRATION BY CONVERGENT WIND-FIELDS

Storm-outflow cold fronts

These are produced at the onset of rainfall in a storm-centre when a core of cold air which has been quickly cooled by evaporating rain plunges downwards and outwards ahead of

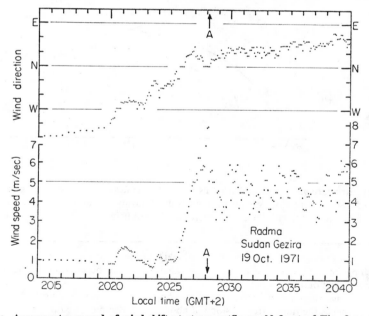

FIG. 8.32. Anemometer record of wind-shift at storm-outflow cold front of FIG. 8.35. Time of passage of leading edge of insect concentration (FIG. 8.35C) marked at A; note how closely this corresponds in time with the wind-shift, recorded at 2 m. Maximum rate of wind-convergence at 2 m is 0·007 per sec, which on the simplest of assumptions would double the area-density of airborne insects in about 2½ minutes.

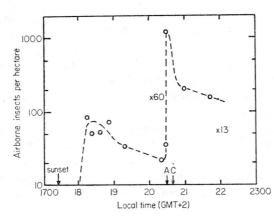

FIG. 8.33. **Concentrating effect of storm-outflow cold front on airborne insects.** Area-density (integrated over all heights) before and after passage of cold front of Figs. 8.32 and 8.35; note 60-fold concentration at the front, followed by resumption of usual exponential decline but with numbers 13 times greater than ahead of front. Front has collected insects for about an hour, bringing new infestations into the Gezira from outside.

the storm. They have been recorded previously as radar line-echoes, attributed to birds [16] or atmospheric echoes [2], and not to insects. Field studies of a number of such fronts have now been made, particularly in the Sudan, illustrated by Fig. 8.35; this shows a rain-storm centred to the NE with a line-echo moving towards the SW several

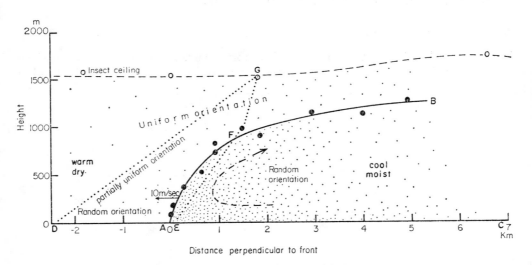

FIG. 8.34. **Structure of head of cold front as inferred from discontinuity in insect densities.** Uniformity of insect orientation was lost at low levels in advance of and within the cold current. Data of Figs 8.32–35, with DG, EFG corresponding to height/density profiles of Fig. 8.37.

B, 2026; range-rings 450 m apart and elevation $1\frac{1}{2}°$; note dense leading edge of cold outflow approaching from NE and now 850 m away; canal-bank also shown, running SE/NW to SW of radar site, and outlines of fields.

c, 2028; range-rings 450 m apart; looking up at 30° elevation at very sharply defined leading edge as it reaches the radar; frontal slope of about 1 in 2 demonstrated by shape of sector of high-density echoes.

D, 2041; range-rings 7·5 km apart and elevation c. $1\frac{1}{2}°$; cold outflow now 8 km away, receding to SW, and with a visible length of at least 60 km.

E, 2050; range-rings 7·5 km apart and elevation $1\frac{1}{2}°$; cold outflow 13 km away, still receding to SW, and visible to at least 45 km; storm collapsing.

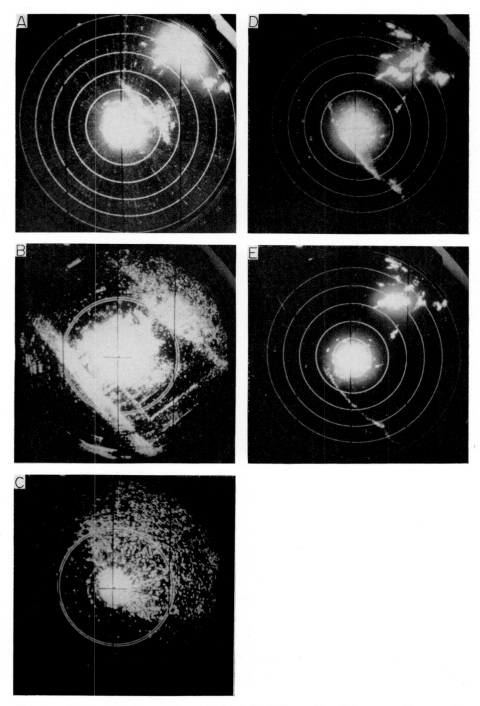

FIG. 8.35. **Insects at a storm-outflow cold front.** Probably mainly *Aiolopus;* see Fig. 5.23 for probable genesis of such a front, there made visible by the neighbouring locust swarm.

Radma, Sudan; 19th October, 1971; single-exposures.

A, 2013; range-rings 7·5 km (8000 yd) apart and elevation 3°; rain-storm centred 35 km away to NE and cold outflow at 8 km, also approaching from NE, undercutting warm SW wind.

[*facing page* 182]

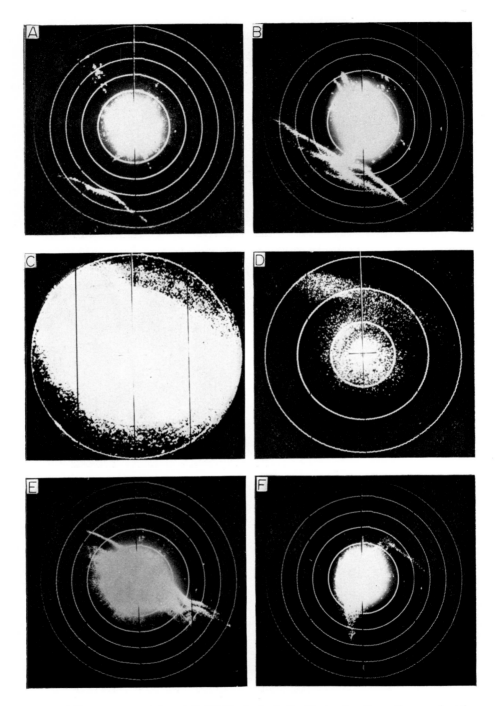

FIG. 8.36. **Insects at an evening wind-shift line in Australia.** Predominantly moths, seen by tele-scope against the moon as the line passed. Coonamble, New South Wales; 15–16th March, 1971; single-exposures.

A, 2305 15th; range-rings 7·5 km apart and elevation $\frac{1}{2}°$; line 30 km away and approaching from SW.

B, 0002 16th; range-rings 3·8 km (4000 yd) apart and elevation $\frac{1}{2}°$; line approaching, at 8 km.

C, 0027 16th; range-rings 450 m apart and elevation 13°; line passing over radar.

[*facing page* 183]

miles ahead of the storm. Insect echoes were responsible for these radar lines on all the occasions studied. Figure 8.32 shows the wind recorded at ground level during the passage of this front. Note the reversal of wind direction from SSW to ENE, and the increase in wind-speed from 1 m/s to 6 m/s in about 90 sec, representing intense convergence, with effects on the density of airborne insects illustrated by Fig. 8.33. From the sequence of time-lapse photographs it was possible to reconstruct the shape A B of the advancing atmospheric density current (Fig. 8.34). This shape has also been found characteristic of the saline density currents used in tank experiments, with a similar slope, half way up the front, of about 1 in 2 [25, 26]. Figure 8.37 shows the effects of the front on the insect height/density profile; such storms are likely to be a major transporting system for insects in the Gezira and in eastern Australia.

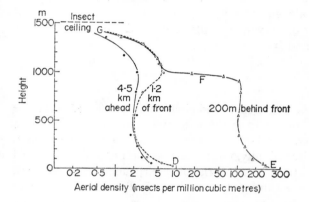

FIG. 8.37. **Influence of advancing front on height/density profile.** Cf. Figs 8.32–8.35; note no accumulation in advance of nose, a 3-fold increase in volume-density above the nose below the ceiling (unlike effect to be expected with smoke), and the 50-fold increase behind the nose.

Nocturnal wind-shift lines

A new type of radar line-echo occurred every few nights in the Sahara, arriving almost invariably from the SE, and one to three times per night in Australia, from a variety of directions including NE, NW and SW. The first appearance was about three hours after sundown (2100 local time); they were more common within an hour of midnight, and were observed as late as 0300 local time; the line-echo was often single, sometimes double and occasionally triple, moving at ground-speeds in the range of 5 m/s to 15 m/s. Their approach was marked by a sudden stilling of the preceding wind (which was often opposed to their direction of motion) and then by a sudden fresh wind as in a storm cold front. Figure 8.36 presents in five photographs (A–E) the history of one such line during its displacement over a distance of nearly 40 km within a period of 100 minutes. The line was about 2 km wide (C), with a double or triple structure (E); with the front edge defined to within a matter of 50 m (as with all these types of line-echo); and with very high volume-densities, here approaching 1 insect per 100 m³ near the front edge—illustrated in (D), taken at the same time as (C) but with the radar gain turned down by a factor of 100. The density contrast before and inside the line was about 500 to 1. A small hill, rising

D, 0028 16th; as C but PPI detection threshold set to 35 dB above noise level instead of 15 dB.
E, 0044 16th; range-rings and elevation as A; line at 8 km, receding to NE, 85 km long.
F, 0015 15th; range-rings 7·5 km apart and elevation 2°; two intersecting fronts, aligned 030°/210° and 135°/315°, appearing to cancel each other out, if not hydrodynamically perhaps by insect deposition. [Cf. Rider & Simpson 1968 'Two crossing fronts on radar' *Met. Mag.* **97** : 24–30—Ed.]

105 m from the almost flat plain to the ESE from the Coonamble radar site, was apparently sufficient to halt these travelling lines, as in (E), and perhaps to cause their enormous insect populations to be deposited. Both in the Sahara and in New South Wales these wind-shifts occurred in a shallow drainage bowl, with slopes of about 1/200 extending over hundreds of kilometres and down which cold air could flow to produce a density current (flowing over stable air near the ground) which could later collapse and deposit its insects, possibly on the opposite slope. This may be suggested as a mechanism in the Sahara for the production from scattered locusts of fairly dense populations perhaps capable of providing nuclei for swarm formation. Two lines of this type are shown meeting in (F), apparently cancelling out each other as they cross; it is suggested that this would leave a line of insects deposited on the plain—a phenomenon which has been reported on a number of occasions in open country, and could enhance the concentration of locusts envisaged. Such katabatic effects may also explain the 'morning-glory' phenomenon observed frequently at dawn with a similar shallow drainage system near the Gulf of Carpentaria in Queensland [5].

Inter-Tropical Front

The Inter-Tropical Front is a feature of the global wind-circulation. In the Sudan it is the meeting place for NE trade-winds coming from the hot dry desert and SW monsoon winds which are moister and cooler. The latter are denser and undercut the former, probably in the form of a density discontinuity, the surface position of which is termed the Inter-Tropical Front (ITF), while the zone of overlapping fairly unmixed air-streams, possibly 100–200 km wide from north to south, is termed the Inter-Tropical Convergence Zone (ITCZ). In the late summer this lies to the north of the Gezira, over which it passes towards the south, with several oscillations, during October.

Over the Gezira the alignment of the front is probably influenced by the mountains of

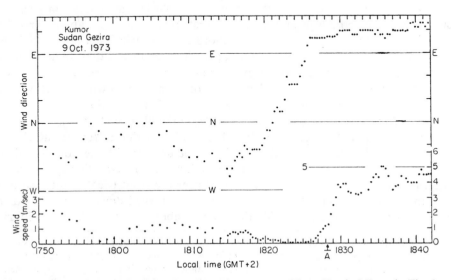

FIG. 8.38. **Anemometer record of wind-shift during passage of Inter-Tropical Front** in Fig. 8.39. Time of passage of leading edge of insect concentration marked at A; note how closely this corresponds in time with the wind-shift recorded at 2 m.

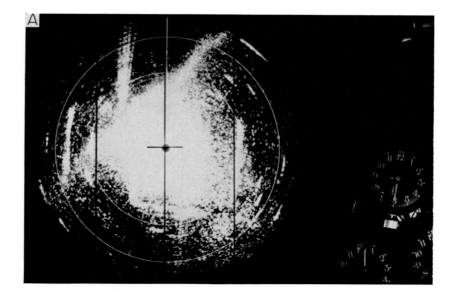

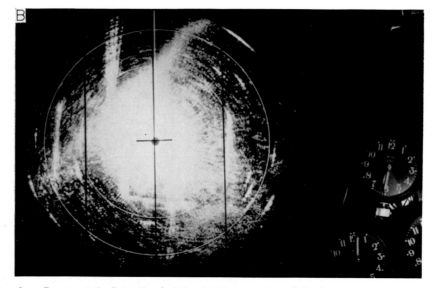

FIG. 8.39. **Insects at the Inter-Tropical Front.** Many moths, off Gezira crops, at high density along front which is aligned SW/NE and has passed overhead from the SE two minutes previously; to be distinguished from the permanent echoes, e.g. from nearer edges of the *Sorghum* fields (aligned N/S), as in Figs. 8.21 & 8.23; corresponding wind-shifts at 2 m and at higher levels shown in Figs 8.38 and 8.40.

Kumor, Sudan; 9th October, 1973*; range-rings 450 m apart and elevation 3°.

A, 1830 single-exposure.

B, 1830 triple exposure; note difference in track-directions between insect echoes ahead of front and those behind it.

* Subsequent to date of symposium but included as particularly well-documented case—Ed.

Ethiopia, and is generally from SW to NE, with predominantly SW winds to the S of it and NE winds to the N. Concentrations of insects associated with meteorological evidence of the surface passage of the ITF during the time of insect flight have been seen on the radar on three occasions during the 1971 and 1973 observations, and echo layers have been observed at the frontal interface of the ITF overhead (p. 189) on other occasions. The degree of concentration has in general been less than that due to storm cold fronts. The radar evidence is illustrated in Fig. 8.39. Here the surface position is clearly defined,

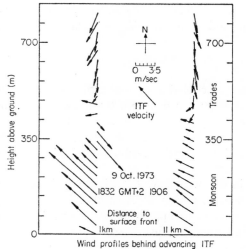

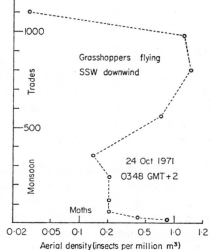

FIG. 8.40. **Wind-profiles through the Inter-Tropical Front following its passage at ground-level.** Data from pilot-balloon ascents accompanying observations of Figs. 8.38 and 8.39.

Note advancing SE monsoon undercutting NNE trade-winds. Around 200 m the monsoon is overtaking the advancing front, with a return flow around 350 m, suggesting vortex flow inside head (Fig. 8.34).

FIG. 8.41. **Insect height/density profile through Inter-Tropical Front.** With the ITF away to the north at ground level, there were grasshoppers (echo signatures indicating entirely *Aiolopus simulatrix*) at 600–1000 m flying southwards with the over-riding northerly trade-winds, and moths below 50 m moving northwards with the under-cutting southerly monsoon. See also Figs. 8.30 and 8.31 for same occasion.

again to within about 50 m, by a dense line of insects approximately 200 m in width, with the appropriate alignment; the line has just passed the radar and is moving NW. The time, 1830, is just at peak take-off period and the front has been collecting and concentrating insects for 20 minutes at most. Figure 8.38 shows the anemograph record of this passage of the ITF. A pilot-balloon, released about 1 km behind the front, gave the winds shown to the left of Fig. 8.40. It will be seen that in the bottom 300 m there were strong winds approaching and overtaking the front, whose direction and speed (NW-wards at $5 \cdot 0 \pm 1 \cdot 0$ m/s) is indicated by the solitary arrow in the centre, while as usual higher up, above the ITF roof, intersected on this occasion at about 500 m, the winds were from the NE. A second pilot-balloon sent up 11 km behind the surface front gave winds shown to the right in Fig. 8.40 and just overtaking the front at the lower levels. Some uniformity of insect orientation persisted at the front, as may be seen from Fig. 8.39. The western part of the 'dumb-bell' of orientation ahead of the front is just visible to the west of the radar, the other parts of the 'dumb-bell' having been swallowed up by the advancing front. Behind the front the 'dumb-bell' shows insect orientation to the south-west, persisting in the SE winds. Evidence that the moving front was converging insects, as well as winds, is provided by the triple exposure, Fig. 8.39B, showing that insects in advance were moving roughly parallel to the front while those in the rear were rapidly approaching the front at right angles. The insects at low level on this occasion were primarily moths, off the Gezira crops; on another occasion (Fig. 8.41) it was found that in the northerly air-stream above the ITF roof there were grasshoppers, mainly *A. simulatrix*, moving towards winter quarters, while close to the ground behind the ITF was a layer of moths.

Bénard convection cells

When a layer of fluid is heated from below and/or cooled from above, buoyant energy is released to transfer the heat aloft and this transfer is usually not by random turbulent mixing, but by organized flow-patterns known as Bénard cells. These cells, often observed in the laboratory, exist also in the atmosphere; they have been observed by long-range radars on a few occasions, with moderate resolution, and the echoes ascribed to insects [14].

This type of pattern has been observed in the present field studies on 8 days, when the day-flying insects were sufficiently numerous, mainly in Australia. In Fig. 8.42A polygonal cells are clearly visible, with diameters of approximately 4 km. The cells have walls which tend to be aligned down-wind, on this occasion N/S, and small cumulus clouds sometimes formed on the tops of these walls. The expected circulation pattern of the winds within a cell is from the centre of the cell along the ground surface towards the walls where similar winds are met from other cells; they then rise together up the walls and return back towards the centres of the cells, there to descend to ground level and repeat the circulation, while each cell as a whole moves along with the wind. These cells were observed on the radar to be translated with the average wind and to maintain their identity for at least 20 km, with their wall up-currents facilitating insect transportation.

In Fig. 8.42B the passage of two portions of a cell wall is seen at close range, with a PPI screen diameter of 6 km. Here a wall running roughly N/S meets at right angles a roughly E/W wall just to the north of the radar; the line of echo seen to the SW and aligned NW/SE is the near edge of a woodland. The sharp definition of these walls, the precise form of the corner, and the high degree of insect concentration within the wall with about 1 insect per 10,000 m³, are striking. The echoes could be seen to be due to insects as the passage of the wall was made visible in the atmosphere by dense clouds of dragonflies and butterflies, appearing and then disappearing within a matter of minutes. As these walls of insects passed over, the anemograph showed a change of wind through a dead calm, confirming the circulation pattern, and the radar echoes were seen to be moving towards the walls from both sides, illustrating the process of concentration. The relative concentration of insects in the walls compared to the surrounds was of the order of several hundred-fold.

An example of such a cell in a Desert Locust swarm is shown in Fig. 8.43, just beyond the central ground clutter; the cell is largely devoid of echo near its centre, and shows a ring of echo around the outside of a polygonal pattern, as in Fig. 8.42. Reference has already been made (p. 163) to the down-wind orientations unexpectedly found in

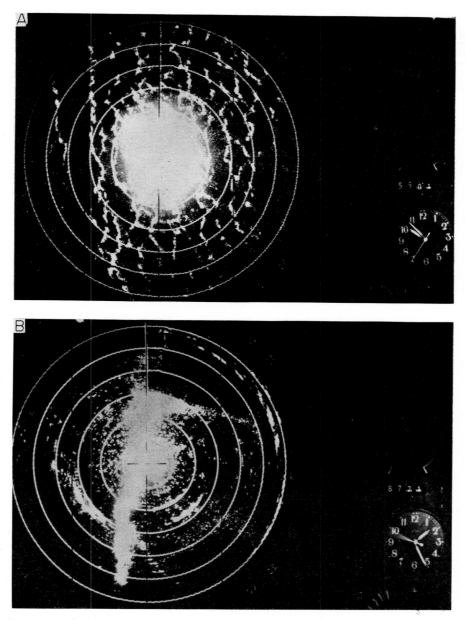

FIG. 8.42. **Distribution of airborne insects reflecting cellular pattern of convection currents.** Mainly dragonflies and butterflies; Coonamble, New South Wales; 14th March, 1971; single-exposures.

A, 1053; range-rings 3·8 km apart and elevation 0°. Convection cells were also marked by 'streets' of small cumulus clouds.

B, 1326; range-rings 450 m apart and elevation 6·3°; intersecting walls of convection cells, aligned 020°/200° and 110°/290°, sharply defined by high radar resolution. Line-echo in SW is from trees—pp. 186–187.

[facing page 186]

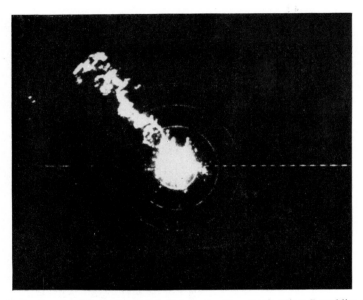

FIG. 8.43. **Flying Swarm of Desert Locusts in Sahara.** In Abangharit, Niger Republic; 1210 15th October, 1968.

Range-rings 3·8 km apart and elevation c. 1°; single-exposure.

Swarm moving towards NW; evidence of several convection cells, e.g. near 2nd range-ring, of diameter about 4 km. Aircraft flying through such cells would encounter intermittent bursts of insects, as has indeed been recorded (in p. 336 in ref. [39], cited on p. 111).

[This swarm probably formed part of the trans-African migration described on pp. 116–119 and in Figs. 6.3 and 6.4—Ed.]

[*facing page* 187]

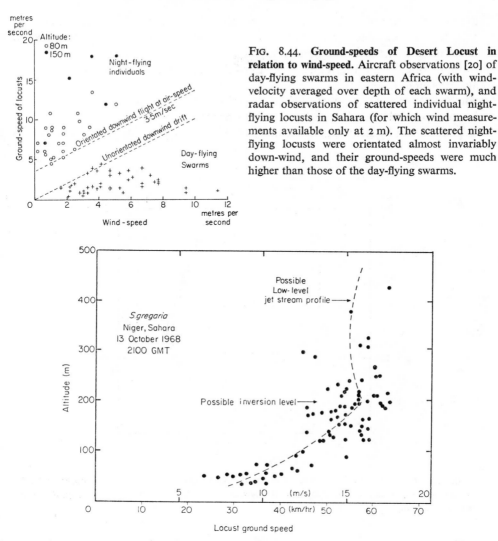

FIG. 8.44. **Ground-speeds of Desert Locust in relation to wind-speed.** Aircraft observations [20] of day-flying swarms in eastern Africa (with wind-velocity averaged over depth of each swarm), and radar observations of scattered individual night-flying locusts in Sahara (for which wind measurements available only at 2 m). The scattered night-flying locusts were orientated almost invariably down-wind, and their ground-speeds were much higher than those of the day-flying swarms.

FIG. 8.45. **Ground-speeds of individual night-flying Desert Locusts in relation to height.** In Abangharit, Niger Republic.

No upper-wind observations were available; dashed curve indicates a possible jet-stream profile (adapted from Izumi [16] and consistent with subsequent observations under roughly comparable conditions in Sudan, e.g. in Fig. 8.48) with 3·5 m/s added as locust air-speed.

scattered night-flying Desert Locusts, giving the high ground-speeds illustrated in Figs. 8.44 and 8.45.

ECHO LAYERS

The radar screen often shows a relatively dense layer of echoes, shallow in depth but large in horizontal extent, and separated from lesser densities above and below. On some occasions several layers—up to five—have been observed simultaneously. Similar layering, both discrete and continuous, has been observed with many radar systems [2].

In the present series of studies it has been concluded that insects have been the main, or only, targets responsible for the layers, since the intense continuous layers have been seen to build up from lower density layers of discrete individual targets which have shown insect-like signature frequencies. All of the main insect species studied have been found in layered distributions.

On most evenings the 'bullet-like' height profile of spruce budworm moths (Fig. 8.29) had a pronounced high-density layer, of the order of 50 m deep, associated with the 'nose', and shown by aircraft temperature profile in Fig. 8.46 to have been at the warmest

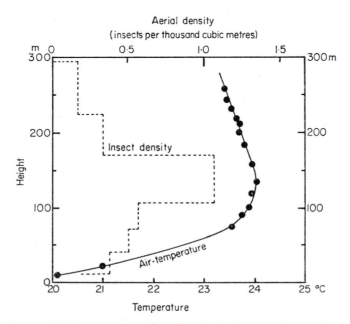

FIG. 8.46. **Characteristic layer concentration of spruce budworm moths at the warmest level above the forest canopy.** Chipman, New Brunswick; 2200 6th July, 1973; see also **Fig. 8.29** for same occasion.

level. Figure 8.9A shows an orientated layer of moths forming at the time of take-off, with a plume rising from the local source to join the incipient layer.

Most dense, shallow layers observed have been at levels of marked wind-shift in the vertical, i.e. strong shear interfaces, associated with frontal systems. Figure 8.9B shows such a layer of moths some 30 m thick during the passage of the minor front illustrated in Fig. 5.21. The layer occurred just above the wind-speed maximum in the lower layer at the inversion height; the layering may well have been influenced by horizontal convergence along the sloping frontal interface.

Well-defined thin layers also occurred some kilometres behind the passage of nearly every storm-outflow cold front or nocturnal wind-shift line, both being density currents. At the sloping shear face between the under-cutting colder air and the over-riding warmer and less dense air, local horizontal convergence aids the formation of the layer. A very good example of this type of layering is shown in Fig. 8.47. This layer formed just after the passage of the line and the illustration was taken two hours later, still in the presence of an under-cutting cold air-stream. A radar-tracked pilot-balloon sent up through this

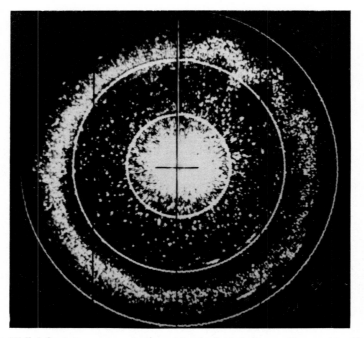

FIG. 8.47. **Well-defined layer concentration at level of wind-change.** Coonamble, New South Wales; 0106 14th March, 1971; range-rings 450 m apart and elevation 19·3°; single-exposure.

Layer 45 m thick, at height of 330 m, at wind-shear zone 2 hours after passage of wind-shift line similar to Fig. 8.36; evidence of waves in layer to NE, of wavelength 410 m; no uniformity of insect orientation—effect of turbulence, or mixture of species?

In layers, densities often reached 1 or more insects per 10,000 m³, giving radar volume reflectivities* of about 10^{-10} cm⁻¹, two orders of magnitude greater than previously recorded [14] at a wavelength of 3 cm. Densities in the line-echoes were still greater, up to 1 insect per 100 m³ (Fig. 8.36D), which gave a volume reflectivity of 10^{-8} cm⁻¹, a hundred times the maximum previously recorded in fronts [14].

* [Radar volume reflectivity may be envisaged as equivalent area of perfect reflector per unit volume (e.g. cm² per cm³, i.e. cm⁻¹).]

layer showed an up-draught just below the layer and down-draughts above. A high-resolution illustration of micro-scale gravity currents, probably of the Kelvin-Helmholtz type (K-H), with a wavelength of 400 m, is also seen, together with 'black holes'.

Layers have often been associated with the sloping interface of the ITCZ in the Sudan, between the southerly air-streams below and the northerly air-streams above. The layers are usually just above the shear calm, and may be dense and shallow, or more diffuse and deeper; a good example of the latter is given in Fig. 8.41.

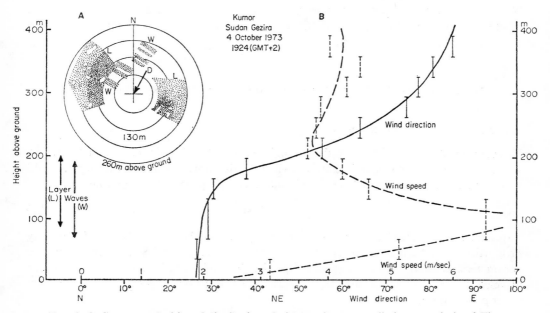

FIG. 8.48. **Some complexities of distribution of airborne insects:** preliminary analysis of Fig. 8.8B. Mainly *Aiolopus* and *Gryllus;* orientation towards SSW (down-wind, direction D); layer L at level of marked change in wind with height (wind-shear), perforated by 'black holes' attributed to turbulence, and with train of travelling waves (W) manifested by successive photographs.

Some layers have occurred away from areas of strong horizontal wind-convergence; during the growth of low-level jet-streams, for example, in the Sudan and in New South Wales, layers of insects have formed just above the height of maximum wind-speed.

A good example of a relatively dense layer forming in these conditions is presented in Fig. 8.8B; the pilot-balloon released at the same time gave the winds shown in Fig. 8.48. This situation was far removed from strong horizontal convergence and the layer formed just below the shear interface, in the presence of travelling (K-H?) waves. There was very uniform orientation, exactly down-wind at the levels shown.

ACKNOWLEDGEMENTS

I am deeply indebted to the Directors of the Centre for Overseas Pest Research, London, the Division of Entomology of C.S.I.R.O., Canberra, the Agricultural Aviation Research Unit of Ciba-Geigy Ltd., and the Maritimes Forest Research Centre of the Canadian

Forestry Service for their foresight and financial provisions which made the field research possible; and to the Organization Commune de Lutte Antiacridienne et de Lutte Anti-aviare of West Africa and the Sudan Gezira Board for generous assistance with the field operations. I am grateful to the Nuffield Foundation, the Canadian Forestry Service and Ciba-Geigy Ltd. for their continuing financial support of the data analysis. I have learned a great deal from the project entomologists Dr D.P.Clark, Dr D.Greenbank, J. Roffey and N.Russell-Smith. The field studies would have been impossible without the untiring technical assistance of J.Sewell and L.Larrad. I wish to express my debt to I. Norton for his assistance with the analysis and preparation of drawings for this paper. Finally, I am especially grateful to R.J.V.Joyce, Dr R.C.Rainey and Dr C.G.Johnson for their unfailing encouragement.

APPENDIX: SOME QUANTITATIVE ASPECTS OF RADAR ENTOMOLOGY

Radar cross-section—definition

The echo-producing ability of a radar target is expressed as its radar cross-section (RCS), which is the area of an isotropic scatterer which, in the same position as the target, would scatter an equal radio power back to the radar. The simplest radar equation

$$P_r = k(\text{RCS})r^{-4} \tag{A1}$$

states that the received echo power is proportional to the target RCS and inversely proportional to the fourth power of the target range, r. The RCS will depend on five parameters: dielectric constants; size (specifically body-length: L) relative to radar wavelength λ; body shape (length/width ratio: LWR); body aspect as seen by radar; radar polarization.

Beam directivity pattern

For the calculation of absolute aerial density, beam shape must be known, and was measured by small airborne spheres, with results shown in Fig. 8.49. The results, which refer only to PPI detection in the surveillance mode and with the visual and photographic threshold set 10 dB above noise, show for example that an insect with an RCS of 2 cm² is detectable to 1·7 km when on beam centre, and to a range of 1 km at 16 m off beam centre.

Radar cross-sections—theory [22]

Organs with high free-water content (muscles, blood, etc.) reflect strongly, almost as intensely as if made of metal. The theoretical RCS of equivalent spheres of the homo-geneous biological dielectric material (unit density; refractive index 6·7; absorption constant 1·4; $\lambda = 3·18$ cm) considered in the equivalent detailed analysis of radar echoes from birds [22] is shown in Fig. 8.50 versus insect weight, W, considering the weight of each sphere as 80 per cent of the corresponding insect weight. The RCS is minute for small insects (Rayleigh region; $L/\lambda \ll 1$), varying as

$$\text{RCS}_R = KW^2/\lambda^4 \tag{A2}$$

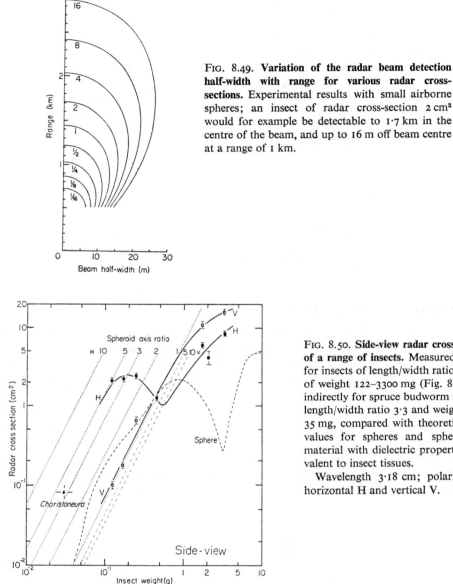

FIG. 8.49. **Variation of the radar beam detection half-width with range for various radar cross-sections.** Experimental results with small airborne spheres; an insect of radar cross-section 2 cm² would for example be detectable to 1·7 km in the centre of the beam, and up to 16 m off beam centre at a range of 1 km.

FIG. 8.50. **Side-view radar cross-sections of a range of insects.** Measured directly for insects of length/width ratio 4–5 and of weight 122–3300 mg (Fig. 8.54), and indirectly for spruce budworm moths of length/width ratio 3·3 and weight about 35 mg, compared with theoretical RCS values for spheres and spheroids of material with dielectric properties equivalent to insect tissues.

Wavelength 3·18 cm; polarization—horizontal H and vertical V.

but oscillates about 1 cm² for the larger insects (interference region; $L/\lambda \sim 1$). In the latter region two interfering waves are sent back to the radar—a specular reflection from the illuminated face, and a diffracted wave which has crept around the 'shadow' side—making the RCS very sensitive to variations in size or shape.

A more realistic model is a prolate spheroid of the same material. Calculations have been carried out only for the Rayleigh region, with LWR values of 1 (sphere), 2, 3, 5 and 10. K now depends on LWR, viewing aspect and radar polarization. The results are presented in Figs. 8.50 to 8.52. Longer spheroids of the same weight have larger side-view RCS values and a dipole-like aspect-dependence with horizontal polarization.

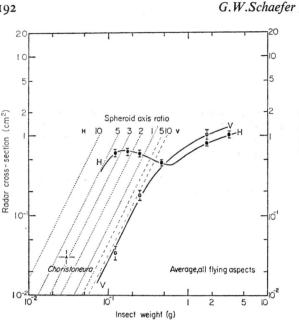

FIG. 8.51. **Radar cross-sections of a range of insects, averaged over all flying aspects.** Measured and theoretical values for insects of Fig. 8.50.

Radar cross-sections—measurements

Further theoretical treatment is difficult. During 1970 RCS measurements were made at X-band on a series of insect species similar in bodily proportions (LWR = 4–5) to pest species under consideration, with L/λ values between 0·5 and 1·7, viewed from all aspects, with H and V polarizations. Fresh specimens in flight posture were used on a radar range along an airfield runway. The results are presented in Figs. 8.50, 8.51 and 8.54; the experimental error was approximately $\frac{1}{2}$ dB. The spruce budworm ($L/\lambda = 0·3$) results of Figs. 8.50–8.52 were obtained by analysis of orientation 'dumb-bells' (p. 162) using a calibrated radar.

These measurements reveal some noteworthy features. In the case of V polarization, the side-view RCS follows the Rayleigh law (A2) reasonably well for all species ($\lambda \geqslant 3$ cm), while the polar diagram is of dipole type for $\frac{1}{2} < L/\lambda < 1$. For H polarization: (i) in the

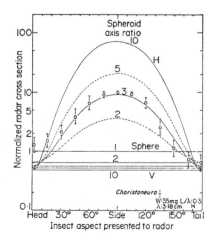

FIG. 8.52. **Aspect-dependence of radar cross-section of spruce budworm moth.** Relative to side-view, normalized to a spheroid of LWR 3·3, and compared with theoretical values for Rayleigh spheroids, normalized to a sphere of equal weight.

Rayleigh region ($L/\lambda \leqslant 0.4$), theory and experiment agree well (Figs. 8.50–52); (ii) in the dipole region ($L/\lambda \sim \frac{1}{2}$), the polar diagram resembles that of a dipole wire (Figs. 8.50–52) and the side-view RCS reaches a maximum; (iii) at the beginning of the interference region ($L/\lambda \sim \frac{3}{4}$), the polar diagram (Fig. 8.54C, D) is similar to that of thin wires; (iv) the side-view RCS greatly exceeds that of the equivalent sphere for $L/\lambda < \frac{3}{4}$, as predicted. The largest insects ($L/\lambda \sim 1\frac{1}{2}$) have polar diagrams which are very sensitive to aspect angle or body distortion. The side-view RCS exceeds the end-view RCS by a factor between 10 and 1000. Figure 8.54A shows that the wings made a negligible contribution to the RCS [23].

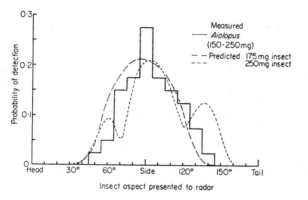

FIG. 8.53. **Aspect-dependence of relative probability of detection of** *Aiolopus simulatrix:* **measured and theoretical.** Observations between 900 and 1350 m, with theoretical values for insects of comparable weight; total probability expressed as unity.

On occasion, insects have random headings, and an RCS averaged over all aspects is required; measured and theoretical values are shown in Fig. 8.51, where a noteworthy feature is the nearly constant RCS, $\frac{3}{4}$ cm^2, for a wide range of insect weights. A useful method for observing larger insects, like locusts, and suppressing echoes from smaller insects is the use of V polarization near the maximum detection range. The results show that insect echoes cannot be suppressed by the use of circular polarization, and so can be distinguished from rain echoes [24].

RCS values for other wavelengths ($L/\lambda > \frac{1}{2}$) may be obtained from the 3.18 cm measurements by dimensional scaling. It is then found that (i) the maximum RCS for a species occurs near $L/\lambda = \frac{1}{2}$ with H polarization, which provides guidance in the choice of radar wavelength; (ii) the largest RCS for any insect is near 20 cm^2; (iii) the use of the multi-wavelength radar technique to prove the presence of insect targets [13] can be misleading if larger insects are present, unless they are flying side-view and V polarization is used; (iv) Desert Locusts have strongest echoes at wavelengths of 3.2 and 8.5 cm, using V and H polarizations respectively.

Orientation pattern and detection aspects

The maximum range for individual detection varies as the fourth root of the RCS, according to (A1); the measured RCS values of Figs. 8.52 and 8.54 have been used to produce the orientation patterns of Fig. 8.11. This figure indicates that only those insects with aspects near side-view will be detected. For example, the probability of PPI detection at a given aspect between 900 and 1350 m (approximately the range of the circle) is proportional to the mean beam detection width for the RCS of that aspect, as given by Fig. 8.49. Applying this to Fig. 8.54B & C gives the curves of Fig. 8.53, which are in good

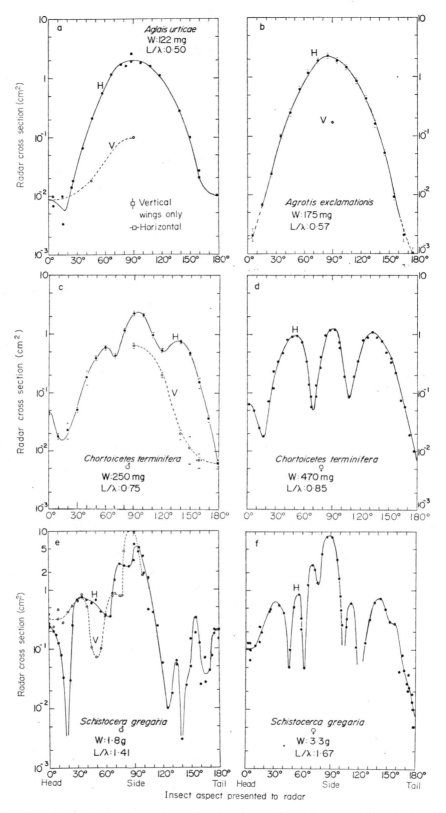

Fig. 8.54. **Measured aspect-dependence of radar cross-sections of a range of insects.** Length/ width ratios 4–5; wavelength $\lambda = 3\cdot18$ cm; polarization H and V; insect weight W; body-length L cm.

agreement with the distribution obtained by measuring the angle between azimuth and heading for a series of ground-tracks chosen equally from all azimuths; simultaneously, signatures and light-trap catches showed the dominant species was *Aiolopus simulatrix*, with the weights indicated.

The degree of orientation will affect the echo count, N. With strong orientation, N was counted at the end of the 'dumb-bell', and no correction was applied. With random headings, N was increased by a factor of 3, i.e. the ratio of side-view to average detection probability (Fig. 8.53).

DISCUSSION (COMMUNICATED)

Mr M.V.Venkatesh: One of the earliest radar studies on locust swarms was made during July 1962 by the Rain and Cloud Physics Research Unit, National Physical Laboratory, New Delhi, using a 3·2 cm model JRC NMB 451 A radar [21]. From the photographic and other data on swarm area, height and density given by these radar observations, it was estimated (Rainey 1967, *Science*, **157** : 98–99) that there were as many as 100,000 million locusts within a radius of 100 km of Delhi on 27th–28th July. However, all the swarm reports received from the usual ground sources for this same area and period account for only about half of the total area of the swarms shown by the radar photographs. Moreover, the radar must have missed low-flying locusts in the more distant parts of its field of view, thus increasing the gap between the radar and ground report estimates. In the past, anti-locust organizations have made considerable use of estimates based on ground reports of this kind, and the anomalies shown by the radar high-light the need for caution in utilizing such estimates and conclusions. At the same time there appears to be further need for improving radar techniques so that a still larger proportion of the locusts can be recorded.

Schaefer: It is not uncommon to have a large discrepancy between ground reports and radar observations, both for birds and insects. For example, radar showed dispersed Desert Locusts taking off uniformly over large areas in the Sahara night after night, and yet these populations were missed by the usual ground observations; again, large populations of pest moths were hidden in groundnut fields until the radar observed their localized mass take-off. Regarding the radar detection of locusts, I have carried out a design study for a locust detection radar [24], and have shown that a network of inexpensive radars separated by about 250 km could keep constant watch in important breeding and gregarizing areas, as well as direct spray aircraft from a central base to the air-to-air spraying of specific swarm targets, whose size and density had been assessed by radar.

Dr G.Jackson: Does Dr Schaefer think that night-flying locusts might orient themselves by perceiving infra-red radiation emitted from the ground, and that they might use infra-red radiation reflected from vegetation at night to locate areas of vegetation and land in them? This latter end could be achieved by a fairly simple mechanism of inhibition of locomotory activity by infra-red after a certain period of flight.

Schaefer: Desert Locusts and the other pest species orientate just as well on the darkest of nights, when it would appear that there would be insufficient ambient light to show up ground patterns at altitudes suitable for wind-drift assessment. Under the same conditions it is most unlikely that there would be sufficient reflected or radiated infrared energy to do the job, except in bright moonlight. However, strong radiations from soil and vegetation exist day or night at far infra-red wavelengths in the region of 10 μm, where the atmosphere is particularly transparent. Competing radiations from the lower atmosphere are unlikely to obscure the ground radiation patterns significantly. The insect receptors for the far infra-red would not have to be particularly sensitive. In the case of night-flying locusts the goal might be the detection of moist soil rather than vegetation. This subject deserves further research.

Dr R.C.Rainey: A leading radar meteorologist recently stated, 'Radar is indeed a versatile tool for observing many aspects of the atmosphere, but . . . the interpretation of radar measurements is subject to many ambiguities, so that its full potential is realized only when used in conjunction with other kinds of observations' (Browning, 1973, *Q. Jl R. met. S.*, **99** : 793). Would Dr Schaefer care to comment?

Schaefer: Dr Browning's statement remains true when applied to the radar observation of the aerial fauna. It is a remote sensing technique, rather analogous to nuclear physics, sensing its targets by the scattering of electro-magnetic waves. In this respect the ambiguities are not about number and density, but about target identification, and the subject must rely on complementary techniques, such as aerial- and light-trapping, moon watching, search-lights and direct observations of wing-beat frequencies. Other ambiguities arise when attempting to relate insect densities and motions to wind-fields. At present there are many unknowns regarding insect flight behaviour, so that on the one hand insects should not be assumed to be perfect tracers for studying wind-fields, and on the other hand a complete knowledge of wind-fields is not sufficient for the complete explanation of the observed insect concentrations. Progress will only be possible by the inter-play of high-quality meteorological and radar observations with direct field and laboratory observations of insect flight behaviour.

Professor B. Hocking: C.B.Williams estimated the insect population of the world at 10^{18} on the basis of suction-trap data. How soon does Dr Schaefer expect to be able to revise this estimate on the basis of radar observations?

Schaefer: Up to the present time radar has been concerned primarily with the larger insects, and only with those which fly at least once during their adult life above an altitude of 10 m. A rough approximation of the population of this minority class might be 10^{15}, the majority of these being forest moths, followed by grasshoppers. But the vast majority of flying insects are small and occur at sufficiently high concentration to make suction-trapping an efficient technique. Even when radar is perfected to deal with these insects it is unlikely to be superior as a technique for estimating the world insect population. In the combined radar and aircraft-trapping field studies, the aerial density of the smaller insects exceeded that of the radar targets on average by a factor of about 10^3, in rough agreement with Williams' estimate. Radar is likely to make a contribution to the assessment of the total biomass of insects which fly, since the combined field studies showed that the major contribution came from the larger insects (p. 181). A very rough estimate would be several times 10^{13} g. In this matter radar cannot claim superiority over thorough ground surveys.

REFERENCES

[1] ATLAS, D., HARRIS, F.I. & RICHTER, J.H. (1970). Measurement of point target speeds with incoherent non-tracking radar: insect speeds in atmospheric waves. *J. geophys. Res.*, **75** : 7588–7595.

[2] BATTAN, L.J. (1973). *Radar observation of the atmosphere.* University of Chicago Press, Chicago and London: x + 324 pp.

[3] BRUDERER, B. (1969). Zur Registrierung und Interpretation von Echosignaturen an einem 3-cm-Zielverfolgungsradar. *Der Ornithologische Beobachter,* **66** : 77–88.

[4] CAMHI, J.M. (1971). Flight orientation in locusts. *Scientific American,* **225** : 74–81.

[5] CLARKE, R.H. (1972). The morning glory: an atmospheric hydraulic jump. *J. appl. Met.,* **11** : 304–311.

[6] CRAWFORD, A.B. (1949). Radar reflections in the lower atmosphere. *Proc. Inst. Radio Eng.,* **37** : 404–405.

[7] DOWNING, J.D. & FROST, E.L. (1972). Recent radar observations of diurnal insect behaviour. *Proc. N.J. Mosq. Ext. Assoc.,* **58** : 114–131.

[8] EASTWOOD, D. (1967). *Radar ornithology.* Methuen, London : xii + 277 pp.

[9] FORSYTH, B.M. & CHAPANIS, A. (1958). Observer under-estimation of the frequency of fluctuating light stimuli. *J. exp. Psych.,* **56** : 385–391.

[10] GEHRING, W. (1967). Analyse der Radarechos von Vögeln und Insekten. *Der Ornithologische Beobachter,* **64** : 145–151.

[11] GLOVER, K.M., HARDY, K.R., KONRAD, T.G., SULLIVAN, W.N. & MICHAELS, A.S. (1966). Radar observations of insects in free flight. *Science,* **154** : 967–972.

[12] GREENEWALT, C.H. (1962). Dimensional relationships for flying animals. *Smithsonian Miscellaneous Collections, Vol.* 144, *Number* 2. Smithsonian Institution, Washington.

[13] HARDY, K.R., ATLAS, D., GLOVER, K.M. (1966). Multi-wavelength backscatter from the clear atmosphere. *J. geophys. Res.,* **71** : 1537–1552.

[14] HARDY, K.R., KATZ, I. (1969). Probing the clear atmosphere with high-power, high-resolution radars. *Proc. I.E.E.E.,* **57** : 468–480.

[15] HARDY, K.R. & OTTERSTEN, H. (1969). Radar investigations of convective patterns in the clear atmosphere. *J. atmos. Sci.,* **26** : 666–672.

[16] HARPER, W.G. (1958). An unusual indicator of convection. *Proc. Seventh Wea. Radar Conf.,* D9–D16. Amer. Meteor. Soc. Boston.

[17] IZUMI, Y. (1964). The evolution of temperature and velocity profiles during breakdown of a nocturnal inversion and a low-level jet. *J. appl. Met.,* **3** : 70–82.

[18] KUTSCH, W. (1973). The influence of age and culture-temperature on the wing-beat frequency of the Migratory Locust, *Locusta migratoria. J. Insect Physiol.,* **19** : 763–772.

[19] LAMBERT, M.R.K. (1970). *General flight characteristics of the Australian plague locust* (Chortoicetes terminifera *Walker*) *in a laboratory wind-tunnel.* Report, Anti-Locust Research Centre, London.

[20] RAINEY, R.C. (1963). *Meteorology and the migration of Desert Locusts.* Tech. Notes Wld met. Org., No. 54 : v + 115 pp.

[21] RAMANA MURTY, BH. V., ROY, A.K., BISWAS, K.R. & KHEMANI, L.T. (1964). Observations on flying locusts by radar. *J. sci. industr. Res., Delhi,* **23** : 289–296.

[22] SCHAEFER, G.W. (1968). Bird recognition by radar: a study in quantitative radar ornithology. In Murton, R.K. & Wright, E.N. (eds.). *The problems of birds as pests:* 53–86. Academic Press, London and New York.

[23] SCHAEFER, G.W. (1969). Radar studies of locust, moth and butterfly migration in the Sahara. *Proc. R. ent. Soc. Lond., Ser. C.,* **34** : 33, 39–40.

[24] SCHAEFER, G.W. (1972). Radar detection of individual locusts and swarms. *Proc. Int. Conf. Acridology, London.* 1970 : 379–380.

[25] SIMPSON, J.E. (1969). A comparison between laboratory and atmospheric density currents. *Q. Jl R. met. Soc.,* **95** : 758–765.

[26] SIMPSON, J.E. (1972). The effect of the lower boundary on the head of a gravity current. *J. Fluid Mech.,* **53** : 759–768.

[27] SOTAVALTA, O. (1947). The flight-tone (wing-stroke frequency) of insects. *Acta. Ent. Fenn.,* **4** : 5–117.

[28] WEIS-FOGH, T. (1956). Flight performance of the Desert Locust. *Phil. Trans. Roy. Soc. B.,* **239** : 459–510.

[29] WEIS-FOGH, T. (1967). Respiration and tracheal ventilation in locusts and other flying insects. *J. exp. Biol.,* **47** : 561–587.

9 · Foraging and homing flight of the honey-bee: some general problems of orientation*

MARTIN LINDAUER

Institut für Vergleichende Physiologie,
Universität Würzburg

When a forager-bee is flying towards a particular individual feeding-site its navigation*
involves two phases:
(i) Teleorientation (*Fernorientierung**) by which the animal reaches the neighbourhood
of the goal [and for which 'long-range navigation' is perhaps the closest English equiva-
lent].
(ii) Orientation immediately at the goal (*Nahorientierung**) by which the animal locates
and identifies the goal itself [and suggested as equivalent to 'short-range navigation'].

I. TELEORIENTATION
(long-range navigation)

In the first phase the bees must be aware of the distance and the direction from the hive
to the feeding-place. This information is gained either from their own experience as
scout-bees or—indirectly—from communication with other bees by the 'waggle-dance'
[5, 6]. The initial acquisition of this information poses some fundamental problems.

ESTIMATION OF DISTANCE

How do the bees measure the distance?—a question which has not so far been formulated
in relation to other animals, not even for migrating insects, fishes, amphibians or birds.
Nevertheless, such an ability has to be postulated as a basic component of navigation
for any swimming, walking or flying animal which is capable of finding a specific
distant goal.

In honey-bees measuring distance is not only of importance to the individual; this
information has to be transferred to hive-mates, encoded in the rhythm of the dance. The
physiology of distance-measurement is still not fully understood, but there are some

---

* '*Orientierung*' appears to have more of a connotation of 'guidance' than does 'orientation', which in
relation to behaviour tends to have the more restricted meaning of 'heading', in terms of compass-
direction, without consideration of distance. On the other hand, 'navigation', particularly in relation to
aircraft, specifically includes visual navigation as well as e.g. astro-navigation.

experimental data to support the hypothesis that *energy consumption* may be involved in providing such information:

(1) *Flight-time* can be excluded as a measure of distance for the following reasons:

(*a*) It is known [7] that only the information gained in the *outward* flight is reported in the dances; and although in head-winds the dances indicate greater distances and in tail-winds smaller ones than under calm conditions, the bee compensates for the effects of the wind in these two cases by keeping approximately the same speed relative to the ground, and energy consumption, not flight-time, shows the best correlation with distance [7, 9].

(*b*) Bees flying *uphill* towards a crop similarly indicate a greater distance than when flying downhill [9].

(*c*) If we force the bees to *walk* towards a feeding-place, through a glass-covered gallery, the distance indicated by dances for a 4 m walk corresponds to a flight of about 100 m [1]. Scholze, Pichler and Heran [26] have quantitatively measured the consumption of sucrose during flight and during walking, and there is a correlation between the amounts used in a 4 m walk and in a 100 m flight.

(*d*) Besides or in addition to energy consumption, information from the optical senses may be used as distance parameters. Flying at constant heights ($2\frac{1}{2}$ m in calm air, 1 m to $\frac{1}{2}$ m in strong winds), the bee can acquire information, by summing the *Reizwechsel* (frequency of stimulus-changes) from the ground, which correlates with the distance. However, Neese (unpublished data) has observed that forager-bees even when flying over a flat and apparently uniform field indicate the same distance as in a landscape which is conspicuously marked by bushes, roads, fences etc.

(*e*) Furthermore, in control experiments with a mutant strain *Cordovan* in which the isolating pigment in the ommatidia is reduced, so that optomotor reactions and pattern recognition are heavily affected, no difference at all has been found in ability to measure distance, although flight-time did increase [25].

(*f*) A similar result was reported by v. Frisch in 1962: when the bees were trained to fly over a lake, distance-measuring remained unchanged [7], although flight-time has been found to be increased over a lake [11].

(2) Even if we accept the theory of distance-measurement by energy consumption, the question remains how an animal is able to measure its energy consumption so exactly. We hope that further research on this problem may be stimulated by our findings that the distance-measuring mechanism is species- and race-specific. Fig. 9.1 shows that an equal dance-rhythm (7 waggle runs/$\frac{1}{4}$ min) corresponds to distances of approximately:

400 m for *Apis mellifica carnica;*
300 m for *A. mellifica ligustica;*
150 m for *A. indica;*
50 m for *A. florea* [18].

It would be of great interest to know whether there is a correlation between differences in energy consumption during flight and in the curves shown in Fig. 9.1.

In relation to insect flight in general it may be worth mentioning extreme foraging ranges and flight performances of bees. Eckert (3) has reported a range of 13·5 km, and we have trained bees to fly to a feeding-site 12·0 km away from the hive. Our bees loaded up with 55 mg nectar (i.e. one third of their body weight) and came again and again, at least every 75 minutes, without interruption from morning to evening. In recent experiments on the Arctic Circle (Messaure Station, Sweden) we induced bees to forage on a feeding-table 400 m away from the hive; on 20th and 21st June they collected for 24 hours without rest, thereby covering a distance of 150 km per day!

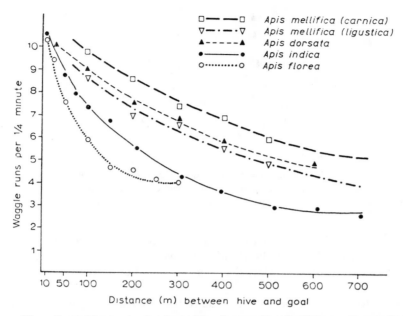

FIG. 9.1. **The code used by a returning forager-bee for reporting the distance of a new feeding-site.** The code, signalled in the 'tail-waggling' dance, is species- and race-specific. Curves are shown for: *Apis mellifica*—the Austrian bee (*Krainer-Biene*); *A. mellifica ligustica*—the Italian bee; *A. indica*—the Indian house-bee; *A. dorsata*—the Indian rock-bee; *A. florea*—the dwarf Indian honey-bee.

DIRECTIONAL ORIENTATION

The route and direction towards a goal can be determined straightforwardly by memorizing landmarks along the route which had been passed in previous excursions. It should be realized, however, that landmarks like bushes, trees and rocks are not as conspicuous for flying insects as they are for animals on the ground, unless they extend above the horizon (i.e. above the level of the flying insect). On the other hand, landmarks which provide a continuous line, like a coast, river, road or forest edge, can be of immense directional value—as every pilot can confirm.

Since it is known [6] that bees use the sun as a compass, a first question to be posed is how landmarks and sun-compass compete with each other. To answer this question we transported a colony of bees to an area unknown to them and trained a group of twenty forager-bees on a feeding-place 180 m to the north-west (Fig. 9.2). Up to the evening the bees (individually marked by coloured spots) collected the sucrose solution again and again, and thus had opportunity to learn either by conspicuous landmarks or by the sun-compass to reach their goal in the north-west. To find out how the sun-compass would operate in the absence of familiar landmarks, we sealed the hive in the evening and moved it to another area unknown to the bees, 23 km away, where there were no familiar landmarks to guide the bees but several unfamiliar landmarks instead. Four feeding-tables were set up, to the NW, SW, SE and NE.

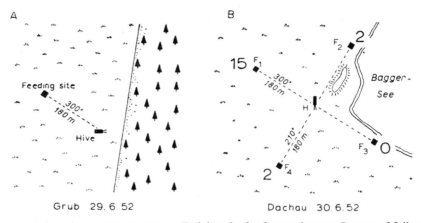

FIG. 9.2. **Orientation by sun-compass.** A. Training site for forager-bees; B. Layout of following day's experiment; large figures denote number of visits made to each of the four alternative feeding-sites F_{1-4}; H = hive; compass-bearings in degrees from north.

It was a big surprise to see that the bees seemed to be in no way disorientated in this unknown area. They flew straight ahead north-west from the hive and appeared at the north-western feeding-station. Only the open sky could offer them precise orientational cues.

However, to prove that bees use the sun to provide true compass guidance—and not simply as a reference direction—we had to change our experiment: the bees were trained to a southern feeding-table in the afternoon only, with the entrance of the hive blocked until noon. Their outward flight-route was accordingly always to the left of the position of the sun.

The next morning after transportation, the bees had to keep to the right of the position of the sun to orientate correctly to the south, i.e. they had to allow for the apparent movement of the sun. If they had orientated at a constant angle to the direction

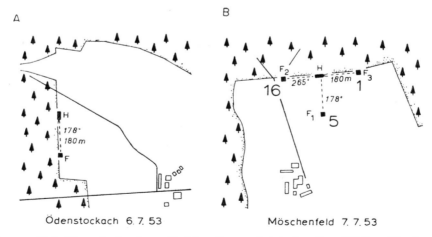

FIG. 9.3. **Use of linear landmarks.** A. Training-site near forest edge aligned N–S; B. Number of bees observed at each feeding-site next day, after hive transferred to vicinity of forest edge aligned E–W.

of the sun (*winkeltreu* or angle-constant orientation), again keeping their flight-route to the left of the sun, they would have come to the feeding-table in the east. All bees, however, appeared in the south, thus establishing a true sun-compass orientation.

What then is the value of landmarks in comparison with the sun-compass? We repeated our shifting experiment: this time we placed the hive on a forest edge and trained the bees to fly southwards along this edge, with the forest to their right (Fig. 9.3). The next morning, 20 km away, the bees found a similar landscape, but the forest edge now led from east to west. This time, most of the south-trained bees followed the guide-line of the forest, and appeared on the western feeding-station. Similar results were obtained when we offered the shore of a lake or a road as guide-line. However, when merely a single conspicuous tree had to be passed on the flight-route, the sun-compass orientation prevailed; if a similar tree was seen next morning in the east, instead of in the south, it did not divert the forager group; without exception the bees flew southwards [21].

Is the sun-compass orientation innate?

The apparent movement of the sun (its speed of change of bearing or azimuth—*Azimutwinkelgeschwindigkeit*) is not uniform from morning till evening, being faster at noon. These changes in azimuth differ during the year in a manner depending on the latitude of the observer; between the Tropics of Capricorn and Cancer the sun moves either clockwise or anti-clockwise according to time of year. The question arises whether the sun-compass orientation in bees is fully innate, or whether at least some of its components—like daily, seasonal and geographical variations—have to be learned.

We reared bees in an artificially illuminated room, where the sun and sky were not visible. After three weeks the colony was brought into the open air and tested in the usual shifting experiment. This time the result was different: the south-trained bees had learned in their single afternoon of training merely that the flight-route had to be kept to the left of the sun; the next morning, in a new, unknown area, they only remembered this angle between the direction of goal and sun, and they appeared at the eastern feeding table. We had to continue the south-training for three afternoons (≈ 500 foraging flights) before the bees were able to take the sun's movement into account in establishing the correct compass-direction instead of maintaining a constant angle to the sun, i.e. to orientate *kompasstreu* instead of *winkeltreu*. Even so, our bees were able to learn from only a section of the sun-arc enough of the 'azimuth-speed' to be able to estimate a further section of the arc.

A colony of bees which was transported from Ceylon to Munich at a time of the year when the sun in Ceylon moved anti-clockwise (from east through north to west) required forty days in Munich to adapt to the new situation, before the forager-bees were familiar with the clockwise sun-arc in Europe [19, 20].

Compensating for side-wind drift

In contrast to the flight behaviour of migrating Desert Locusts, African armyworm moths etc., of which we have been hearing, honey-bees are never diverted away from their flight-route by a cross-wind. Keeping the body oblique to the wind they compensate for the lateral drift, the direction and strength of which are determined by processing optical and mechanical data.

In calm air the pattern of the ground below passes from the front to the rear; as soon as a side-wind begins to drift the bee, this pattern-dynamic changes in direction and the bee turns towards the wind direction. If the ground offers no optical pattern, e.g. when we train a group of bees to fly across a lake, the bees fail to make this allowance for drift [11]. However, as soon as we laid out a line of floating markers (wooden planks 4 m long and spaced 3 m apart), the bees tried to follow these markers, in a somewhat scalloped course (Figs. 9.4 and 9.5), but it should be emphasized that their

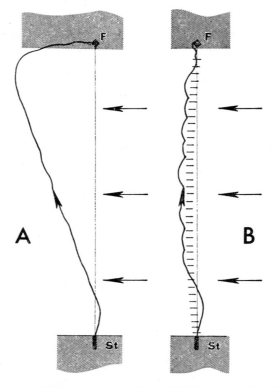

FIG. 9.5. **Use of markers in flight over water.** A. Lateral drift by cross-wind when bees are flying over rippling water without markers; B. Flight-path of bees using visual markers (Fig. 9.4). St = hive; F = feeding-site.

flight behaviour was quite different from that on a patterned landscape. The bees were drifted off course again and again, because the widely spaced markers did not enable them to compensate continuously for the drift by keeping the body-axis continuously at the appropriate angle to the side-wind. They could use the planks merely as a series of separate landmarks leading to the goal. Hence it follows that the angular direction of movement of the optomotor ground-pattern, with respect to the body-axis, is of great value for cross-wind compensation.

Optomotor reactions can also be inferred from the observation that with increasing wind-speed the bees fly closer to the ground; in calm air they fly at about 2·5 m, in winds of 2 m/s at about 1·5 m and in winds of 4 m/s at about 0·5 m above the ground. In this way, the change of stimulus intensity in a single ommatidium increases markedly [11].

FIG. 9.4. **Floating markers for guidance in flight over water.** A. Viewed from feeding-table; B. Viewed from hive on other side of lake.

In addition, the air-flow against the antennae and the bristles on the eye-surface provides some information for orientation; in a cross-wind, fixation of the antennae or removal of the eye-bristles results [24] in very marked drift, i.e. movement relative to the ground at an angle to the longitudinal axis of the insect [in this case likely to involve yaw, i.e. movement relative to the air at an angle to the axis of the insect—Ed.]. Johnston's organ in the scapus and distinct receptor cells on the base of the eye-bristles can therefore be expected to provide information on varying speeds and directions of air-flow relative to the bee [10]. How this information from mechanical senses is combined with optical information has still to be resolved.

Up till now we have discussed only one part of the problem. The forager-bee returning to the hive has to indicate the direction of her route in her subsequent dances. Will she refer to the direct route towards the goal, or to the angle between the orientation of her own flight and the sun (since in a cross-wind she must view the sun at an angle different from that in a calm situation)? In extensive experiments [8] we came to the conclusion that forager-bees estimate the angle they would have seen in calm weather from the oblique position of their bodies and the direction of their track which they see relative to the ground, i.e. their dances indicate the true azimuth of the goal. Only in this way will their nest-mates never be led astray, since the direction and strength of the wind may change over short intervals of time or of distance. This performance requires remarkable sensory and neuronal capacities. In measuring the angle of drift due to the side-wind, the bee is able to refer to the simultaneous position of the sun; it is known that bees can memorize the sun's azimuth for each time of the day [19]. When the sun and the sky are covered by clouds, conspicuous landmarks which are on the route may be taken as reference points.

How this oblique position of the body-axis is used by the brain to establish the correct line to the goal is still an open question; it cannot be solved by the assumption that the forager simply averages outward and homeward angles of drift. A similar feat of integration by the bees is demonstrated by detour experiments, in which a group of forager-bees is obliged to fly around a rock ridge or a high building (Fig. 9.6). Although their own flight-path had included marked changes in direction, starting westwards and

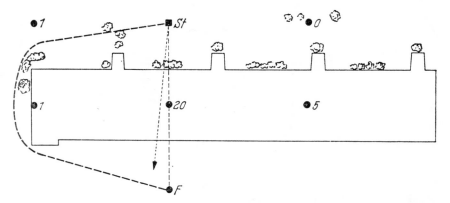

Fig. 9.6. **Detour experiment around a high building.** Initial forager-bees had followed dashed route; numbers record other bees observed at series of scent-plates while navigating on information provided by initial foragers in the 'waggle-dance'. Most of the newcomers took a direct route over the building, rather than the path followed by the initial foragers.

finishing eastwards, their dances unanimously indicated the direct air line from hive to food. Following this information from the dancers, most of the alerted novices accordingly flew directly *over* the building, rather than following the route of the initial foragers.

II. ORIENTATION NEAR THE GOAL
(short-range navigation)

On approaching the goal, both on the outward route and on the homeward flight, the bees are faced with navigational problems of a different kind, for which they are able to utilize different forms of guidance.

(*a*) Honey-bees are *flower-constant*, i.e. they visit the same species for hours and days, as long as the blossoms open their calices; only in this way can the bees serve as pollinators. Odour, colour and pattern of the flower are used as signals for identification of the species, after the bees have reached the general area of the feeding-site.

(*b*) Returning to the hive, the forager has to find the exact mother colony, and there may be a dozen or even hundreds of similar hives, one next to the other, as in the large 'bee-houses' in Germany. The guard-bees will only allow their own nest-mates to enter the hive, recognizing them by the *colony-specific odour* which clings to the hairy body of the returning forager.

OLFACTORY ORIENTATION

The bees can recognize a specific flower scent among the multiplicity of fragrances in a meadow, by the olfactory receptors of the 'plate organs' on the antennae. Electrophysiological analysis of this system has shown that a single sensory neurone responds to a limited spectrum of odorants, but this spectrum is never identical for two different

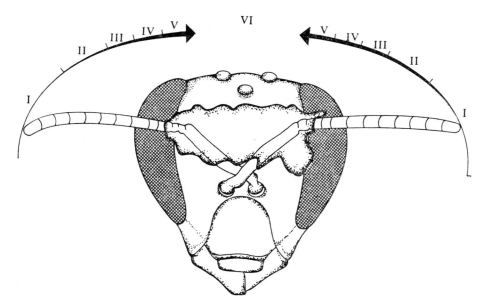

FIG. 9.7. **Olfactory tropotaxis** I. Bee with antennae crossed but distally free to move. (After Martin [22].)

receptors; as a consequence, there must exist an almost unlimited number of possible responses to different odours [15, 16].

The source of an odour can be located in principle either *klinotactically*, i.e. by the successive comparison of different concentrations emanating radially from the centre, or *tropotactically*, i.e. by the simultaneous comparison of different concentrations with the two antennae. Using bees with crossed antennae, as in Fig. 9.7, Martin [22] could determine the minimum difference in concentration required to evoke a tropotactic reaction, shown by the bee turning to the wrong side at a Y-junction; another tropotactic test procedure is illustrated in Figs. 9.8 and 9.9. It should be realized, however, that in normal flight *anemotaxis*, i.e. up-wind orientation, will result in an approach towards the fragrant flowers, since the wind is then transporting the odour molecules towards the flying bee. Approaching closer to the goal the bee may be helped by tropotactic orientation to find the actual nectar-source.

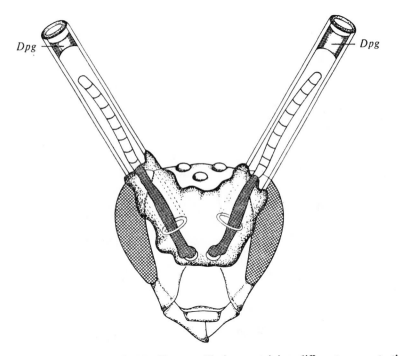

FIG. 9.8. **Olfactory tropotaxis** II. Glass capillaries containing different concentrations of fragrance (*Dpg*) and fitted over antennae. (After Martin [22].)

PATTERN RECOGNITION

In pattern recognition two parameters are of decisive value: *figural quality* and *figural intensity*. V.Frisch in 1914/15 had already found that although bees cannot distinguish a closed circle from a closed square or triangle, solid figures are easily distinguished from open or divided ones even if they are similar in basic geometric form [4]. In her classical investigations, M. Hertz [12 to 14] emphasized the *figural intensity*—the extent to which a figure is divided up, and hence the extent of outline and the amount of contrast which

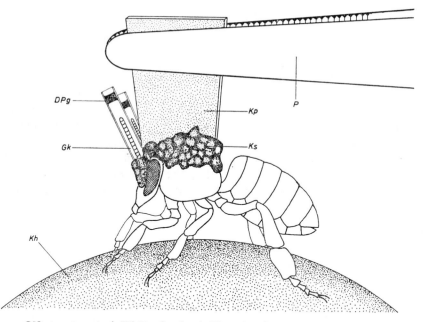

FIG. 9.9. **Olfactory tropotaxis** III. Bee fitted as in Fig. 9.8 and suspended onto a cork hemisphere which is readily rotated by the legs. The trained bee will run continuously to the side on which the higher odour concentration is presented in the capillaries over its antennae. (After Martin [22].)

it presents—as a decisive parameter in pattern attractiveness and discrimination. Cruse [2] expresses this parameter as a 'function of pattern recognition':

$$U = C_1 \frac{R^+ + R^-}{G} F^+ + C_2 (\log K^+ - \log K^-)$$

where U = pattern recognition,
C_1, C_2 = value of the geometric pattern, which is genetically fixed,
G = common part of the two figures,
F^+ = part of the positive figures,
R^+, R^- = remaining parts,
K^+, K^- = length of outline of each figure.

Figure 9.10 illustrates the application of this function to discrimination between a circle and a rectangle.

LEARNING PROCESSES IN APPROACHING AND IDENTIFYING THE GOAL

At first glance it seems as if such evaluation of figural intensity and figural quality by the central nervous system is innate. It is however essential to know if and how subjective experience and learning processes may take part in this phase of approaching and identifying a goal. Let me explain the problem by describing a simple experiment.

A 2-molar sucrose solution is offered in the centre of a circular table, above a black star for the bees to use as a feeding mark. The black star, however, represents only one part of a complex pattern, being a mark in the centre of a circular table positioned e.g. in

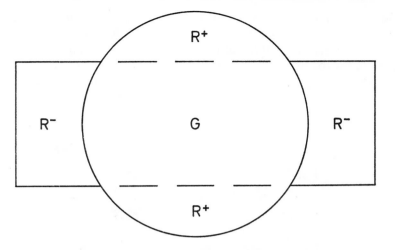

FIG. 9.10. **Pattern discrimination:** rectangle (−) and circle (+). Cruse's function of pattern recognition [2] takes into account the extent of the area peculiar to each pattern (relative to the area common to both), together with the relative lengths of outline of the two patterns.

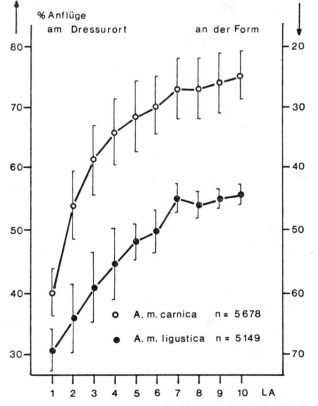

FIG. 9.11. **Localization experiment** to compare significance of a single visual marker and of position of training-place relative to its other surroundings. LA = learning act.

%Anflüge:	%flying towards:
am Dressurort	training-place
an der Form	visual marker (black star)

an orchard. Let us now test which is of prime importance: the *pattern* of the star or the central position on the table—the training *place*. The result is demonstrated in Fig. 9.11: at the first learning step the *pattern* of the star proves more important than its relative position, but in the following learning acts progressively more approaches are recorded towards the centre of the table, i.e. to the *place* the bees were trained on.

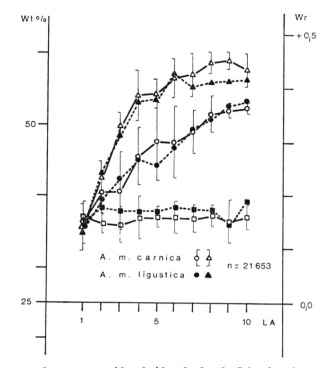

FIG. 9.12. **Local use of sun-compass with and without landmarks.** LA = learning act.
Upper curves (triangular points)—feeding-site in an orchard, though without conspicuous landmarks in its immediate vicinity.
Middle curves (circular points)—feeding-site in a large open field.
Lower curves (square points)—in a closed area, with only the sky visible, and the food placed at the southern point of a black star marker.

These learning curves were surprisingly different, when we tested the different races *Apis mellifica carnica* and *ligustica;* the choice frequency for 'relative position' (*Anflüge am Dressurort*) rose in *carnica* from 40 per cent to 75 per cent, but in *ligustica* from 32 per cent to only 55 per cent. These results refer to tests in which the surroundings of the feeding-site presented many other landmarks, like trees, bushes, and roads; as soon as we moved the feeding-table to a large clearing, we obtained nearly identical learning curves for both races.

These observations demonstrate that:

(1) The visual structure both of nearby and of more distant surroundings is taken into account, when bees learn to use figural patterns as feeding marks. Thus, approaching the feeding-site, the bees locate themselves not only by one particular visual sign, but make use of the surrounding spatial complex of landmarks.

(2) *A. mellifica carnica* learns to use preferentially the more distant of the surrounding features; *ligustica* tends to remain dominated by the more immediately adjacent figural patterns. Thus, in the initial location of the goal from a distance *carnica* proves superior to *ligustica*.

I would like to refer now to a series of experiments, which present more difficult orientation problems for honey-bees. We restrict the bees to the use of only the sun-compass in orientating near the goal, by placing the circular table in an arena with black walls, allowing perception of the sky, but not of auxiliary landmarks. The feeding jar is no longer placed in the centre, but at the southern tip of the black star. The learning curves reach only a 37 per cent level, with no further improvement in orientation after the first learning step (Fig. 9.12 lower curves).

The results of other experiments confirm that the sun-compass is still of importance in this short-range navigation; thus under a heavily overcast sky, when sun-orientation becomes impossible, the learning curves remain at a level of 25 per cent, which represents the statistical distribution of random approaches to the four points of the star, in the

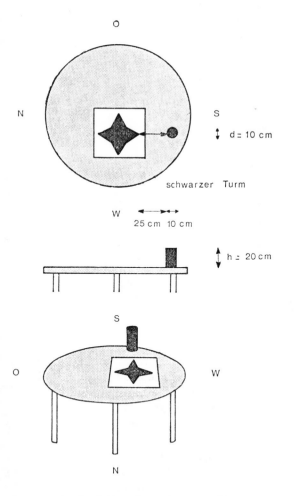

FIG. 9.13. **Layout of localization experiment within arena.** Only the view to the sky is open; sucrose solution is placed only at the southern tip of the black star, with a black cylinder as an auxiliary landmark which improves the orientation performance of the bees.

Dressurort : S—Spitze des schwarzen Sternes

15

south, east, west and north. This behaviour alters when we offer additional visual markers such as the black cylinder in Fig. 9.13; under these conditions the orientation performance immediately improves, but *carnica* again learns the position of the goal better than *ligustica*.

GENERALIZATION AND RE-LEARNING

In one aspect our experimental set-up is not similar to natural conditions: in nature identical signals rarely recur from one learning situation to the next, as they do in our training procedure. The great problem of re-orientation in a changing environment is whether the bees are able to generalize, on which I refer to the experiments of Wehner [27] and Mazochin-Porschnjakov [23].

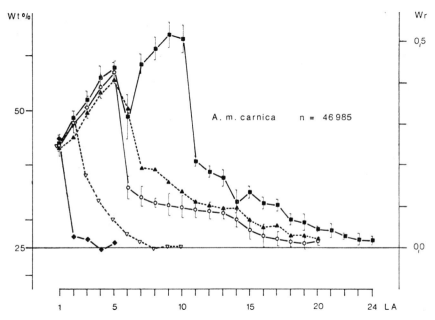

FIG. 9.14. **Reactions to changing landmarks.** In 1, 3, 6 and/or 10 training steps the bees are assisted in locating the feeding-site in Fig. 9.13 by the auxiliary landmark of the black cylinder, which is subsequently removed. The bees are unable to re-learn their approach procedure in the new situation. (From Höfer, J.—unpublished.)

In nature the situation near the goal may change drastically from one trip to the next, so that important factors (e.g. food signals) cannot be recognized by generalization or transfer. The bees then have to locate themselves by new cues; they have to *re-learn*.

In the following experiments food was offered, as before, at the southern tip of a black star on a circular table, with the black cylinder as a further landmark (Fig. 9.13). After the sixth learning act the black cylinder was removed, so that the bee had to locate itself in a new situation, with the southern tip of the black star no longer associated with a black cylinder. Figure 9.14 shows that the learning curve falls rapidly, from 54 per cent to 35 per cent. It was still more surprising that when one rewarded by providing food in this new situation (without the cylinder), the curve nevertheless continued to fall, until the average value for a random statistical distribution—25 per cent—was reached after the 24th approach. Thus, in spite of rewards, the bees did not re-learn nor did they grasp the new situation. Similar behaviour was shown when the cylinder was removed after the first, third or tenth learning act (Fig. 9.14).

Let us now ask, whether the old situation had been entirely forgotten, or on the other hand whether the normal learning process had been disturbed by the removal of the cylinder so that re-learning was not shown. After the sixth approach we replaced the black cylinder, and the reaction showed that the bees remembered the previous situation quite well; the learning curve rose to the former level, although the progress of the original learning curve was delayed by the equivalent of one learning act. Thus the earlier learning steps were obviously recalled.

Of the further analysis of these complex orientation performances I will quote only the reverse experiment: we trained the bees first without the cylinder and then tested them with it, so that the bees had to re-learn in a situation which provided additional orientation cues. In this case re-learning occurred immediately, but, in comparison with the original learning curve, the re-learning curve levelled out at a lower percentage of correct choices.

It seems to me that this particular series of experiments represents a new approach to some still unsolved problems of learning during foraging flight.

DISCUSSION

Professor Southwood (Chairman) contrasted the precise navigation of the honey-bee, as described in Professor Lindauer's fascinating paper, with the other end of the behavioural spectrum as represented by insects travelling with the wind in the manner discussed in the previous papers.

Dr Rainey enquired as to the maximum range over which bees can locate a food-source and return to the hive, and the maximum wind-speed in which they can operate successfully. In reply, Professor **Lindauer** began by quoting the observation recorded by Eckert [3] in California, which had given a value of 13·5 km, and then described a series of observations of his own at Graz in Austria, in which, under particularly favourable conditions during one September with no nearer supply of food, the bees used a food-source 12 km from the hive. He added the further finding that the bees announced a food-source by waggle-dance only up to a distance of 10 km, which appeared to represent the limit of profitable foraging because of the increase in energy consumption with foraging distance. On the second point Professor Lindauer quoted unpublished observations on bees flying in a really strong wind with gusts of about 6–9 m/s, in which they could just glide to the feeding-table. They had no obvious difficulty in taking off, nor in maintaining flight against the wind, but in attempting to land were often blown away. Professor **Pringle** recalled Heran's figure of about 6 m/s as a maximum wind-speed for foraging, which Professor **Lindauer** confirmed for steady winds.

Professor Southwood enquired further about the experiments with bees flying over water in a cross-wind, and asked whether the bees zig-zagged to each successive plank of the 'bridge' (Fig. 9.4) instead of flying in a complete arc as he understood they did over land. Professor **Lindauer** confirmed the zig-zag or rather scalloped flight of the bees (Fig. 9.5), apparently only intermittently in visual contact with the 'bridge'—which he felt was not an ideal substrate for this experiment. Owing to the relatively low rate of optical stimulus-change (*Reizwechsel*) provided by this ground-pattern, the bees appear to have been unable to recognize the continuity of the line which is suggested to us by the ends of successive planks of the 'bridge'.

Dr J.A.Kefuss (communicated): Is there any information on possible differences in learning or code between colonies within each race? If only a single colony was tested from each race, is it possible that the differences recorded as between races may in fact have been only differences between ecotypes?

Lindauer: You refer to our experiments on the orientation of bees near the goal, where they had to code specific patterns to signal a source of food. This coding process is surely race-specific, i.e. genetically fixed: *Apis mellifica carnica* locates itself preferentially by the general surroundings of the goal, *A. mellifica ligustica* by the patterns in its more immediate vicinity. This conclusion has now been confirmed by my students Fr. Lauer and Fr. Höfer on *different* colonies of both races, in different seasons over three years, and the results, based on more than 800,000 test-data, are consistent on this point.

Professor W. Nachtigall (communicated): Is there experimental evidence for any alternative to the theory that bees can measure their distance flown by their consumption of fuel?

Lindauer: Theoretically there is only one alternative: the forager-bee, flying at a constant height, may sum up all the stimulus-changes (*Reizwechsel*) originating from the optical ground-pattern and given by lighted and shaded objects like grass-stems, stones etc. My collaborator V.Neese has tried to test this alternative experimentally but with negative results.

Dr R.C.Rainey (communicated): About the navigational problems so impressively solved by foraging bees, such as those which Professor Lindauer trained to use a feeding-site at a distance of 12 km: had they flown at the usual height of 2–3 m (p. 198), the extent of the visual field they traversed (and learned) could be regarded as very roughly equivalent (p. 28) to that observed by a human navigator flying at 1000 m for a distance of 4800 km, to a destination he may have visited only once before—which I understand is probably approaching the limits of unaided human visual aircraft navigation. Professor Lindauer has shown that the bees allow for the effects of cross-wind drift in establishing and reporting the position of a food-site; in doing so they must face a further major problem arising from short-period wind fluctuations (due to the thermal convection characteristic of fair weather with sunshine), which must cause large variations in the angle of drift (between the heading of the insect and its direction of displacement relative to the ground), perhaps of $\pm 45°$ or more within periods of seconds at gust-fronts. Is the high flicker-fusion-frequency of bees (about 200 per sec) perhaps important in making available visual information (e.g. *Reizwechsel*) on the continuously changing direction and speed of movement of the bee in relation to the ground, of sufficient accuracy to be capable of integration to provide the kind of assessment of geographical displacement which bees achieve? Finally, for communicating such information to other bees, a feeding-site at 10 km would appear to require some extension of the waggle-dance code shown up to 500 m in Fig. 9.1; does the method of coding change at greater distances?

Lindauer: (1) Dr Rainey raises a serious problem and I must confess that we are far from understanding how bees perceive exact information on changes in drift due to fluctuations in wind. I agree fully with Dr Rainey's hypothesis that bees try to maintain direction and speed along the optical pattern of the passing ground. Contrast and direction of shadow may play an important role; relating this to the sun-position by the ommatidia of the upper part of the eye, combining this procedure with the information delivered by Johnston's organ (see p. 203), and furthermore aligning the body-axis in relation to distant landmarks—all these data may be integrated in the central nervous system into a constant 'direction pattern'.

(2) We could record dances from bees collecting up to 10 km, and the 'dance tempo' decreases in an asymptotic curve from 6 circuits per 15 sec for 500 m (Fig. 9.1) down to $1\frac{1}{2}$ circuits per 15 sec in 10 km distance; see v.Frisch, 1967, p. 69 [6].

REFERENCES

[1]   BISETZKY, A.R. (1957). Die Tänze der Bienen nach einem Fußweg zum Futterplatz. *Z. vergl. Physiol.*, **40** : 264–288.

[2]   CRUSE, H. (1972). Versuch einer quantitativen Beschreibung des Formensehens der Honigbiene. *Kybernetik*, **11** : 185–200.

[3]   ECKERT, J.E. (1933). The flight range of the honey bee. *J. agric. Res.*, **47** : 257–285.

[4]   FRISCH, K.VON. (1914/1915). Der Farbensinn und Formensinn der Bienen. *Zool. Jb. (Physiol.)*, **35** : 1–188.

[5]   FRISCH, K.VON (1948). Gelöste und ungelöste Rätsel der Bienensprache. *Naturwissenschaften*, **35** : 12–23, 38–43.

[6]   FRISCH, K.VON. (1967). *The dance language and orientation of bees*. The Belknap Press of Harvard University Press, Cambridge, Massachusetts.

[7]   FRISCH, K.VON. & KRATKY, O. (1962). Über die Beziehung zwischen Flugweite und Tanztempo bei der Entfernungsmeldung der Bienen. *Naturwissenschaften*, **49** : 409–417.

[8]   FRISCH, K.VON. & LINDAUER, M. (1955). Über die Fluggeschwindigkeit der Bienen und ihre Richtungsweisung bei Seitenwind. *Naturwissenschaften*, **42** : 377–385.

[9]   HERAN, H. (1956). Ein Beitrag zur Frage nach der Wahrnehmunggrundlage der Entfernungsweisung der Bienen. *Z. vergl. Physiol.*, **38** : 168–218.

[10] HERAN, H. (1959). Wahrnehmung und Regelung der Flugeigengeschwindigkeit bei *Apis mellifica* L. *Z. vergl. Physiol.*, **42** : 103–163.

[11] HERAN, H. & LINDAUER, M. (1963). Windkompensation und Seitenwindkorrektur der Bienen beim Flug über Wasser. *Z. vergl. Physiol.*, **47** : 39–55.

[12] HERTZ, M. (1929). Die Organisation des optischen Feldes bei der Biene, I. *Z. vergl. Physiol.*, **8** : 693–748.

[13] HERTZ, M. (1930). Die Organisation des optischen Feldes bei der Biene, II. *Z. vergl. Physiol.*, **11** : 107–145.

[14] HERTZ, M. (1931). Die Organisation des optischen Feldes bei der Biene, III. *Z. vergl. Physiol.*, **14** : 629–674.

[15] LACHER, V. (1964). Elektrophysiologische Untersuchungen an einzelnen Rezeptoren für Geruch, Kohlendioxyd, Luftfeuchtigkeit und Temperatur auf den Antennen der Arbeitsbiene und der Drohne (*Apis mellifica* L.). *Z. vergl. Physiol.*, **48** : 587–623.

[16] LACHER, V. & SCHNEIDER, D. (1963). Elektrophysiologischer Nachweis der Riechfunktion von Porenplatten (Sensilla Placodea) auf den Antennen der Drohne und der Arbeitsbiene (*Apis mellifica* L.). *Z. vergl. Physiol.*, **47** : 274–278.

[17] LAUER, J. & LINDAUER, M. (1971). *Genetisch fixierte Lerndispositionen bei der Honigbiene*. Akademie der Wissenschaften und der Literatur, Mainz.

[18] LINDAUER, M. (1956). Über die Verständigung bei indischen Bienen. *Z. vergl. Physiol.*, **38** : 521–557.

[19] LINDAUER, M. (1957). Sonnenorientierung der Bienen unter der Äquatorsonne und zur Nachtzeit, *Naturwissenschaften*, **44** : 1–6.

[20] LINDAUER, M. (1963). Kompaßorientierung. *Ergebn. Biol.*, **26** : 158–181.

[21] LINDAUER, M. (1972). *Communication among social bees*. 2nd edition. Harvard University Press, Cambridge, Mass.

[22] MARTIN, H. (1964). Zur Nahorientierung der Biene im Duftfeld, zugleich ein Nachweis für die Osmotropotaxis bei Insekten. *Z. vergl. Physiol.*, **48** : 481–533.

[23] MAZOCHIN-PORSCHNJAKOV, G.A. (1969). Die Fähigkeit der Bienen, visuelle Reize zu generalisieren. *Z. vergl. Physiol.*, **65** : 15–28.

[24] NEESE, V. (1965). Zur Funktion der Augenborsten bei der Honigbiene. *Z. vergl. Physiol.*, **49** : 543–585.

[25] NEESE, V. (1968). Zum optischen Orientierung der augenmutante Chartreuse von *Apis mellifica* L. I Teil. *Z. vergl. Physiol.*, **60** : 41–62.

[26] SCHOLZE, E., PICHLER, H. & HERAN, H. (1964). Zur Entfernungsschätzung der Bienen nach dem Kraftaufwand. *Naturwissenschaften*, **51** : 69–70.

[27] WEHNER, R. (1967). Zur Physiologie des Formensehens bei der Honigbiene. *Z. vergl. Physiol.*, **55** : 145–166.

10 · Lability of the flight system: a context for functional adaptation

C.G.JOHNSON

O.D.A. Lethal Yellowing Research Team,
Coconut Industry Board Research Laboratories,
Kingston, Jamaica*

Within a species, the capacity for prolonged flight varies; this variation makes possible adaptive change in the dispersal of the species. The variation is obvious when, within a species, wings may be absent, short, or full length; much has been published on this. However, the variation in innate flight capacity of fully winged adults is not so obvious and has been little studied, so that its causes are little understood.

Many fully winged adults either cannot fly or can fly for a few seconds only; others that are flight-worthy can dispense with the activity, temporarily or permanently—can opt out, as it were—either by quiescence or by histolysis of the flight muscles. Because of this individual lability, within a population the proportions of individuals with various flight capacities change, often quickly.

Thus, though alary polymorphism must be mentioned, my main concern is with the variable intrinsic flight duration of macropters (which is a component of the migratory condition) and with some factors that affect it. Lability in the nervous control of flight movements, and appetitive responses, are outside the scope of this paper.

DIFFERENCES IN SUSTAINED FLIGHT BETWEEN INDIVIDUALS OF A SPECIES

Apart from free-flying aphids in a flight chamber, where flight thresholds have free play, individual variation in intrinsic flight duration has been measured mostly with tethered insects, with all the associated limitations; in that context especially, the flightlessness of macropters is usually mentioned only casually.

Thus different individuals of *Aphis fabae* Scop. made tethered flights to exhaustion that ranged from two to twelve hours [26] and in a flight chamber from seven minutes to more than three hours [53]; however many alatae never fly [79]. Some moths *Trichoplusia ni* (Hb) and *Mythimna separata* Walk. flew, tethered, for up to seven and 36 hours non-stop respectively; other individuals, though winged, did not fly [49, 54]; the related *Spodoptera exigua* Hb has been inferred to fly for up to five days [52]. Some

* Now 42 Luton Road, Harpenden, Herts, England.

Oncopeltus fasciatus Dallas flew for nearly 8 hours, *Schistocerca gregaria* up to 23 hours and *Dysdercus intermedius* Distant up to 25 hours; all produced adults which did not fly and some that flew only briefly [31, 58, 82].

Where known, the frequency distribution of such durations in a species is skew, many individuals making short flights, few making long ones. For example, 40 to 50 per cent of the bug *Lygaeus kalmii* Stål flew, in five successive flights, for less than 5 minutes, about 30 per cent for 5–15 minutes and only 5 to 10 per cent for more than 150 minutes

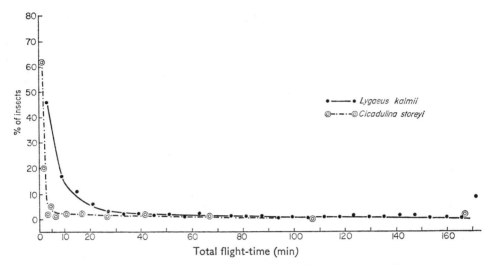

FIG. 10.1. **Differences in flight-time within species** (total of five consecutive flights). *Lygaeus kalmii* (Heteroptera: Lygaeidae), after [19]; *Cicadulina storeyi* (Homoptera: Cicadellidae), after [73].

[19] (Fig. 10.1). About 47 % of free-flying *Aphis fabae* flew less than 1 hr in the laboratory and only about 10 per cent exceeded 2 hrs [53]. Of three species of *Cicadulina* most flew for a few seconds or not at all, a few for over three hours and one for nine hours (p. 223–224); two forms were involved, with different wing-loadings, with mean durations of eight to fifteen seconds and about eight minutes respectively; for each form the distribution was less skew than when both were combined [73].

CHANGES IN FLIGHT DURATION THROUGHOUT LIFE

The above examples are of flight durations of individuals fairly early in life. But before possible causes of individual differences can be considered, variation of flight duration with age must be discussed and this has always been expressed by averages; it would be interesting to know how individuals vary in this respect also, and although the data must exist they have not been published. Nevertheless the average change in duration with age could help to elucidate the individual differences mentioned above.

Tethered flight durations of adults at different ages have been published for about a dozen taxa. Duration generally increases after emergence to a maximum relatively early in adult life and diminishes with age and, in females, as ovaries develop. Where measured, wing-beat frequency follows a similar course. Examples are: *Drosophila funebris* (F.) [106], *D. melanogaster* Mg. [104], *Oscinella frit* (L.) [76], *Calliphora vicina* R.-D. [90],

Musca domestica L. [70], *Aedes aegypti* (L.) [75], *Schistocerca gregaria* Forsk. [58], *Oncopeltus fasciatus* Dallas [31], *Dysdercus intermedius* Distant [82], *Cicadulina* spp. [73], *Mythimna separata* (Walk.) [49] (Fig. 10.2). All are of insects that either do not diapause as adults or were flown soon after imaginal ecdysis; flight in relation to diapause is discussed briefly on p. 224. It has been supposed that the period of maximum flight duration is the migratory period and that the flights represent migratory flights [31, 52, 73].

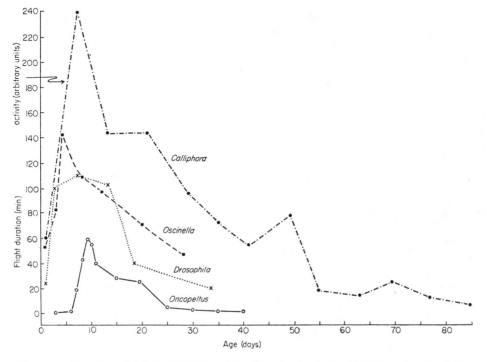

FIG. 10.2. **Duration of tethered flight in relation to age.** *Oscinella frit* (Diptera), after [76]; *Drosophila funebris* (Diptera), after [106]; *Oncopeltus fasciatus* (Heteroptera), after [31]; activity of free *Calliphora vicina* during 24 hours, after [90].

Many other examples, from the field, of prolonged flight relatively early in adult life and its subsequent shortening are summarized elsewhere [52] but the change in duration was not then measured accurately.

This change with age can be considered in relation to the biochemical maturation of the flight apparatus, fuel, oogenesis and the endocrine balance, and ageing *per se*: their treatment here must be superficial and perhaps naive; also it is their integrated effect which is important, but it is possibly too early for anyone to attempt to assess this.

SOME IMMEDIATELY POST-EMERGENCE CHANGES RELATING TO FLIGHT DURATION

After emergence, but before maximal flight activity, the flight muscles or associated enzymes, or both, may continue to develop; flight may thereafter be short or absent.

These have been studied together in detail mainly in locusts and in the Colorado Beetle representing hemi- and holometabolous insects respectively.

Briefly, in *Locusta migratoria migratorioides* (R. & F.) flight muscles become structurally complete after emergence: some muscle enzymes lessen and others increase in activity. For example, a key enzyme in flight muscle activity, α-glycerophosphate dehydrogenase (GDH), concerned with the release of energy, increased 20-fold by the third day of adult life at 30°C, when the wings can move, and doubled again in the next five days, as did all cellular constituents [15]. In *Schistocerca vaga*, also a migrant, the enzyme increased in activity five to eight times after five days [13], while in the flightless grasshopper, *Romalea microptera* it did not increase appreciably; lactate dehydrogenase, important for jumping muscles, was very active but not in the flying species. The metabolisms of the two species flying and flightless differed fundamentally [14].

In *Locusta* the development of the structural components and enzyme patterns were correlated; several enzymes from different metabolic pathways behaved as 'constant proportion' groups [17], the activity of one enzyme representing that of several others at different stages of development. It was suggested that GDH activity itself might control flight potential at the enzyme level [14].

In the Colorado Beetle some enzymes developed as constant proportion groups but others did not and the view that one enzyme might control flight potential was disputed [30]. As with many other Coleoptera [52] the indirect flight muscles of *Leptinotarsa* are also incompletely developed at emergence, when the beetles cannot fly; in long-day beetles they and the enzyme systems develop to flight-worthiness during the next two weeks. In short-day beetles and in allatectomized beetles, myofibrils and mitochondria begin to develop normally but enzymes decrease after nine days and the muscles histolyze before they have completed their development; such beetles never fly.

By contrast, in other holometabolous insects (e.g. *Musca domestica*, *Apis mellifica*, *Tenebrio molitor* L.) flight muscles were complete at emergence (though in view of the lability of the system perhaps this is not always so) but the enzymes were not, and they attained their characteristic pattern later [30].

Sustained flight depends also on enzymes in the nervous system. For example, brain cholinesterase activity becomes maximal seven to ten days after emergence in the honeybee and in one day in *M. domestica*, that is, when flight begins. β-glycerophosphatase and ATPase activity, associated with muscle contraction, becomes maximal in seven to ten days and one hour respectively in the two species [52]. In the Colorado Beetle brain cholinesterase doubles in the ten days after emergence [30]. Motor efficiency and efficient muscle contraction develop more or less together, though at different rates in different species.

There are many instances of macropterous insects with flight muscles differentiated but incompletely developed at emergence, although some can fly and others cannot. Tsetse flies fly to feed but flight muscle volume increases greatly afterwards [55]. Muscles develop more slowly and flight is less in laboratory bred flies than in wild ones and flight hastens development [18, 27]; but there seems to be no information on these effects on flight in the field. The indirect muscles in teneral Scolytidae and in *Sitona regensteinensis* increase greatly in size before flight [10, 28] and variation in the process affects flight performance (p. 226). A halt in muscle development at any of several stages causes various forms of flight muscle polymorphism in a wide variety of insects and has been studied particularly by Jackson and Young [52] in Coleoptera and Corixidae, as has the flight-muscle ultrastructure [84, 77, 1]. In both groups incomplete musculature may be a temporary or permanent condition resulting from differences in photoperiod, temperature, or food, either during larval development or in the teneral stage, so producing flightless macropters seasonally. A block in muscle development, not a general arrest, evidently involving mRNA is a cause in *Cenocorixa bifida* (Hung.) and can be removed by suitable temperatures [78].

Thus there are many causes of flightlessness in macropters but the problem is to know if a hiatus in the development of muscles or enzymes can also produce brief fliers. For example, in honey-bees the mitochondrial α-glycerophosphate p-oxidase system is active soon after emergence and limited flight is possible; but the pyruvate metabolizing system is not complete until 16 to 20 days and it has been suggested that sustained flight depends on the Krebs cycle becoming fully functional [3].

This superficial account is given to make the following point: flightlessness in macropters, for which some of the causes are known, grades into brief flight and that into longer flight between individuals in a species. Ontogeny admits a hiatus at several points, possibly with a graded outcome. The literature on the biochemistry, ultrastructure and physiology of flight muscles is voluminous but mostly about muscle function itself, or diapause. Flight is not so commonly considered and, if it is, usually as an all-or-none effect regardless of its variability between individuals of the same species, which can be as great as between species and is, for ecologists, perhaps the more important aspect.

FLIGHT DURATION AND MUSCULAR EFFORT

FUEL

Colorado Beetles cannot fly until several days after emergence, possibly, as de Kort suggested, because the flight muscles and enzyme systems are immature. However, beetles are then also short of glycogen, and full flight ability is not attained till fuel has been replenished by feeding. *Eurygaster integriceps* Put. also migrates successfully only when enough food in the larval stages provides enough fuel for the newly-emerged adult [52]. Cultured *Prodenia eridania* (Cram.) can lack similarly [88]. Post-emergence delay in maturation of enzymes is not the only factor that could produce adults flying only briefly or not at all.

Many Diptera emerge without a fuel supply and need nectar for flight [47]. Of the many papers on the biochemistry of fuel metabolism relatively few relate to needs in nature but the following is an example of one that does: *Aedes sollicitans* (Wlk.) and *Ae. taeniorhynchus* (Wied.) did not use their glycogen store as long as sugar meals lasted; but flight after resting from exhaustion used glycogen, which was replaced by synthesis from the blood-meal. Evidently enough glycogen is carried over from the pupa for the post-emergence exodus but further migration depends on replenishment. Nectar provides an immediately available supply but a long rest after feeding leads to an accumulation of triglycerides, useless as fuel, which limit flight potential. Also a long rest after a blood-meal may cause ovaries to compete for glycogen whose storage is limited by a hormone which suppresses its synthesis [65].

Differences between sugars may also affect flight in nature. *Ae. taeniorhynchus* flew fastest on glucose, followed by other mono- and disaccharides, but pentoses did not allow flight [67]. Exhausted houseflies flew after meals of glucose, fructose, maltose, sucrose and trehalose but not after galactose, mannose or cellobiose, probably because flight muscles lacked particular enzymes or other biochemical mechanisms or because only some sugars can pass membranes to reach the muscles [85]. Honey-bees used fructose rapidly, sorbitol slowly and flew less rapidly and less far on the latter [56]. Maltose and sucrose cannot maintain flight muscle activity in *S. gregaria* but glucose can [20, 21]. Nectar can vary greatly in its sugars as shown by its effect on the developing ovaries of *Erioischia brassicae* [39] and it has been postulated that the effect of its vitamin E content, speeding or delaying ovarian development, decreases or increases the amount of migratory flight in moths [95].

Culex tarsalis Coq. flew only two hours on its glycogen and longer flights are evidently limited by the rate of mobilization of carbohydrate. Glycogen reaches its maximum early in life in *Ae. aegypti* but its use also appears to be limited, for flights to exhaustion are brief [74, 75]; failure to utilize this fuel could be one reason why *Ae. aegypti* is not a long-distance flier. *Lucilia cuprina* (Wied.) uses fat for development, carbohydrate for flight; fatty acid oxidation provides most of the energy for the developing enzyme systems, sarcosomes and labile organophosphates in the thorax, carbohydrate for the exoskeleton and flight [29]. The rate at which carbohydrate is utilized also affects flight duration. In *Phormia regina* (Mg.) the fuel trehalose is synthesized from fat almost as fast as the muscles can use it and its rate of use might control flight duration [25].

Many migratory Lepidoptera fly on fat but feed en route on nectar; but carbohydrate cannot be used directly because the necessary enzymes are absent or deficient. The mode of conversion of carbohydrate to fat and of its utilization is the subject of many papers [6, 87, 88, 89] which suggest factors in biochemical pathways that may limit flight. For example, if fat-burning insects transport fatty acids from fat-body to flight muscles as diglyceride the muscles must have enough lipolytic enzymes to produce free fatty acids fast enough to sustain flight. Several species of moth contain enough monoglyceride lipases but little di- or triglyceride lipases or esterases. For flight to be sustained by the oxidation of fatty acids derived from glycerides, flight muscle enzymes must be able, presumably, to hydrolyze the latter faster in strong fliers than, for example, in the weakly flying Saturniidae. With them the monoglyceride lipases were sufficient but the di- and triglyceride lipases could be limiting [87]. But the effect of ancillary systems may also be important and it has been suggested [89] that juvenile hormone (JH) might control fatty acid and lipid transport in *Hyalophora cecropia*.

As mentioned above, partitioning of food between ovaries and flight muscles can limit flight, though not necessarily from lack of fuel (p. 224). Partitioning between somatic growth processes and fat-body has been studied for male *S. gregaria* [97, 98] but it is difficult to link the information with problems of natural needs. Perhaps it is too early to expect a study of any variation in these metabolic pathways, between long and short fliers of the same species, though different taxa have been compared.

COMPARISONS BETWEEN SPECIES

Locusts, Syrphids, fleshflies and different races of honey-bee, all allegedly strong fliers, were compared [16]. GDH activity increased about five times in the flight muscles of honey-bees one week after emergence and was particularly great in strong fliers; it also declined as flight activity lessened with age [52], but no comparisons have, I think, been made between weak and strong fliers of the same species.

Flight muscles of species that use both fat and carbohydrate as fuel (*Locusta*, *Pieris*, *Agrotis*) and some of whose individuals fly long distances, had a much higher glyceraldehydephosphate dehydrogenase (GAPDH) activity than Lepidoptera that do not feed as adults [5], and it has been postulated that butterflies known to migrate, e.g. species of Danaids and Nymphalids, have muscle fibres smaller in diameter than do Pierids and Papillionids which, it is said, have less sustained flight [40]. But muscles similar in speeds of mechanical and electrical response may differ cytologically and in their enzymes; a distinction has been made between 'sprinter' and 'stayer' muscles whose different rates of fatigue may depend on intracellular carbohydrate substrate and enzyme composition, the metabolic pathways preceding the Krebs cycle possibly being more inhibited by accumulation of intermediate products in sprinters than in stayers. The flight muscles of locusts were able to oxidize fatty acids more intensively than the corresponding muscles of the cockroach, the metabolism of the muscles being more aerobic. Stayers had more post-synaptic vesicles than had sprinters and therefore possibly a more efficient re-synthesis of transmitter substance and longer lasting neuromuscular transmission [83].

Differences in cytology and metabolic systems are great from species to species; even such closely related taxa as *Glossina* and *Sarcophaga* differ greatly [68], as do different species of mosquito [67]. But, after reading many papers on these subjects, and not being interested in biochemical mechanisms for their own sakes, I feel that experimenters have been satisfied with the assumption that their chosen individual insects, often of different species, were safely classified as 'strong' or 'weak' fliers. As shown above, *the range of flight duration within a species can be as great as from one species to another;* also flight performance can quickly suffer by laboratory selection. A particular grade of flight potential cannot be safely assumed without flight tests on the experimental insects.

WING-LOADING

The effects of fuel, wing-loading and related factors are often difficult to disentangle. For example in *Cicadulina* presumably wing-loading is lessened by small body size and most long-bodied forms fly briefly, if at all; but the correlation is incomplete and many factors could contribute to the relative lack of strong fliers among long-bodied forms [73]. In *Aphis fabae* the correlation coefficient of fat content and initial live weight is 0·97 but the range of size of all aphids was 0·45 to 0·90 mg (i.e. × 2) while the range of fat content was 0·01 to 0·09 mg/aphid (i.e. × 9). For the same mean live weight a field aphid could have two to six times the amount of fat as a cultured aphid. Aphids from culture flew from three to eight hours (av. 5·3) with 4 per cent average fat/live wt; wild aphids flew seven to twelve hours (av. 8·9) with a 10 per cent ratio [26]. Thus though big tethered aphids flew on average further than small ones on account of the extra fat, they flew relatively less long than fat would have allowed had that been the only factor.

FLIGHT, OVARY DEVELOPMENT AND ENDOCRINES

Females of many taxa make more sustained flights before ovaries ripen than afterwards. However there are exceptions. For example *Aegeria apiformis* (Clerch) (Lep.) makes long mass flights soon after emergence but not before many eggs have been laid [91], so possibly lessening wing-loading. The spruce budworm *Choristoneura fumiferana* (Clem.) behaves similarly, and most aphids have well-developed embryos during migration. Other examples have been reviewed in detail elsewhere [52].

Though the general relation of prolonged flight with ovarian immaturity cannot be doubted, the system is obviously labile, though the degree of lability has rarely been measured, as it has in *Cicadulina*, where about 60 per cent of all three species studied did not fly. The proportions of long fliers however varied being 27, 19 and 27 per cent among sexually immature insects and 20, 9 and 14 per cent respectively among gravid ones for the three species. However, other factors blurred the relation. There were more long fliers among short-bodied forms (42 to 69 per cent) and more were produced after feeding on drying wheat than on fresh (12 compared with 9 per cent of gravid females of all species combined, and 28 compared with 6 per cent for non-gravid ones). The inter-correlation made it impossible to say whether ovarian immaturity would have been more or less strongly correlated with long flight had the effects of body length and host plant been allowed for [73]. This looseness in such correlations is another aspect of lability.

Various factors affect ovary development and its correlative, flight duration. As long ago as 1911 Lepidoptera were classified as those that emerged with immature ovaries and flew long distances, those that emerged with ripe eggs and did not, and those that were sometimes gravid, sometimes not, on emergence. The delayed ovarian development was attributed to high temperatures during the larval stage [52]. Since then other factors have been found that delay the growth of ovaries, for example a short photoperiod, or the lack of water-soluble vitamin E [38], on which a theory of migration has been based [95]; absence of host plants or their constituents or isolation from males in the adult stage are others [45].

Many of these factors affect corpus allatum (CA) activity and the association of ovarian immaturity and post-teneral prolongation of flight suggests that the hormone balance may affect flight directly. It is well known that prothoracic gland homogenates and ecdysone lessen activity in the leg muscles of locusts through a direct effect on the central nervous system (22, 42, 43], and that the prothoracic glands disappear and CA activity begins to increase early in adult life when migratory flight is often most evident. It was therefore suggested very tentatively indeed that the balance between CA and prothoracic gland hormones affects flight duration [52], an idea which has found some support in experiments with *Locusta migratoria migratorioides* whose flight speed and duration, in males, lessened when the CA was removed [96], as also occurred when prothoracic gland extracts were injected into *S. gregaria* [62]. Allatectomy is also known to decrease, and CA implantation to increase, general activity (not flight) in male *S. gregaria* [69]. If CA has an effect in regulating flight duration its presence is not necessary for flight and individual locusts that normally fly longest are least upset by allatectomy [61]. Crowding however, which is a token stimulus for increased CA activity, has a pro-found effect not only on wing production generally but on flight itself. Maximum flight durations by *S. gregaria* kept in crowds before testing were over ten times more than of those kept in isolation; and even those isolated as nymphs and crowded only after imaginal ecdysis flew longer than adults isolated during and after development. Rearing for one generation in isolation eliminated all flight activity of a kind regarded as migra-tory. Limited, continuous flight occurred only in the first generation of these laboratory *solitaria* [59, 60].

There are many papers on the 'corpus allatum deficiency syndrome' as it affects diapause and fat metabolism, but surprisingly few that consider flight. Yet migratory flight ceases at diapause (though limited flight often still occurs, weather permitting) and begins again when diapause ends, and when CA again starts to become active. The fact that winter temperatures tend to inhibit prolonged flight has probably drawn attention from the more direct effect of a depressed physiological state on flight. But to the insect flight is as important as its ovaries, which it cannot exploit to advantage unless it finds, by flight, a suitable place to recommence breeding. The ultrastructure of flight muscles in those insects which do not lose them by histolysis during hibernation would repay study, if linked with flight tests.

Flight also tends to be self limiting where, as in *Locusta migratoria* and in *S. gregaria*, daily flights to exhaustion stimulated the release of neurosecretory material which hastened growth of ovaries. The male pheromone and crowding also caused a similar effect and may be the reason why some flying swarms mature more quickly than would be expected from environmental effects [46]. Similar effects have been noted in *Oscinella frit* and *Plutella maculipennis* and in some aphids [52].

THE DEGENERATION AND REGENERATION OF FLIGHT MUSCLES

Flight muscles are now known to histolyze in more than 50 species of more than 20 families and 8 Orders (Table 10.1), in some of which muscles also regenerate later. This suggests that insects generally have a potential for structural lability in the flight apparatus even after reaching the adult stage. Histolysis, timed in relation to ecological needs, ensures that a habitat or hibernation site is not relinquished too soon, though the means seems very drastic. The flight muscles of many aphids histolyze a day or two after the first flight, though in some this can be delayed for more than two weeks, for example, by

TABLE 10.1. **Insects in which histolysis of the indirect flight muscles has been recorded**

Homoptera	Aphids; at least 15 species.	
Heteroptera	*Corixa* spp.	(Corixidae)
	Dysdercus intermedius	(Pyrrhocoridae)
	Nepa cinerea	(Nepidae)
	Ranatra spp.	,,
	Gerris spp.	(Gerridae)
	Ilycoris cimicoides	(Naucoridae)
	Belostoma malkini	(Belostomatidae)
	Lethocerus maximus	,,
Coleoptera	*Leptinotarsa decemlineata*	(Chrysomelidae)
	Psylliodes chrysocephala	,,
	Several spp. of Scolytidae	see text
	Trachyostus ghanaensis	(Platypodidae)
	Hylobius abietus	(Curculionidae)
	Sitona hispidula	,,
	S. lineatus	,,
	S. regensteinensis	,,
	Hydroporus palustris	(Hydroporidae)
	Lucanus cervus	(Lucanidae)
Diptera	*Lipopterna cervi*	(Hippoboscidae)
	Carnus hemapterus	(Milichiidae)
	Aedes communis	(Culicidae)
	Mochlonyx culiciformis	(Chaoborinae)
Hymenoptera	*Lasius niger*	(Formicidae)
	Formica fusca	,,
Isoptera	Termites generally	
Dermaptera	*Forficula auricularia*	(Forficulidae)
Orthoptera	*Acheta domestica*	(Gryllidae)

darkness which stops flight [44, 50]. In some Scolytids the muscles histolyze in the brood chamber as the ovaries ripen (*Dendroctonus monticolae*) and in some individuals, but not in all, they regenerate later and the cycle can be repeated up to three times (*D. pseudotsugae*) [71, 2]. Muscles can also regenerate as ovaries develop, as in the Colorado Beetle after diapause [30]. Evidently in the Scolytidae, as in *A. fabae* and other insects, some individuals deposit eggs, or larvae, before flight and subsequent histolysis, and others fly first and oviposit and histolyze later [79, 80, 81] and it is in the autogenous not in the anautogenous race of *Ae. communis* that flight muscles histolyze [4, 48]. Histolysis sometimes occurs prematurely, before flight (*Leptinotarsa* [30], *Dysdercus intermedius* [36], *A. fabae* [80, 81], *Nepa cinerea* [41]) and either before (*Leptinotarsa*) or after diapause (*Sitona regensteinensis* [28]).

The stimuli that trigger histolysis in the flight muscles seem to vary. In aphids the sequence of flight, settling, feeding and larviposition is not necessary, as once thought, for the muscles can histolyze before flight and in both fed and unfed aphids. Interference with locomotion seemed to be the common factor in several tests [50, 51]. In *Acheta domestica* L., the house cricket, the histolysis is said to be regulated by the intensity of afferent impulses from the working wing, via the CNS, whose cessation after flight coincides with an inhibition of JH production [8, 24]. Cessation of wing-beat, in contrast to its absence, is not however a general cause, for autolysis can precede flight. Thus histolysis begins so soon in newly emerged, mated females of *Dysdercus intermedius* that they never fly, though starved virgin females fly vigorously for many days and rarely lose the flight muscles. Mating seems to be the trigger, but unfed males are reluctant to mate, so muscle histolysis in females usually follows feeding by males and mating, both of which are enhanced by crowding; males rarely lose the ability to fly. There is, however, much variation in the process. Migration of females in nature is probably mostly by virgins and the proportions of migrants and non-fliers probably depends on the availability of seeds as food and on the sex ratio in the small aggregations that develop from single egg batches [36, 37, 82].

A blood-borne factor evidently controls flight muscle histolysis in aphids [51] though other profound, concomittant changes also occur, such as hypertrophy of the fat body and renewed growth of embryos, and are possibly associated with the same humoral factor. As so far studied, histolysis does not occur because the ovaries grow, for the muscles degenerate in both ovariectomized and intact mated females of *D. intermedius* [37], and when ovarian growth ceases in *Leptinotarsa* or continues in *D. pseudotsugae*. As in aphids, the CA in *Dysdercus* was not markedly different in size between newly histolyzed and non-histolyzed insects. Flight muscles degenerate when CA activity is minimal at diapause or after allatectomy in *Leptinotarsa* [86], but after topical applications of synthetic JH (or when the insects burrow or a male is introduced into the brood chamber) in *Ips paraconfusus* Lanier [9]. Nor is the histolysis a general prerequisite for ovary development [7] for ovaries grow as muscles regenerate in *Leptinotarsa*.

After diapause, flight muscles in many insects regenerate, for example as days lengthen for *Leptinotarsa* [30, 86] and as moisture in the bark drops below a critical value for *D. monticolae* [72].

In view of the current controversy on the role of humoral and nervous factors in the control of muscle growth, maintainance and degeneration, in different parts of the insect body, a discussion on possible mechanisms of histolysis and regeneration of flight muscles is inappropriate here.

THE COLLECTIVE EXPRESSION OF INDIVIDUAL VARIATION

Variation in individual flight capacity is reflected collectively in the speed at which the proportions of adults with different flight capacities change and in the relative effects of environment and heredity on the change.

As populations of A. *fabae* became crowded, the percentage of flightless alatae dropped from 100 to zero, fliers and migrants each increasing to about 50 per cent. Thereafter about 90 per cent born were migrants. These proportions are not general but change with external factors; but they show what can happen in about six weeks in successive broods. Variation in wing production and flight capacity in clonal parthenogenetic aphids is controlled ontogenetically by crowding, diet, or, experimentally by JH application [52, 63, 102, 103], not by selection through mortality and by genetic reassembly. But ontogenetic plasticity is manifested also in bisexual populations. Thus in *Ae. taeniorhynchus* larval crowding produced migrant and non-migrant forms that had the same morphological characters, glycogen content and degree of autogeny; but the migrants on average had less dry body-weight and a smaller proportion of lipids than non-migrants [66].

Even where flight characters are strongly inherited they can be modified by the environment. In Britain *Gerris odontogaster* Zett. has homozygous macropters and heterozygous brachypters but temperature probably modifies the response to JH so that those with a genetical constitution for long wings develop short ones. In Finland a photoperiod longer than 18 hours during larval development produces brachypters, and a shorter photoperiod in the 4th instar, macropters, from the heterozygote [11, 12, 94]. Variation in the flight of *Gerris* macropters is not known but in the beetles *Callosobruchus maculatus* (F.) and *C. chinensis* L. the proportion of active and sluggish fliers which had, respectively, immature and mature ovaries varied with population density and food supply [23, 64, 92, 93]. Adult *Oncopeltus fasciatus* that flew longest at first also flew longest subsequently and the proportion of 'migrants' changed by selective breeding from 20 and 30 per cent in males and females respectively to over 60 per cent in the F_3 generation; however, more migrants were produced at 23°C than at 27°C [32, 33]. The progeny of non-flying *Cicadulina* spp. were mostly but not all non-fliers, and those of long-fliers mostly long-fliers, but flight duration was correlated with body length which is inherited; the gene complexes for flight are probably inherited in a block and protected by chromosomal inversions [73]. Nevertheless variation within the forms is probably ontogenetically controlled to some extent, as the effect of host plant (p. 224) suggests.

Flight capacity may be selected in populations by mortality or by redistribution. In *Malacosoma pluviale* (Dyar) up to nearly 40 per cent of larvae from single egg batches walked vigorously and became active fliers. The rest wandered aimlessly or little, many died as they competed for food and they produced sluggish fliers. The active moths migrated; their larvae were better fed and active. As these more distant infestations aged they became smaller and more sluggish but, when minimal, again increased in size from the hardier survivors [99–101].

CONCLUSION

In an earlier Symposium Sir Vincent Wigglesworth [105] said that the first and greatest polymorphism in insects was that in which all parts of the body changed during ontogeny. Post-emergence changes come into this class and by differential development of the parts evidently produce a variable intrinsic flight duration which has been called (for want of a better term) behavioural polymorphism. A graded variation would not, by some, be called a polymorphism; but it is the basis of that extreme state which would be called behavioural dimorphism in which nearly all individuals of a generation seem to be migrants and others not, whether that effect is produced environmentally and so seasonally by ontogenetic effects, or genetically. There are many factors that limit flight duration in which they can be selected from time to time.

I have not considered thresholds . . . a major omission . . . and difficult to fit into an ontogenetic context. Migratory flight in nature, as distinct from on a pin, seems to be analogous to a manic condition and the biochemistry of even mammalian moods, let alone those of insects, is a new, difficult and controversial subject [34].

DISCUSSION

Professor Southwood (Chairman) commented that Dr Johnson had high-lighted the point that whereas the physiologist is primarily concerned with establishing the mechanism of a process, the ecologist, confronted by the massive variance in all his data, is always interested in the variation in physiological processes shown between individuals, within and between populations.

Dr J.Brady: Dr Johnson has pointed to the large differences in flight activity level that can occur between individual insects. This is conspicuously shown by actograph records, but it is worth emphasizing that these differences in the *amount* of activity generally do not extend to its circadian *pattern*. In the tsetse fly, for example, the shape of individuals' circadian rhythms is very similar, even though the amount of activity they show varies by a factor of about ten.

Johnson: I do not know how general such a result may be; for example, both the diurnal rhythm and the amount of flight activity of *Plusia gamma* were found to change during adult life. Activity increased to a maximum up to the third day after emergence, tending to become continuous in a light/dark regime during this period and so to obliterate the rhythm which was established later when activity lessened (McCauley, E.D.M., 1972, *Entomologia exp. appl.*, **15** : 387–391); and this tendency for continued flight, day and night, has been noticed in other migrants [52]. But the pattern varied between different moths in the *Plusia* culture, as it does also in migrants and non-migrants of other species. It would be interesting to know if it also differs intrinsically between fliers of long and short duration, if these exist in tsetse and other insects.

Mr J.Hargrove: With tsetse it is more apt to talk of short-fliers and non-fliers. While many species have individuals which fly for hours, the longest continuous flight that I have recorded from about 2,000 tsetse is 11 minutes, and I would estimate that about 50 per cent of these flies could not be induced to fly for more than 5 seconds. Of those which would fly, the mean flight duration increased during the first hunger cycle from 78 to 135 seconds; and this increase was paralleled by increases in flight muscle and thoracic proline.

Johnson: This increase in flight activity as flight muscles mature is, I think, general in insects. Presumably flight in tsetse is very local during this early period. But it would be interesting to know if flight activity stabilizes at the higher level, when I suspect most dispersal might occur, and later diminishes again to more local flight.

Hargrove: We have no data on the effect of age on the flight performance of mature testse but Bursell has unpublished evidence that sarcosomes from 45-day-old flies maintain good oxidative capacity.

Neville: I have made field measurements on teneral dragonflies (*Sympetrum striolatum*), and found them to fly at about half the speed of fully matured adults (Neville, 1960, Ph.D. Thesis). Few studies of this kind seem to have been made. [Further information on this point has recently been provided by field data on the air-speeds of Desert Locusts, over a considerable range of conditions (Waloff, Z., 1972, *Anim. Behav.* **20** : 367–372), and evidence of a new kind is now becoming available from radar (Schaefer, G.W., pp. 161–162)]. In those adult insects which feed as carnivores or bloodsuckers, the food-supply is liable to be more erratic than in herbivores, so that their muscles, cuticle and fuel deposits are not likely to grow as steadily during the teneral period. Could this contribute to the variability of flight performance shown by such insects as tsetse?

Johnson: Yes. On this general question of variations in growth-rate, I think there may sometimes be a discrepancy between the timing of the formation of the layers of cuticle, especially of the wing-hinge, and the onset of flight. Thus, among herbivores, although there are eight or nine wing-hinge layers in the Milkweed Bug (*Oncopeltus*) in which the maximum duration of flight becomes possible at 8–10 days, and there is only one such layer in *Vanessa*, which can migrate on the day of emergence, yet in locusts, also herbivorous, wing-hinge layers are added daily for some three weeks after fledging (under laboratory conditions), suggesting that they may do quite a lot of flying before they have a fully developed wing-hinge (though differences in temperature regime may be involved). [There is indeed repeated field evidence of young flying Desert Locust swarms moving out of the area in which they had fledged, and starting on long-range migration, at about one week after fledging.—Ed.] Again some dragonflies are known to migrate while the cuticle is still very soft. I think there is growing evidence that prolonged (migratory) flight sometimes begins before all the systems involved in flight have developed to their maximum.

Weis-Fogh: May I comment briefly on an 'ecological' experiment I did on locusts in the laboratory, which illustrates the refined interplay there must be between the act of flying and the general ecology of a species? After sexual maturity, the males will still fly at any time, but the females will fly only between periods of egg-laying. In the experiment, using four groups of locusts which were fully-grown (in terms of weight) but not yet sexually mature, the locusts were flown every day for long periods, say four hours or so of continuous flight, and the sexes were kept separate. After each period of flight, each locust was given just sufficient food to bring it up to its pre-flight weight, and no more. Under these conditions, the males matured sexually in the normal way, but the females failed to mature, and instead retained their ability to fly over an extended period of time. As soon as the females were provided with more food than just that necessary to bring them up to their pre-flight weight, they matured rapidly— almost overnight—and went into the usual rhythmical on-and-off ability to fly for long periods. And there we have a very good ecological parallel in the field, where if they reach a really good situation where they can feed well they will mature and lay eggs. [While this is true in general terms, the sexual maturation of Desert Locusts in the field can be delayed, sometimes for months, without obvious shortage of food or other simple explanation—and can then occur almost simultaneously over a wide area, possibly triggered off by bud-burst of aromatic desert shrubs (Carlisle, D.B., Ellis, P.E. & Betts, E., 1965, *J. Insect Physiol.*, **11** : 1541–1558)—Ed.]

Johnson: The case of the *Cicadulina* I mentioned is perhaps somewhat similar, with the short-winged form having a much more rapid maturation than the long-winged form, and also showing effects of feeding regime as well as of other factors.

REFERENCES

[1]   ACTON, A.B. & SCUDDER, G.G.E. (1969). The ultrastructure of the flight muscle polymorphism in *Cenocorixa bifida* (Hung.) (Hem., Het., Corixidae). *Z. Morphol. Tiere*, **65** : 327–335.

[2]   ATKINS, M.D. & FARRIS, S.H. (1962). A contribution to the knowledge of flight muscle changes in the Scolytidae (Coleoptera). *Can. Ent.*, **94** : 25–32.

[3]   BALBONI, E.R. (1967). The respiratory metabolism of insect flight muscle during adult maturation. *J. Insect Physiol.*, **13** : 1849–1856.

[4]   BECKEL, W.E. (1954). The lack of autolysis of the flight muscles of *Aedes communis* de Geer (Culicidae) in the laboratory. *Mosquito News*, **14** : 124–127.

[5]   BEENAKKERS A.M.T. (1969). Carbohydrate and fat as fuel for insect flight. A comparative study. *J. Insect Physiol.*, **15** : 353–361.

[6]   BHAKTHAN, N.M.G. & GILBERT, L.I. (1970). Studies on lipid transport in *Manduca sexta* (Insecta). *Comp. Biochemistry and Physiology*, **33** : 705–706.

[7]   BHAKTHAN, N.M.G., BORDEN, J.H. & NAIR, K.K. (1970). Fine structure of degenerating and regenerating flight muscles in a bark beetle, *Ips confusus*. I. Degeneration. *J. Cell. Sci.*, **6** : 807–811.

[8]   BOCHAROVA-MESSNER, O.M., CHUDAKOVA, I.V. & NOVAK, V.J.A. (1969). The effect of castration on the process of wing muscle degeneration in the domestic cricket (*Acheta domestica* L.). Transl. from *Doklady Akademii Nauk SSSR.*, **188** : 242–244.

[9]   BORDEN, J.H. & SLATER, C.E. (1968). Induction of flight muscle degeneration by synthetic juvenile hormone in *Ips confusus* (Coleoptera: Scolytidae). *Z. vergl. Physiol.*, **61** : 366–368.

[10] BORDEN, J.H. & SLATER, C.E. (1969). Flight muscle volume change in *Ips confusus* (Coleoptera: Scolytidae). *Can. J. Zool.*, **47** : 29–31.

[11] BRINKHURST, R.O. (1959). Alary polymorphism in the Gerroidea (Hemiptera–Heteroptera). *J. Anim. Ecol.*, **27** : 211–230.

[12] BRINKHURST, R.O. (1963). Observations on wing-polymorphism in the Heteroptera. *Proc. R. ent. Soc. Lond.* (A), **38** : 15–22.

[13] BROSEMER, R.W. (1965). Changes in glycerophosphate dehydrogenase activity during development of the grasshopper *Schistocerca vaga*. *Biochim. biophys. Acta*, **96** : 61–65.

[14] BROSEMER, R.W. (1967). The levels of extramitochondrial glycerophosphate dehydrogenase in the wing muscle of a flightless grasshopper. *J. Insect Physiol.*, **13** : 685–690.

[15] BROSEMER, R.W., VOGELL, W. & BÜCHER, TH. (1963). Morphologische und enzymatische Muster bei der Entwicklung indirekter Flugmuskeln von *Locusta migratoria*. *Biochem. Z.*, **338** : 854–910.

[16] BROSEMER, R.W., GROSSO, D.S., ESTES, G. & CARLSON, C.W. (1967). Quantitative immuno-chemical and electrophoretic comparison of glycerophosphate dehydrogenases in several insects. *J. Insect Physiol.*, **13** : 1757–1767.

[17] BÜCHER, TH. (1965). Formation of the specific structural and enzyme pattern of the insect flight muscle. In Goodwin, T.W. (ed.) *Aspects of Insect Biochemistry:* 15–28. Academic Press, New York and London.

[18] BURSELL, E., SLACK, E. & KUWENGA, T. (1972). Aspects of the development of flight musculature in the tsetse fly (*Glossina morsitans*). *Trans. R. Soc. Trop. Med. & Hyg.*, **66** : 305–323.

[19] CALDWELL, R.L. & HEGMANN, J.P. (1969). Heritability of flight-duration in the milkweed bug *Lygaeus kalmii*. *Nature, Lond.*, **223** : 91–92.

[20] CANDY, D.J. (1969). Carbohydrate utilization by locust thoracic muscle *in vitro*. *Proc. Biochem. Soc., Biochem. J.*, **108** : 12 pp.

[21] CANDY, D.J. (1970). Metabolic studies on locust flight muscle using a new perfusion technique. *J. Insect Physiol.*, **16** : 531–543.

[22] CARLISLE, D.B. & ELLIS, P.E. (1963). Prothoracic gland and gregarious behaviour in locusts. *Nature, Lond.*, **200** : 603–604.

[23] CASWELL, G.H. (1956). Observations on the biology of *Callosobruchus maculutus* F. (Coleoptera Bruchidae). *Divl. Rep. Fac. Agric. Ibadan Univ. Coll. (Ent.)*, no. 3 : 9 pp.

[24] CHUDAKOVA, I.V. & BOCHAROVA-MESSNER, O.M. (1968). Endocrine regulation of the condition of the wing musculature in the imago of the house cricket (*Acheta domestica* L.) Transl. from *Doklady Akademii Nauk SSSR.*, **179** : 489–492.

[25] CLEGG, J.S. & EVANS, D.R. (1961). The physiology of blood trehalose and its function during flight in the blowfly. *J. exp. Biol.*, **38** : 771–792.

[26] COCKBAIN, A.J. (1961). Fuel utilization and duration of tethered flight in *Aphis fabae* Scop. *J. exp. Biol.*, **38** : 163–174.

[27] DAME, D.A., BIRKENMEYER, D.R. & BURSELL, E. (1969). Development of the thoracic muscle and flight-behaviour of *Glossina morsitans orientalis* Vanderplank. *Bull. ent. Res.*, **59** : 345–350.

[28] DANTHANARAYANA, W. (1970). Studies on the dispersal and migrations of *Sitona regensteinensis* (Coleoptera: Curculionidae). *Entomologia exp. appl.*, **13** : 236–246.

[29] D'COSTA, M.A. & BIRT, L.M. (1969). Mitochondrial oxidations of fatty acids in the blowfly, *Lucilia*. *J. Insect Physiol.*, **15** : 1959–1968.

[30] DE KORT, C.A.D. (1969). Hormones and the structural and biochemical properties of the flight muscles in the Colorado Beetle. *Med. Lab. Entom., Wageningen*, no. 159 : 63 pp.

[31] DINGLE, H. (1965). The relation between age and flight activity in the Milkweed Bug, *Oncopeltus*. *J. exp. Biol.*, **42** : 269–283.

[32] DINGLE, H. (1968). The influence of environment and heredity on flight activity in the Milkweed Bug *Oncopeltus*. *J. exp. Biol.*, **48** : 175–184.

[33] DINGLE, H. (1968). Life history and population consequences of density, photoperiod and temperature in a migrant insect, the Milkweed Bug *Oncopeltus*. *Am. Nat.*, **102** : 149–163.

[34] ECCLESTON, D. (1973). The biochemistry of human moods. *New Scientist*, **57** : 18–19.

[35] EDWARDS, F.J. (1969). Development and histology of the indirect flight muscles in *Dysdercus intermedius*. *J. Insect. Physiol.*, **15** : 1591–1599.

[36] EDWARDS, F.J. (1969). Environmental control of flight muscle histolysis in the bug *Dysdercus intermedius*. *J. Insect. Physiol.*, **15** : 2013–2020.

[37] EDWARDS, F.J. (1970). Endocrine control of flight muscle histolysis in *Dysdercus intermedius*. *J. Insect. Physiol.*, **16** : 2027–2031.

[38] EICHLER, F. (1970). Studies in migratory Lepidoptera, V. Ovarial maturation in *Herse convolvuli* L. after feeding with vitamin E. *Entomol. Nachr.*, **13** : 132–136.

[39] FINCH, S. & COAKER, T.H. (1969). Comparison of the nutritive values of carbohydrates and related compounds to *Erioischia brassicae*. *Entomologia exp. appl.*, **12** : 441–453.

[40] GEORGE, J.C. & BHAKTHAN, N.M.G. (1960). A study on the fibre diameter and certain enzyme concentrations in the flight muscles of some butterflies. *J. exp. Biol.*, **37** : 308–315.

[41] HAMILTON, M.A. (1931). The morphology of the Water Scorpion, *Nepa cinerea* Linn. (Rhynchota, Heteroptera). *Proc. zool. Soc. Lond.* (1931): 1067–1136.

[42] HASKELL, P.T. & MOOREHOUSE, J.E. (1963). A blood-borne factor influencing the activity of the central nervous system of the Desert Locust. *Nature, Lond.*, **197** : 56–58.

[43] HASKELL, P.T., CARLISLE, D.B., ELLIS, P.E. & MOOREHOUSE, J.E. (1965). Hormonal influences in locust marching behaviour. *Proc. XII Int. Congr. Ent.* (London, 1964) : 290–291.

[44] HEATHCOTE, G.D. & COCKBAIN, A.J. (1966). Aphids from mangold clamps and their importance as vectors of beet viruses. *Ann. appl. Biol.*, **57** : 321–336.

[45] HILLYER, R.J. & THORSTEINSON, A.J. (1969). The influence of the host plant or males on ovarian development or oviposition in the diamondback moth, *Plutella maculipennis* (Curt.). *Can. J. Zool.*, **47** : 805–816.

[46] HIGHNAM, K.C. & HASKELL, P.T. (1964). The endocrine systems of isolated and crowded *Locusta* and *Schistocerca* in relation to oocyte growth, and the effects of flying upon maturation. *J. Insect Physiol.*, **10** : 849–864.

[47] HOCKING, B. (1953). The intrinsic range and speed of flight of insects. *Trans. R. ent. Soc. Lond.*, **104** : 223–245.

[48] HOCKING, B. (1954). Flight muscle autolysis of *Aedes communis* (de Geer). *Mosquito News*, **14** : 121–123.

[49] HWANG, GUAN-HUEI & HOW, WU-WEI (1966). Studies on the flight of the Army-worm moth (*Leucania separata* Walker). I. Flight duration and wingbeat frequency. *Acta ent. sin.*, **15** : 96–104. [*Rev. appl. Ent.* (A), **55** : 88–89.]

[50] JOHNSON, B. (1957). Studies on the degeneration of the flight muscles of alate aphids.—I. A comparative study of the occurrence of muscle breakdown in relation to reproduction in several species. *J. Insect Physiol.*, **1** : 248–256.

[51] JOHNSON, B. (1959). Studies on the degeneration of the flight muscles of alate aphids.—II. Histology and control of muscle breakdown. *J. Insect Physiol.*, **3** : 356–377.

[52] JOHNSON, C.G. (1969). *Migration and dispersal of insects by flight.* Methuen, London: 763 pp.

[53] KENNEDY, J.S. & BOOTH, C.O. (1963). Free flight of aphids in the laboratory. *J. exp. Biol.*, **40** : 67–85.

[54] KISHABA, A.N., HENNEBERRY, T.J., HANCOCK, P.J. & TOBA, H.H. (1967). Laboratory technique for studying flight of Cabbage Looper moths and the effects of age, sex, food, and tepa on flight characteristics. *J. econ. Ent.*, **60** : 359–366.

[55] LANGLEY, P.A. (1970). Post-teneral development of thoracic musculature in the tsetse flies *Glossina austeni* and *G. morsitans*. *Entomologia exp. appl.*, **13** : 133–140.

[56] LOH, W. & HERAN, H. (1970). How well can bees utilise sucrose, glucose, fructose and sorbitol in flight metabolism? *Z. Vergl. Physiol.*, **67** : 436–452.

[57] McCAMBRIDGE, W.F. & MATA, S.A. (Jr.) (1969). Flight muscle changes in Black Hills beetles, *Dendroctonus ponderosae* (Coleoptera: Scolytidae), during emergence and egg laying. *Can. Ent.*, **191** : 507–512.

[58] MICHEL, R. (1969). Étude expérimentale des variations de la tendence au vol au cours de vieillissement chez le criquet pélerin *Schistocerca gregaria* (Forsk.). *Revue du Comportement Animal*, **3** : 46–65.

[59] MICHEL, R. (1970). Étude expérimentale de l'activité maximum de vol journalière du criquet pélerin (*Schistocerca gregaria* Forsk.) élevé en groupe on en isolement. *Behaviour*, **36** : 286–299.

[60] MICHEL, R. (1970). Experimental study of the variations in flight tendency in the migratory locust, *Schistocerca gregaria* (Forsk.) reared in isolation for several generations. *Insectes. Soc.*, **17** : 21–38.

[61] MICHEL, R. (1972). Influence de l'ablation des corpora allata sur l'activité du vol soutenu du criquet pélerin *Schistocerca gregaria*. *Ann. Soc. ent. Fr.*, **8** : 729–734.

[62] MICHEL, R. (1972). Étude expérimentale de l'influence des glandes prothoraciques sur l'activité de vol du criquet pélerin *Schistocerca gregaria*. *Gen. comp. Endocr.*, **19** : 96–101.

[63] MITTLER, T.E. & SUTHERLAND, O.R.W. (1969). Dietary influences on aphid polymorphism. *Entomologia exp. appl.*, **12** : 703–713.

[64] NAKAMURA, H. (1969). Geographical variation of ecological characters in *Callosobruchus chinensis* L. *Jap. J. Ecol.*, **19** : 127–131.

[65] NAYAR, J.K. & VAN HANDEL, E. (1971). The fuel for sustained mosquito flight. *J. Insect Physiol.*, **17** : 471–481.

[66] NAYAR, J.K. & SAUERMAN, D.M. Jr. (1969). Flight behaviour and phase polymorphism in the mosquito *Aedes taeniorhynchus*. *Entomogia exp. appl.*, **12** : 365–375.

[67] NAYAR, J.K. & SAUERMAN, D.H. Jr. (1971). Physiological effects of carbohydrate on survival, metabolism and flight potential of female *Aedes taeniorhynchus*. *J. Insect Physiol.*, **17** : 2221–2233.

[68] NORDEN, D.A. & PATERSON, D.J. (1969). Carbohydrate metabolism in flight muscle of the tsetse fly (*Glossina*) and the blowfly (*Sarcophaga*). *Comp. Biochem. Physiol.*, **31** : 819–827.

[69] ODHIAMBO, TH.R. (1966). The metabolic effects of the corpus allatum hormone in the male desert locust: II Spontaneous locomotor activity. *J. exp. Biol.*, **45** : 51–63.

[70] PATTERSON, R.S. (1957). On the causes of broken wings of the House Fly. *J. econ. Ent.*, **50** : 104–105.

[71] REID, R.W. (1958). Internal changes in the female Mountain Pine Beetle, *Dendroctonus monticolae* Hopk., associated with egg laying and flight. *Can. Ent.*, **90** : 464–468.

[72] REID, R.W. (1962). Biology of the Mountain Pine Beetle, *Dendroctonus monticolae* Hopkins, in the East Kootenay Region of British Colombia. I. Life cycle, brood development, and flight periods. *Can. Ent.*, **94** : 531–538.

[73] ROSE, D.J.W. (1972). Dispersal and quality in populations of *Cicadulina* species (Cicadellidae). *J. Anim. Ecol.*, **41** : 589–609.

[74] ROWLEY, W.A. (1970). Interval flights and glycogen utilization by the mosquito, *Culex tarsalis*. *J. Insect Physiol.*, **16** : 1839–1844.

[75] ROWLEY, W.A. & GRAHAM, C.L. (1968). The effect of age on the flight performance of female *Aedes aegypti* mosquitoes. *J. Insect Physiol.*, **14** : 719–728.

[76] RYGG, T.D. (1966). Flight of *Oscinella frit*. L. (Diptera, Chloropidae) females in relation to age and ovary development. *Entomologia exp. appl.*, **9** : 74–84.

[77] SCUDDER, G.G.E. (1971). The post-embryonic development of the indirect flight muscles in *Cenocorixa bifida* (Hung.) (Hemiptera; Corixidae). *Can. J. Zool.*, **49** : 1387–1398.

[78] SCUDDER, G.G.E. and MEREDITH, J. (1972). Temperature-induced development in the indirect flight muscle of adult *Cenocorixa* (Heteroptera: Corixidae). *Developmental Biology*, **29** : 330–336.

[79] SHAW, M.J.P. (1970). Effects of population density on alienicolae of *Aphis fabae* Scop. I. The effect of crowding on the production of alatae in the laboratory. *Ann. appl. Biol.*, **65** : 191–196.

[80] SHAW, M.J.P. (1970). Effects of population density on alienicolae of *Aphis fabae* Scop. II. The effects of crowding on the expression of the migratory urge among alatae in the laboratory. *Ann. appl. Biol.*, **65** : 193–203.

[81] SHAW, M.J.P. (1971). Effects of population density on alienicolae of *Aphis fabae* Scop. III. The effects of isolation on the development of form and behaviour of alatae in a laboratory clone. *Ann. appl. Biol.*, **65** : 205–212.

[82] SHERRARD, J. & JOHNSON, C.G. (MS).

[83] SMIT, W.A., BECHT, G. & BEENAKKERS, A.M.TH. (1967). Structure, fatigue and enzyme activities in 'fast' insect muscles. *J. Insect Physiol.*, **13** : 1857–1868.

[84] SMITH, D.S. (1964). The structure and development of flightless Coleoptera: a light and electron microscopic study of the wings, thoracic exoskeleton and rudimentary flight musculature. *J. Morph.*, **114** : 107–184.

[85] SRIVASTAVA, P.N. & ROCKSTEIN, M. (1969). The utilization of trehalose during flight by the housefly, *Musca domestica*. *J. Insect Physiol.*, **15** : 1181–1186.

[86] STEGWEE, D. (1964). Respiratory chain metabolism in the Colorado Potato Beetle. II. Respiration and oxidative phosphorylation in 'sarcosomes' from diapausing beetles. *J. Insect Physiol.*, **10** : 97–102.

[87] STEPHEN, W.F. & GILBERT, L.I. (1970). Alterations in fatty acid composition during the metamorphosis of *Hyalophora cecropia*: correlations with juvenile hormone titre. *J. Insect Physiol.*, **16** : 851–864.

[88] STEVENSON, E. (1968). Carbohydrate metabolism in the flight muscle of the Southern Armyworm Moth, *Prodenia eridania*. *J. Insect Physiol.*, **14** : 179–198.

[89] STEVENSON, F. (1969). Monoglyceride lipase in moth flight-muscle. *J. Insect Physiol.*, **15** : 1537–1550.

[90] TRIBE, M.W. (1966). Some physiological studies in relation to age in the blowfly, *Calliphora erythrocephala* Meig. *J. Insect Physiol.*, **12** : 1577–1593.

[91] TURUNDAYEVSKAYA, T.M. (1970). On the biology of *Aegeria apiformis* (Lep., Aegeriidae). *Zool. Zh.*, **49** : 746–752.

[92] UTIDA, S. (1969). Photoperiod as a factor inducing the flight-form in the population of the southern cowpea weevil, *Callosobruchus maculatus*. *Jap. J. appl. Entomol. Zool.*, **13** : 129–134.

[93] UTIDA, S. (1970). Secular change of percent emergence of the flight form in the population of southern cowpea weevil, *Callosobruchus maculatus*. *Jap. J. appl. Entomol. Zool.*, **14** : 71–78.

[94] VEPSALAINEN, K. (1971). The role of gradually changing day length in determination of wing length, alary dimorphism and diapause in a *Gerris odontogaster* (Zett.) population (Gerridae, Heteroptera) in South Finland. *Ann. Acad. Sci. fenn.* A, IV *Biologica*, **183** : 25 pp.

[95] VOJNITS, A. (1969). Reproductive biological aspects of the migration of the Gamma moth (*Autographa gamma* L.). *Acta Phytopathol. Acad. Sci. Hung.*, **4** : 163–179.

[96] WAJC, E. & PENER, M.P. (1971). The effect of the corpora allata on the flight activity of the male African Migratory Locust, *Locusta migratoria migratoriodes* (R. & F.). *General and Comparative Endocrinology*, **17** : 327–333.

[97] WALKER, P.R. & BAILEY, E. (1970). Metabolism of glucose, trehalose citrate, acetate, and palmitate by the male desert locust during adult development. *J. Insect. Physiol.*, **16** : 499–509.

[98] WALKER, P.R., HILL, L. & BAILEY, E. (1970). Feeding activity, respiration, and lipid and carbohydrate content of the male Desert Locust during adult development. *J. Insect. Physiol.*, **16** : 1001–1015.

[99] WELLINGTON, W.G. (1957). Individual differences as a factor in population dynamics; the development of a problem. *Can. J. Zool.*, **35** : 293–323.

[100] WELLINGTON, W.G. (1960). Qualitative changes in natural populations during changes in abundance. *Can. J. Zool.*, **38** : 289–314.

[101] WELLINGTON, W.G. (1964). Qualitative changes in populations in unstable environments. *Can. Ent.*, **96** : 436–451.

[102] WHITE, D.F. (1971). Corpus allatum activity associated with development of wingbuds in cabbage aphid embryos and larvae. *J. Insect Physiol.*, **17** : 761–772.

[103] WHITE, D.F. & LAMB, K.P. (1968). Effect of a synthetic juvenile hormone on adult cabbage aphids and their progeny. *J. Insect. Physiol.*, **14** : 395–402.

[104] WIGGLESWORTH, V.B. (1949). The utilization of reserve substances in *Drosophila* during flight. *J. exp. Biol.*, **26** : 150–163.

[105] WIGGLESWORTH, V.B. (1961). Insect Polymorphism: a tentative synthesis. *Symp. R. ent. Soc. London*, No. 1: 103–113.

[106] WILLIAMS, C.M., BARNESS, C.A. & SAWYER, W.H. (1943). The utilization of glycogen by flies during flight and some aspects of the physiological ageing of *Drosophila*. *Biol. Bull.*, **84** : 263–272.

11 · The fossil record and insect flight

R.J.WOOTTON

Department of Biological Sciences, University of Exeter

We have no fossil record of the origin, nor indeed of much of the early evolution of flight in insects. It is probable that insects developed the ability to fly in the Devonian (see Fig. 11.1 for stratigraphic divisions and time-scale); but they are unknown as fossils until well into the Carboniferous (Namurian A) and scarce before the Bashkirian [20]. Upper Carboniferous insects are already diverse, and it is becoming increasingly clear that they represent a highly developed fauna in which extensive adaptive radiation had already taken place; and that many forms which were previously believed to be generally primitive show a variety of relatively specialized (apomorphic) characters. Distinguishing genuinely primitive (plesiomorphic) features is far from straightforward, and several widely-quoted conclusions on the nature of the first Pterygota, drawn half a century ago from inadequately known fossil material, are now in question.

Moreover, and despite the fact that most insect fossils are wings, the modes of flight of Palaeozoic species are not easy to determine. All information on the mechanisms and dynamics of insect flight is necessarily drawn from living forms, and the species which have been studied in detail were chosen less for phylogenetic interest than for convenience, which includes their readiness to fly under experimental conditions, and has focused attention on active, efficient, and therefore probably highly-evolved fliers. Even for these, published investigations, on Polyneopterous and Palaeopterous orders at least, have tended to avoid the aspects of most value to a palaeontologist: the detailed kinematics and functions in flight of the skeletal components of thorax, axilla and wings—the structures which are available to him for study. Some information of this kind is available for the thoracic and axillary sclerites of Odonata [32, 51, 62] and *Ephemera vulgata* [2], but no full account has yet been published for a Polyneopterous species; and the functional significance of most details of the architecture of wings is still effectively unknown in any insect.

In the absence of such information, study of the origin and early evolution of insect flight is still at the level of informed guesswork, to which students of living insects have contributed as much as palaeoentomologists. Guesses are more likely to be correct, however, if they take account of both fossil and living material. Appropriate study of the latter can provide information on the extant insect groups whose flight mechanisms appear to show primitive characteristics; and palaeontology can elucidate the evolutionary history and relationships of these groups, reveal extinct groups displaying relevant structures which are absent from living forms, and clarify the environmental circumstances under which flight may have arisen and developed.

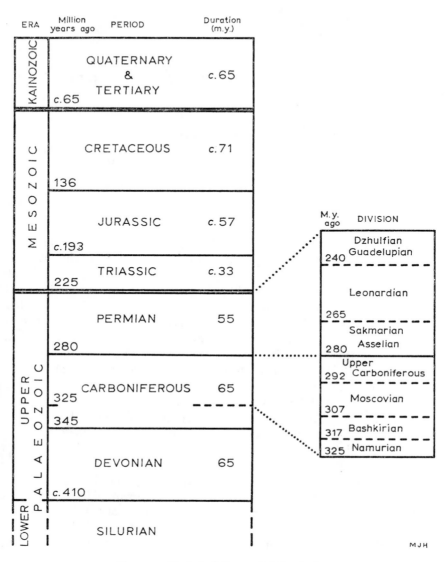

FIG. 11.1. **Geological time-scale** (after [31]).

It is now widely, but not universally accepted (see e.g. p. 256), that insect wings arose from meso- and metathoracic paranotal lobes, homologous with those which occur on the prothorax of members of the Palaeozoic order Palaeodictyoptera, of the problematical Bashkirian *Lithoneura lameerei* [5]; and of some Protodonata, Permian Ephemeroptera, and Protorthoptera (see Table 11.1 for classification used). Such lobes probably also contribute to the prothoracic shields of some other Protorthoptera and of Blattodea. Similar structures are found on the abdomens of some Palaeozoic nymphs of Protorthoptera, Palaeodictyoptera [69] and Blattodea, and of some adults of these orders, but their homology with the wings and prothoracic lobes has been questioned—at least in Palaeodictyoptera [41]. Where they occur elsewhere, for example in Peloridiid and Tingid Hemiptera, Cassidine and larval Elminthid and Silphid Coleoptera and nymphal

Heptageniid Ephemeroptera, paranotal lobes are probably secondary, and serve such functions as elimination of shadow in camouflage, protection against predators, or streamlining, any of which may also have been the initial function served by the primary lobes of Pterygote ancestors.

In these ancestors it is supposed that the lobes later came to serve for delaying descent, as parachutes and then as aerofoils, those on the meso- and metathorax becoming enlarged and modified until, with associated changes in the skeleton, musculature and

TABLE 11.1. **A summary of the classification employed:** emphasizing the groups discussed in detail.

Subclass PTERYGOTA
  Infraclass PALAEOPTERA
    *Superorder **Palaeodictyopteroidea**
      *Order *Palaeodictyoptera*
      *Order *Megasecoptera*
      *Order *Diaphanopterodea*
      *Order *Archodonata*
    Superorder **Ephemeropteroidea**
      Order *Ephemeroptera*
      *Family *Syntonopteridae* incertae sedis
    Superorder **Odonatoidea**
      *Order *Protodonata*
      Order *Odonata*
  Infraclass NEOPTERA
    Cohort **Polyneoptera**
      *Order *Protorthoptera*
      Order *Blattodea*
      *Family *Paoliidae* incertae sedis
      ?*Order *Caloneurodea*
      *Order *Miomoptera*
      Orders *Orthoptera, Plecoptera, Dermaptera* etc.
    Cohort **Paraneoptera**
      Orders *Psocoptera, Hemiptera, Thysanoptera* etc.
    Cohort **Oligoneoptera** (= **Endopterygota**)
      Orders *Megaloptera, Neuroptera, Mecoptera, Coleoptera, Trichoptera,*
        Lepidoptera, Diptera, Hymenoptera etc.

---

*Signifies wholly extinct

nervous supply of the thorax, true flapping flight was evolved. The prothoracic lobes became obsolete and ultimately disappeared, after persisting for a while in a few lines.

The prothoracic paranota appear at their most wing-like in the Palaeodictyoptera, and these insects became the corner-stone of Handlirsch's theory of the origin of insects from trilobites. Although this theory is entirely obsolete, the concept of the Palaeodictyoptera as ancestral Pterygota was for a while accepted by other authorities, and sometimes still is, although it has been untenable for twenty years.

Palaeodictyoptera, and the related Megasecoptera, share with Odonata and Ephemeroptera the inability to fold their wings back over the abdomen; and it was this above all which led Martynov [48] to distinguish these orders, as the division Palaeoptera, from the remainder of winged insects—the Neoptera. It was assumed, and still is by many authors, e.g. Carpenter [14], that the first winged insects had no power of wing-folding in this way, and that Neoptera arose later from Palaeopterous forms.

This, among other factors, points to the importance of the Ephemeroptera, as the more generalized of the two existing Palaeopterous orders. Lameere used the wings of Ephemeroptera and Palaeodictyoptera to derive a hypothetical insect wing archetype, and a general theory of wing venation. Tillyard [64] suggested that the venation of mayflies was in some respects the most primitive of all, and this view was developed by Edmunds and Traver, in a paper which made a rare attempt to relate the details of wing structure in both Recent and fossil forms to flight behaviour. In again attempting to derive an archetype wing from those of early mayflies these authors drew attention to the family Syntonopteridae, treated by Carpenter [8] as Palaeodictyoptera with marked Ephemeropteran characteristics.

Martynov [49] believed however that the wings of Palaeozoic Blattodea were closer to the primitive pattern than were those of any known Palaeoptera; and he proposed that Neoptera and Palaeoptera were both independently derived from unknown ancestors with Blattoid-like venation, rather than the one from the other. Rohdendorf [53] followed this view, but later [54], after the discovery of *Eopterum devonicum*, a Devonian fossil initially identified as an insect, which he believed to be Neopterous, suggested that the first winged insects were Neoptera, from which Palaeoptera later arose. *Eopterum* and the later-described *Eopteridium* have now been acknowledged by Rohdendorf [55] to be Crustacea; but in the meantime Sharov [57] had introduced a new candidate for near-archetypal status—the ancient family Paoliidae. These were believed by Sharov to represent a new order of Pterygota: the Protoptera, which arose from *Eopterum*-like forms, and were characterized by their wings being held diagonally backward when at rest; and which gave rise to both Palaeoptera and Neoptera.

The division of winged insects into Palaeoptera and Neoptera has never been universally accepted. In particular several authors have stressed the many differences between the Odonata, on one hand, and the Ephemeroptera and the wing-folding orders on the other. Lemche [46] derived the Odonata and the related Protodonata, independently of the remainder of the Pterygota, from Palaeodictyopterous stock; chiefly because the wing-pads of Odonata develop with the leading edge mesally situated, whereas in those of Ephemeroptera and exopterygote Neoptera the leading edge forms the lateral margin of the pad. At that time the wing-pads of nymphal Palaeodictyoptera were believed to project perpendicularly to the body axis, a position from which both the Odonatan condition and that of other Pterygota could theoretically have been derived—were it true. The many unique features of modern Odonata have recently been re-emphasized [47] with a further suggestion [50] that wings and flight arose completely independently in this order.

Seven groups, therefore, seem to merit particular attention in relation to the early evolution of insect flight. These are the Ephemeroptera; the Syntonopteridae; the Palaeodictyoptera, together with the related Megasecoptera and Diaphanopterodea; the Odonata and Protodonata; the Blattodea; the Paoliidae; and finally the Protorthoptera, in the sense of Carpenter [12], including a large and inadequately known array of Palaeozoic insects which evidently represent, with the Blattodea, Orthoptera, Plecoptera, and a handful of minor orders, the diverse products of the initial Polyneopterous radiation.

EPHEMEROPTERA

Ephemeroptera clearly have many plesiomorphic features—but it is not wholly clear which aspects of their wings and flight should be so described.

The most remarkable characteristic of the wings of mayflies is their fluting, which has two components. Firstly, there is present a complete series of main longitudinal veins, alternately convex and concave in position, and separate almost from the base of the wing. The media is usually considered [43] to have an anterior convex and a posterior concave component, respectively MA and MP (Fig. 11.2). Many authorities, from Adolph [1] to Hamilton [27], have taken this condition to be primitive; others [19, 49, 57] have disagreed.

Secondly, the fluted condition is extended to the wing margin, despite the branching of the main veins, by the presence of 'triads', wherein an intercalary vein of opposite aspect lies between each pair of branches. This condition occurs elsewhere: in Odonata, Protodonata, the Odonate-like Carboniferous wing *Campyloptera*, Acrididae (Orthoptera), and Osmylidae (Neuroptera); in each of which the intercalary veins are clearly secondary, and formed by the alignment of the edges of cells between the main veins.

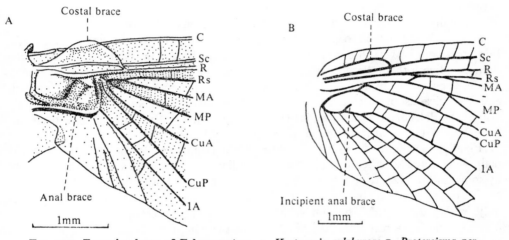

FIG. 11.2. **Fore-wing bases of Ephemeroptera.** A, *Heptagenia sulphurea;* B, *Protereisma permianum,* after Carpenter [4].

Most workers have assumed that intercalary veins in Ephemeroptera are likewise secondary structures, but Edmunds and Traver [21] supposed them to be primitive and a component of the archaic fluting which they believed to have provided the support for the earliest membranous wings, associating it with a hypothetical primitive vertical flight pattern, similar to the nuptial flight of the males of many existing mayfly species. These authors suggest that the limitation of fluting in Neoptera to the anterior and posterior regions of the wing, and the flattening of the area supported by the branches of Rs and M, was consequent on the development of the 'sculling' type of flight characteristic of Neopterous insects. This view is supported by Hamilton [27] and in part by the work of Brodskiy on the flight of *Ephemera vulgata*.

Brodskiy [3] has distinguished two kinds of flight in *Ephemera*, suggested as differing sharply in power: the vertical nuptial flight of males, with a high body angle and the cerci usually held together; and forward flight, in which the body angle is low and the cerci usually spread. He suggests [2] that the vertical nuptial dance may have been the oldest flight regime of the order.

Whether or not the fluted wings and vertical flight of some modern mayflies are primitive for the Pterygote insects as a whole, it is certain that many features of their flight apparatus are not. These include the reduction of the hind-wing, and its frequent linking to the fore-wing in flight; the associated disproportionate development of the mesothoracic flight motor; and the complex structure of the fore-wing base, which has never been properly described in functional terms (Fig. 11.2A). Here C, Sc and R are linked by an arcuate ridge, the costal brace. R arises from a convex domed plate adjacent to the second axillary sclerite, and immediately gives off Rs, which is united basally with MA, so that C, Sc, R, Rs and MA form a group of veins moving as a unit about the pleural wing process. Posteriorly the cubital and anal veins originate from, and are likewise linked into a functional unit by the anal brace—a convex, more or less sclerotized bar which curves anteriorly as a membranous ridge towards the point of separation of MA and R. This brace and ridge enclose an area in which the membrane is deeply but precisely folded, in a way which permits the cubito-anal complex and the region between C and MA some limited independent movement, the concave MP supporting the area between them.

More distally, the concave veins Sc, Rs and sometimes MP are typically interrupted by a line of breaks which form a one-way hinge, allowing the wing to bend downwards. Edmunds and Traver, in drawing attention to these breaks, associated them with the fluting and with their allegedly primitive flight pattern; but this may be an oversimplification, for a one-way hinge line occurs in the same place in the fore-wings of Recent and fossil cicadas, of Siricidae and several other groups of Hymenoptera, and apparently also of the Palaeozoic Blattinopsidae (Protorthoptera), all of which may be expected to have flight patterns far from that of the nuptial flight of mayflies.

Finally, the flight of modern Ephemeroptera is powered by indirect muscles, the main wing depressors being the dorsal longitudinal muscles. This situation is supposed by some authors [22, 25, 61] to be apomorphic to a condition where wing depression is carried out by the direct basalar and subalar muscles, with the dorsal longitudinal muscles only weakly developed. The first subalar muscle of Ephemeroptera is large, and apparently acts [2] in restricting the amplitude of the wing-beat. Brodskiy suggests, however, that this muscle may formerly have been the principal depressor, at a stage when the flight of Ephemeroptera resembled that of the nuptial flight today.

Palaeozoic Ephemeroptera show early stages in the development of several of these characters. Best known are members of the Leonardian superfamily Protereismatoidea [67] from the U.S.A., U.S.S.R. and Czechoslovakia [4, 6, 38, 64, 65, 66]. The body was slender, with long cerci and normally a caudal filament. The adult prothorax, in *Protereisma permianum* at least, shows paranotal expansions [6]. Meso- and metathorax are similar in size, as are fore- and hind-wings, the latter usually being slightly broader. There is no evidence of wing-coupling. At the wing base (Fig. 11.2B) the costal brace is partly developed, as a vein-like structure lying between C and Sc whose line is continued beyond Sc to R by an aligned cross-vein. MA is arched near the base in both wings, so that the anterior vein-grouping of modern forms is already established. The anal brace too is present in *Protereisma* and *Misthodotes* as a ridge linking the proximal parts of the anal veins, without involving CuA and CuP. The bases of the cubital and median veins are distinct, and the folded membranous area which lies between anal brace and R in modern mayflies is not developed in these forms, though it may be in their small contemporary *Eudoter delicatulus* [65].

Triads are present, and the intercalary veins appear to be branches of those adjacent. Despite this, Tillyard [64] had no doubt that they are secondary structures, principally because within a single species the same intercalary could arise impartially from the vein before or behind it, which is suggestive, but not conclusive.

Only one certain older mayfly is known: the Upper Carboniferous *Triplosoba pulchella* from Commentry in France, which has been carefully redescribed by Carpenter [11]. The wings are similar in shape to those of Protereismatoids. Their bases are poorly preserved, so that it is not clear whether costal and anal braces were present. M is fused with the common stem of R and Rs in the fore-wing, but lies adjacent to it in the hind-wing. Rs arises more distally.

Mayflies with fore- and hind-wings approximately equal in size (Mesephemeridae) continued into the Upper Jurassic, and so overlap in time with families of more modern form: Siphlonuridae are recorded from the Lower or Middle Jurassic of Siberia [67].

It is probable that the evolution of the modern mayfly facies, with enlarged mesonotum, reduced metanotum and hind-wing, and the structural refinements particularly of the fore-wing which have been described, accompanied considerable changes in the mode of flight of the group. There is evidence [64] that Permian forms had a subimaginal stage; and that adult Misthodotidae at least had functional mouthparts [66]. It seems therefore that the winged instars of Palaeozoic mayflies lasted longer than do those of existing species. Whether their flight was largely vertical as has been suggested [21], employing direct muscles for wing depression [2], one can only guess.

SYNTONOPTERIDAE

This family is known only from Moscovian beds at Mazon Creek, Illinois, U.S.A., and consists of one nearly complete insect, *Lithoneura lameerei*; three fore-wings belonging to two other species; and a hind-wing of a fourth. Carpenter [5, 8] and Richardson included the Syntonopteridae in the Palaeodictyoptera; but Carpenter [5] drew attention to their resemblence to Ephemeroptera, and suggested that they were related to the stock from which the latter evolved [8]. It is now certain that Ephemeroptera did not arise from Palaeodictyoptera. Edmunds and Traver pointed out that Syntonopteridae could well be true Ephemeroptera, though in some ways exceedingly primitive. This is supported in principle by Hamilton [28] although he follows Laurentiaux [45] in giving the family distinct ordinal status.

Lithoneura lameerei has a slender body; prothoracic paranotal lobes; and broad fore- and hind-wings, more or less equal in length, but the hind being much broader than the fore. The wings have complete peripheral fluting, with triads on Rs, MA, MP and CuA. M is free at the base from R and Rs, which were probably also independent of each other as in the hind-wing *Lithoneura mirifica* [8]; but MA is arched anteriorly near its base to fuse briefly with Rs. The common stem of M is basally fused with CuA in the hind-wing, but not in the fore. No anal or costal braces have been recorded.

The fore-wing of *Lithoneura lameerei* [Fig. 11.3A] is therefore probably the only described insect wing in which C, Sc, R, Rs, and the common stems of MA and MP, and CuA and CuP are wholly independent at the wing base; and hence closest to the archetypal insect wings envisaged by Edmunds and Traver and Hamilton [28]. It is accordingly possible to see *L. lameerei* and the Syntonopteridae as exceedingly archaic forms combining the most primitive features of Palaeodictyoptera and Ephemeroptera, and close to the ancestry of both—or merely as Carboniferous mayflies with unusually broad wings.

ODONATA AND PROTODONATA

The Odonata are a remarkable mixture of primitive and specialized characters. Matsuda lists 14, possibly 15 thoracic muscles exclusive to this order, of which five are employed in flight [62]; and three skeletal characters which he states to be otherwise present only in Machilidae or Japygidae. On this basis he postulates that Odonata arose, entirely independently from other Pterygota, from Machilid or Japygid-like ancestors. The isolation of the Odonata among Pterygotes has been emphasized by Lemche and Mackerras, the former stressing the curious position of the nymphal wing pads, the latter the unique features of sperm transfer and of venation—MP and CuA appear to be entirely absent. It has been shown [27] that the position of the pads can be explained by the superimposition of body flattening—a nymphal adaptation—on the adult condition, anticipated in the nymph, of high and sloping thoracic pleura. The venation can likewise be derived without undue difficulty from a more generalized pattern; and it seems

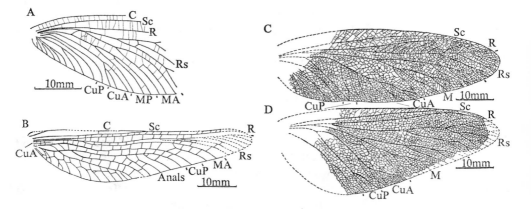

FIG. 11.3. **Some very early fossil insects** (Carboniferous: Namurian to Moscovian).
A, *Lithoneura lameerei* fore-wing, after Carpenter [5]: Syntonopteridae;
B, *Erasipteron larischi* fore-wing, after Kukalová [37]: Odonata;
C, *Zdenekia grandis* fore-wing, after Kukalová [35]: Paoliidae;
D, *Z. grandis* hind-wing, after Kukalová [35]: Paoliidae.

probable that the other unique characters of the group may be explicable in terms partly of the retention of plesiomorphic structures from the unknown ancestors of Pterygota, partly of the extreme specialization of these and other structures for an unusual mode of life involving aerial predation combined with remarkable powers of manœuvrable and sometimes rapid flight. The flight of Odonata has attracted considerable attention, most recently by the discovery [68] of their employment of 'flip' effects (p. 67) in hovering and manœuvring at low air-speeds.

The fossil record shows that these orders, and many of their special characteristics, evolved extremely early. *Erasipteron larischi* (Fig. 11.3B) from the early Bashkirian of Czechoslovakia, is a wing with clear Odonatoid characteristics, but retains a basal rudiment of CuA and lacks the arculus and nodus of typical Odonata. The Protodonata, which also lack arculus and nodus, and include the largest insects as well as many smaller forms, are known from several Upper Carboniferous and Permian localities, and continued into the Lias alongside true Odonata, which developed and radiated in lower Permian times. Most early Odonata resembled Zygoptera, which diversified from the

Permian onwards with Anisozygoptera and Anisoptera appearing in the Triassic and Jurassic respectively.

A number of changes in the venational support of Odonatoid wings can be followed in the evolution of the various lineages, some, like the development of the arculus and nodus, appearing to be structural refinements, others such as the number of branches and cells being evidently correlated with size. This notwithstanding, the overall shapes of the wings of Protodonata and Odonata conform with remarkable constancy to a few fairly distinct types, varying only in minor details of proportion, which seem to have arisen convergently several times. Figure 11.4 shows wing outlines, drawn from several groups, of Permian and recent species. In overall wing form many Protodonata (Fig. 11.4A, B) strikingly resemble the larger Anisoptera today; the Ditaxineuridae (Fig. 11.4D), representatives of a line which possibly did not survive the Permian, recall small extant Libelluloids; and the Protozygoptera (Fig. 11.4F) were an early stage of the Zygopterous

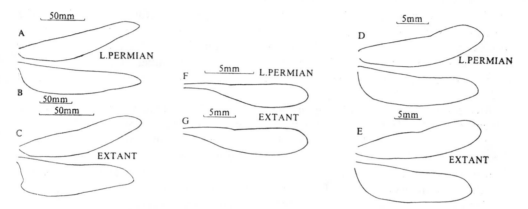

FIG. 11.4. **Wing-outlines of Protodonata (A, B) and Odonata (C–G).** A, *Typus gracilis* fore-wing B, *Megatypus schucherti* hind-wing; C, *Uropetala* sp. (Petaluridae); D, *Ditaxineura anomalostigmata* (Ditaxineuridae, S.O. Protanisoptera); E, *Nannophlebia risi* (Libellulidae); F, *Progoneura nobilis* (Progoneuridae, S.O. Protozygoptera); G, *Coenagrion dyeri* (Coenagrionidae).

Adapted: A, F from Carpenter [9]; B from Carpenter [7]; C, G from Fraser [26]; D after Carpenter, from Laurentiaux [45].

radiation which has produced many similar forms with nearly homonomous, petiolate wings up to the present day.

It is reasonable to guess that these similarities in wing shape may represent similar flight patterns, and probably modes of life. If this is so, it may be that the flight techniques of modern Odonata, including the use of 'flip' effects, are Palaeozoic in origin.

No account of the thorax of Palaeozoic Odonata has been published, but the thoracic structure of Meganeuridae (Protodonata) has been described [17, 18], and appears to correspond in most details with that of modern dragonflies, except that the pronotum is relatively larger, and the unique Odonatan condition of the mesothoracic episterna is not developed. The form of Odonatoid thoraxes appears to be far removed from that of other Pterygote insects but, as will be shown, resembles far more nearly the thorax of some Palaeodictyoptera, which in turn have features of more orthodox Pterygota.

Palaeontological evidence therefore suggests that Odonata and Protodonata represent an evolutionary side-line, diverging in the lower Carboniferous from a point somewhere near the biting ancestors of Palaeodictyoptera and Ephemeroptera, and specialized from the early Upper Carboniferous at least for aerial predation—Commentry Megan-euridae clearly show spiny dragonfly-like legs. It is probable that this specialization involved the development of patterns of flight which, as today, were in some cases powerful and fast, but always precise and manœuvrable to an extent which is unparalleled elsewhere among exopterygote insects, and achieved by quite different means from endopterygotes whose performance is comparable. Wing depression in extant Odonata is powered exclusively by direct musculature which, however, differs in detail from that of other insects. Modern forms show no evidence that indirect depression was ever employed.

PALAEODICTYOPTERA, MEGASECOPTERA, DIAPHANOPTERODEA AND ARCHODONATA

The discovery [44] that *Stenodictya*, the most frequently cited of all Palaeodictyoptera, had haustellate mouthparts disposed finally of the legend that this order included the most primitive known Pterygota, possibly ancestral to all the rest. Since then many new species have been described; and the crucial collection of Upper Carboniferous Palaeo-dictyoptera from Commentry, France, has been entirely restudied [39, 40, 41]. This is hence now the most fully known of all extinct insect orders, and appears as a successful and extremely varied Palaeozoic group.

Megasecoptera and Diaphanopterodea also have haustellate mouthparts, and are evidently closely related to Palaeodictyoptera; and Archodonata, which had only two wings, seem on venational grounds to belong in the same assemblage. All are limited to the upper Carboniferous and Permian; and they together represent a remarkable radia-tion, probably of liquid-feeding forms [14, 41], although it has been suggested [34, 60] that their mouthparts may have been employed in probing for spores and pollen.

The wing-venation of Palaeodictyoptera is in some respects perhaps less primitive than that of Permian mayflies. They retain a complete series of main longitudinal veins, alternately convex and concave; but they usually* lack intercalary veins, so that the distal areas supported by the branched veins are effectively flat. The fluting therefore only operates as a support in the basal part of the wing, and along the leading edge. R and Rs, MA and MP, and CuA and CuP separate near but not at the wing base. At the base of many wings of Palaeodictyoptera are visible structures which resemble those described in Ephemeroptera: a post-costal vein recalling in position and sometimes in appearance the incipient costal brace of Permian mayflies; and a thickened V-shaped ridge in the cubito-anal region [39, 40, 41] linking the cubital and anal veins into a functional unit, as does the mayfly anal brace.

The diversity of the Palaeodictyopteroid complex of orders is particularly evident in their wings. The Diaphanopterodea appear to have developed the ability to fold their wings over the abdomen, but Megasecoptera, Palaeodictyoptera and Archodonata are always preserved with the wings straight out, as in Anisopterous Odonata, and it seems clear that this was their position in life. Figure 11.5 shows to the same scale the outlines, as far as they are known, of a selection of Palaeodictyoptera and Megasecoptera. The area of the wings of Palaeodictyoptera is always large for the size of the body; but some forms (e.g. *Dictyoptilus*, Fig. 11.5H) have a very high aspect ratio, while others (e.g. *Lamproptilia*, Fig. 11.5J) have broad wings of almost Orthopteroid appearance, with an expanded anal area in the hind-wing. Narrow and broad-winged forms can occur within

* Intercalary veins occur in the related families Calvertiellidae and Archemegaptilidae.

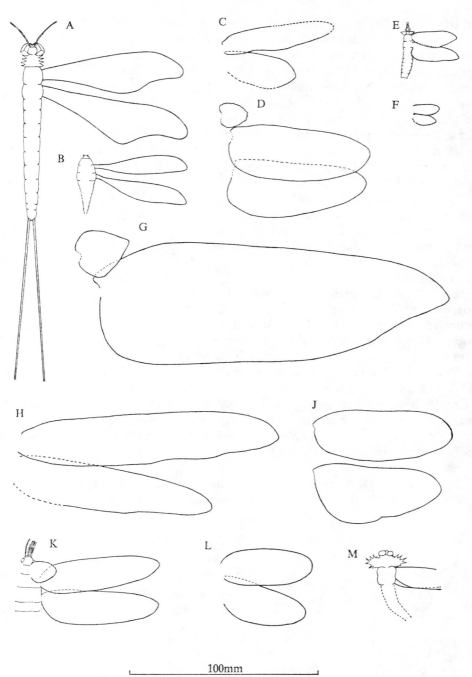

FIG. 11.5. **Outline drawings of Megasecoptera (A, B) and Palaeodictyoptera (C–M).** A, *Mischoptera nigra*, adapted from Carpenter [10]; B, *Sylvohymen sibiricus*, from Kukalová-Peck [42]; C, *Peromaptera filholi;* D, *Homoioptera woodwardi;* E, *Homaloneura lehmanni;* F, *Madera mamayi;* G, *Homoioptera gigantea;* H, *Dictyoptilus peromapteroides* (C, D, G, H, from Kukalová [40]); J, *Lamproptilia grandeuryi;* K, *Stenodictya spinosa,* from Kukalová [41]; L, *Mecynostomata dohrni* (E, J, L, from Kukalová [39]); M, *Notorachis wolfforum,* from Carpenter and Richardson [16].

a single family. In some species, e.g. *Homoioptera woodwardi* (Fig. 11.5D), fore- and hind-wings overlap considerably, and this appears to reach an extreme in *Notorachis wolfforum* [16] (Fig. 55.1M)—whose mode of flight is wholly obscure.

Palaeodictyopteroid insects were probably active flyers. Outstretched wings do not permit mobility on the ground, and except in still air present problems even to a stationary insect. Their varied wing forms suggest a similar variety of flight habits, many of which are unlikely to be particularly primitive. In general the wings of Palaeodictyoptera themselves suggest predominantly steady-state, fairly rapid forward flight; they show no marked weakening of the posterior part of the wing, such as is found in many Odonata and Protodonata, and is perhaps associated with the use of 'flip' effects (p. 67) in hovering and slow flight. Moreover the long cerci would only become effective as stabilizers in forward flight; Brodskiy [3] noted that the cerci of *Ephemera* were spread in forward flight, but closed when the insect was rising vertically.

Megasecoptera and Diaphanopterodea, however, may have been more versatile. Both groups show reduction in the number of vein branches. The Megasecoptera in particular are characterized by marked narrowing of the wing-bases, which in some forms are clearly petiolate, closely recalling Zygopterous Odonata. The small Asthenohymenid Diaphanopterodea show similar basal narrowing, and indeed are closely convergent in form and venation with some Megasecoptera. Moreover these forms and particularly the Protohymenid and Bardohymenid Megasecoptera (Fig. 11.5B) show concentration of the anterior veins along the leading edge with the rest of the wing far more weakly supported. It is quite possible that they were using 'flip' and even 'fling' mechanisms [68 and p. 65], so that hovering and a high degree of control of flight speed and direction may have been available to them. They too had long cerci, which would not hinder hovering, but would probably only function as stabilizers in forward flight. One can envisage these insects moving with ease and precision through vegetation, in the manner of many Zygoptera today, and in contrast to most Palaeodictyoptera, which would probably have been restricted to more open areas of the lowland forests in which many seem to have lived [41].

There are other similarities between Palaeodictyopteroid and Odonatoid species. I have recently examined the thoracic morphology of the Commentry Palaeodictyoptera

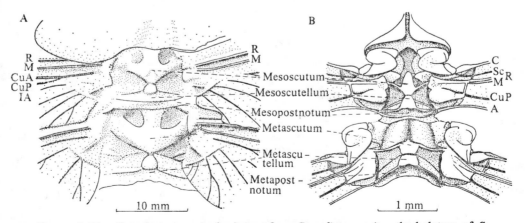

FIG. 11.6. **Pterothoracic terga and wing-bases of:** A, *Stenodictya agnita*—the holotype of *S. fayoli* Meunier (Palaeodictyoptera)—drawn from silicone rubber cast of concave impression, with reference to the original; B, *Pyrrhosoma nymphula* (Odonata—Coenagrionidae).

in the Musée d'Histoire Naturelle, Paris. Two specimens of *Stenodictya* species show clear tergal structure which resembles markedly the pterothorax of Odonata (Fig. 11.6). It has been shown [36] that the axillary structures of *Rochlingia* are reconcilable with those of Odonata. It seems probable that similar thoracic flight mechanisms were operating in the two groups of orders, although the Palaeodictyopteroids clearly lacked many of the Odonatoid thoracic and alar specializations for aerial predation. It is to be presumed that Palaeodictyoptera and their relatives were primarily plant-feeders, although the apparently raptorial fore-limbs of *Mischoptera nigra* (Megasecoptera, Fig. 11.5A) suggest a secondarily insectivorous habit.

The functions of the paranotal lobes are obscure, but it is likely that they played a positive role in the insects which retained them. In a few Palaeodictyoptera they were reduced and thickened, and in one modified into spines; but they were typically membranous, anteroposteriorly symmetrical, and supported by radiating veins which may or may not be homologous in detail with the wing veins. In *Lithomantis carbonarius* they are convex domes. Four possibilities are that they provided accessory lift generators, slots, ailerons, or air-brakes. Their position close to the body would minimize their value in the second and third of these functions, but given some muscular control there would seem to be no reason why they should not to some extent have served all four.

PAOLIIDAE AND PROTORTHOPTERA

Archetypal Palaeoptera would, it seems, have had wings most like those of archaic Ephemeropteroids and the most generalized Palaeodictyoptera, with a complete series of alternating convex and concave veins. In Neoptera, by contrast, the branches of Rs and M are neutral, supporting a flat membrane, although this may become secondarily fluted by the insertion of intercalary veins, as in Acridid Orthoptera. MP is either lost or has become unrecognizable because of the absence of relief.

Although ancestral Paraneoptera and Oligoneoptera were probably established by the end of the Carboniferous, the great majority of early Neoptera are Polyneoptera. Blattodea and true Orthoptera are present in the upper Carboniferous; but many Polyneopterous insects existed whose relationships are still unclear, and which are conveniently lumped into the Protorthoptera [12]. Most of these are only recorded as single wings, but some are more completely known. *Lemmatophora*, a lower Permian genus, has prothoracic paranotal lobes of almost Palaeodictyopteran type, but more broadly attached; and several others have prothoracic shields which may be derived from paranota.

Typically all hind-wings of Protorthoptera have an underfolding vannus. It is reasonable to assume that they flew like modern Orthoptera. Since the underfolding of the vannus is dependent on the folding of the wings and is unlikely to have preceded it, and as Oligoneoptera primitively had homonomous wings, the ancestral Neoptera presumably lacked an underfolding vannus. A very few Carboniferous Neoptera do— the Caloneurodea (which are clearly not archetypal Neoptera, though their relationships are uncertain)—and the Paoliidae.

The Paoliidae are among the oldest insects known, having so far been described only from the Namurian and lowest Bashkirian divisions of the Carboniferous. Their fore-wings alone would not justify their separation from Protorthoptera, but the absence of an underfolding hind-wing vannus would make them unique in that artificial order (Fig. 11.3C, D).

Sharov [57] has credited the Paoliidae with unique significance as sole representatives of an evolutionary grade, ancestral to both Palaeoptera and Neoptera, in which the wings were held obliquely backwards at rest. His evidence comes from a single fossil, *Sustaia impar* [35], in which fore- and hind-wings of one side, preserved incompletely without a body, appear to occupy relative positions which might result from being held obliquely. It is wholly inconclusive; and there is no reason to suppose Paoliidae to be more than early and primitive Neoptera. This does not, of course, disprove the one-time existence of Sharov's oblique-winged grade, which concept has much to recommend it.

BLATTODEA

Cockroaches are present in large numbers in most Upper Carboniferous insectiferous deposits, and are clearly an ancient group. Their interest to students of early flight mechanisms is twofold.

Firstly Martynov [49] believed the fore-wings of Palaeozoic Blattodea to be the most primitive known, principally because of the symmetry of the wing and its venation about the median longitudinal axis, which recalls the paranotal lobes of Palaeodictyoptera. He supposed that the first true wings would be of this kind, before aerodynamic factors brought about the concentration of the more anterior main veins close to the leading edge of the wing. The Blattoid hind-wing is similar, but has a vannus; although the vanni of some Palaeozoic forms were small and probably folded under as a single lobe, rather than fanwise.

The second point of particular interest concerns the flight motor. Tiegs showed that depression of cockroach wings is brought about largely by direct muscles, the first, second and third basalars and the subalar muscle; while the dorsal longitudinal and oblique dorsal muscles are weakly developed and seem to play only a minor role. He took this to be a plesiomorphic state, a view supported by Snodgrass and Ewer. Ewer and Nayler largely confirmed Tiegs' conclusions in a comparative study of the thoracic musculature of winged male and wingless *Deropeltis erythrocephala*. These authors emphasized, moreover, that the flight muscles were also leg muscles, and that their sequence of contraction was the same as in walking—but in phase, not antiphase [23]. Ewer suggested that insect wing flapping was initially powered by such bifunctional muscles, originally developed for walking—with the ingenious corollary that these may at first have operated in antiphase in flight, as in walking. This would account for the early reduction, or failure to develop, of the prothoracic wings: for the forces on the thorax in flight would have been seriously asymmetrical if six wings were functioning.

Whether or not this last proposal is true, the hypothesis that direct flight power by bifunctional leg muscles preceded an indirect motor is plausible. The Blattodea, which have altered remarkably little since their first appearance, may well have retained nearly unchanged a primitive musculature which was widespread among Carboniferous Palaeoptera and Neoptera. Against this must be considered the fact that existing cockroaches are weak fliers, and their soft thoracic terga show much less relief than most known Carboniferous insects, including apparently Blattodea. Some secondary reduction in the flight motor cannot be ruled out.

PALAEOZOIC NYMPHS AND THE ORIGIN OF FLIGHT

The last few years have seen a considerable advance in our knowledge of the nymphs of Palaeozoic insects, with descriptions of juveniles of Permian Protereismatoid Ephemeroptera [38], of Megasecoptera [15], and of Palaeodictyoptera [16, 58, 69]. Palaeozoic Blattoid nymphs and those of a variety of Protorthoptera have been known for some time; so that of the groups under consideration only Paoliidae, Syntonopteridae and the Odonatoids are unknown as nymphs in the Palaeozoic.

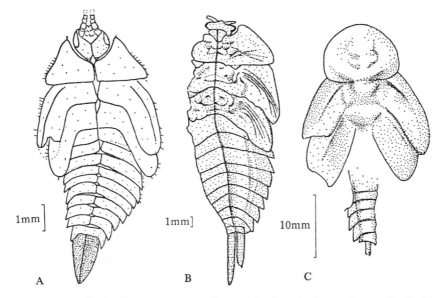

FIG. 11.7. **Upper Carboniferous nymphs.** A, *Euryptilodes horridus* (Protorthoptera?). (Redrawn after Sharov [56]); B, *Rochdalia parkeri* (Palaeodictyoptera). (Redrawn after Wootton [69]); C, *Leptoblattina exilis* (Blattodea).

These apart, it is clear from described material that the wing-pads of early Palaeoptera and Neoptera were directed posterolaterally, as in modern Plecoptera and Blattodea. Known nymphs of Megasecoptera and Protereismatoidea are slender, with prominent wing-pads, as are those of the Siberian and American Palaeodictyoptera discussed by Sharov, and Carpenter and Richardson. However, in some representatives of Palaeodictyoptera [69] and Protorthoptera [56] as well as Blattodea, the pads form part of a series of paranotal lobes extending from the prothorax to the end of the abdomen (Fig. 11.7). They show some resemblence, therefore, to what some authors have considered a plausible ancestor to Pterygota: an onisciform insect with a complete series of paranotal lobes on all segments.

Palaeodictyoptera at least were probably arboreal, and it is likely that the nymphs too fed in the vegetation above the ground. They may well have occupied similar situations to the immediate precursors of Pterygotes, and some of the same selective pressures may have operated to determine their form. The prothoracic lobes of several Palaeozoic nymphs project forward, apparently protecting the sides of the head. In Blattodea they are united in front to form a complete shield overlying the head. The paranota as a whole appear protective, perhaps allowing the insects to flatten against the substrate making it hard for predators, such as the many arachnids which were their contemporaries, to get a grip.

Sharov, drawing attention to deep grooves between the wing-pads and thoracic terga of Siberian nymphal Palaeodictyoptera, has suggested that the pads may have been movable as one might expect the paranota of pterygote ancestors to move [58]. Another Palaeodictyopteran nymph, *Idoptilus onisciforme*, has similar grooves, and also shows precociously incipient axillary plates [69]; but these characters are not conclusive evidence for pad movement. The nymphs of Eustheniidae and some other families of extant Plecoptera have similar grooves, but their pads are certainly not actively movable (Dr P. Zwick, personal communication).

A curious feature of *Idoptilus*, and the similar *Rochdalia parkeri* is the relief of the pronotum, which is similar, though on a smaller scale, to that of the pterothorax, implying similar internal structure. The nymphal pronotal paranota, however, show no sign of articulation, and the pronota of adult Palaeodictyoptera which I have examined show no obvious relief, although the paranotal lobes of these may well have been mobile.

CONCLUSIONS

The several theories of the origin of flight in insects are discussed by Professor Wigglesworth (pp. 255–268). The palaeontological evidence, slender as it is, seems to favour the existence in the Devonian of small—c. 10 mm? [24]—insects, whose initially protective thoracic paranota became enlarged in association with their value first as parachutes, in delaying descent; next as gliding surfaces; then as steering vanes, as they developed the ability slightly to pronate, supinate, elevate and depress the pads by the action of the pleural leg muscles; and ultimately, in the case of the meso and metathoracic lobes, as flapping aerofoils, powered mainly by bifunctional leg muscles including those operating in Blattodea and perhaps supplemented by others now retained only in Odonata. This is the classical view, and differs little from that of Forbes although following Hinton, Flower and Smart in stressing parachuting as a probable precursor to gliding.

As the wings and prothoracic paranota became membranous they were supported by branching blood-filled tubes, the veins, linked by a network of finer veins. It is probable that the membranes were from the first stiffened by fluting, with alternate vein stems occupying ridges and troughs in the basal parts of the wings. It may be that of known wings those of *Triplosoba* and the Syntonopteridae are closest to the ancestral pattern, but whether the fluting extended throughout the archetypal wing is still doubtful.

Equally uncertain is the position of the first wings, both in flight and at rest. Apparent homologues of the flexor muscle have been reported in both Ephemeroptera [50] and Odonata (Neville, personal communication); but it is hard to believe that the Neopterous condition was present from the first. Yet the resting position of the wings found in Palaeodictyopteroids and Anisopterous Odonata renders these latter insects so wholly dependent on aerial locomotion that it seems very improbable in early, weakly-flying insects. A more feasible alternative is that adopted by Ephemeroptera, whose wings are held above the body at rest, but partly remoted.

The position of the wing pads of Palaeozoic nymphs lends weight to Sharov's proposal [57] that insect wings were at first posterolaterally directed; but they may have been flexible enough at the base to allow promotion to the perpendicular position in flight.

From this condition true Neopterous folding could readily have evolved by refinement of the axillary sclerites and musculature.

Neoptera, with the development of the expanded anal hind-wing lobe may from the first have relied mainly on steady-state aerodynamics and flights of relatively short duration. In contrast several lines of Palaeoptera, notably among Odonatoids and Palaeodictyopteroids, seem to have exploited 'flip' mechanisms to a considerable extent, in association with finely-controlled manœuvrable flight, although flight patterns were diverse in both groups. In the Palaeodictyoptera prothoracic paranotal lobes were retained probably as accessory control surfaces. Their persistance in Protorthoptera and Blattodea seems rather to have been due to continued exploitation of their nymphal protective function.

In understanding the early evolution of insect flight information is badly needed in two main areas. The first is palaeontological: we have as yet no direct knowledge of Devonian and lower Carboniferous insects. The second is the significance in flight of some of the visible characters of relevant extinct and extant insects; particularly the mode of operation of fluted wings, and of homonomous wings, both linked and unlinked, in a variety of forms. Until these are resolved the possibly crucial importance of the recent and fossil Ephemeroptera is unknown. Their mosaic of archaic and specialized characters makes it particularly hard to decide whether we are seeing in the nuptial dance a reflection of the flight of the first winged insects, or a wholly apomorphic pattern.

ACKNOWLEDGEMENTS

I am indebted to Professor Torkel Weis-Fogh for valuable discussion, and for the opportunity to read the manuscript of his most recent paper [68] before its publication.

DISCUSSION

Dr N.P.Kristensen: The question whether the Palaeoptera or the Polyneoptera came first necessitates a consideration of whether the two are truly monophyletic in the sense of Hennig, which is hardly the case. It may well be that some sub-groups of the Palaeoptera are more closely related to the Neoptera than are others. Indications that the Odonata are closer to the Neoptera than are the Ephemeroptera are the presence in the Odonata of direct spiracular muscles, the absence of a sub-imago stage, and the direct innervation of the corpora allata from the corpora cardiaca, in contrast with the occurrence of the sub-imago and the isolation of the corpora allata from the corpora cardiaca in the Ephemeroptera.

Wootton: My working principle is to assume a group to be monophyletic unless one has evidence to the contrary—but it is perfectly conceivable that either the Palaeoptera or the Neoptera may be polyphyletic. Dr Kristensen's points about direct spiracular muscles and the innervation of the corpora allata are new to me and interesting, though such features could perhaps appear more than once in the course of evolution. The question of the sub-imago may be complicated by Sharov's claim that at least three distinct winged instars can be identified in a Permian Protorthopteran called *Atactophlebia termitoides*, which is of course a member of the Polyneoptera.

[Communicated]: I am myself less concerned with classification than with discovering what Palaeozoic insects were like and how they lived. With our present limited knowledge, Palaeoptera and Neoptera are useful abstractions; but they may well disappear as formal taxa when we know more about the nature of the first Pterygota.

REFERENCES

[1] ADOLPH, G.E. (1879). Über Insektenflügel. *Nova Acta Leopoldina*, **41** : 215–291.
[2] BRODSKIY, A.K. (1970). Organisation of the flight system of the mayfly *Ephemera vulgata* L. (Ephemeroptera). *Ent. Rev. Wash.*, **49** : 184–188.
[3] BRODSKIY, A.K. (1971). An experimental study of flight in the mayfly *Ephemera vulgata* L. (Ephemeroptera). *Ent. Rev. Wash.*, **50** : 25–29.
[4] CARPENTER, F.M. (1933). The Lower Permian insects of Kansas. Part 6. *Proc. Am. Acad. Arts Sci.*, **68** : 411–503, 1 pl.
[5] CARPENTER, F.M. (1938). Two Carboniferous insects from the vicinity of Mazon Creek, Illinois. *Am. J. Sci.*, **36** : 445–452.
[6] CARPENTER, F.M. (1939). The Lower Permian Insects of Kansas. Part 8. *Proc. Am. Acad. Arts Sci.*, **73** : 29–70, 2 pl.
[7] CARPENTER, F.M. (1943). The Lower Permian insects of Kansas. Part 9. *Proc. Am. Acad. Arts Sci.*, **75** : 55–84.
[8] CARPENTER, F.M. (1943). Carboniferous insects from the vicinity of Mazon Creek, Illinois. *Illinois State Museum, Scientific Papers*, **3** : 9–20, 4 pl.
[9] CARPENTER, F.M. (1947). Lower Permian insects from Oklahoma. Part 1. *Proc. Am. Acad. Arts Sci.*, **76** : 25–54.
[10] CARPENTER, F.M. (1951). Studies on Carboniferous insects from Commentry, France. Part II. The Megasecoptera. *J. Paleont.*, **25** : 336–355.
[11] CARPENTER, F.M. (1963). Studies on Carboniferous insects from Commentry, France: Part IV. The Genus *Triplosoba*. *Psyche, Camb.*, **70** : 120–128.
[12] CARPENTER, F.M. (1966). The Lower Permian insects of Kansas, Part II. *Psyche, Camb.*, **73** : 46–88.
[13] CARPENTER, F.M. (1970). Fossil insects from New Mexico. *Psyche, Camb.*, **77** : 400–412.
[14] CARPENTER, F.M. (1971). Adaptations among Paleozoic insects. *Proc. N. Am. Palaeont. Convention, Chicago*, 1969, **1** : 1236–1251.
[15] CARPENTER, F.M. & RICHARDSON, E.S.Jr. (1968). Megasecopterous nymphs in Pennsylvanian concretions from Illinois. *Psyche, Camb.*, **75** : 295–309.
[16] CARPENTER, F.M. & RICHARDSON, E.S.Jr. (1971). Additional insects in Pennsylvanian concretions from Illinois. *Psyche, Camb.*, **78** : 267–295.
[17] CARPENTIER, F. (1953). Sur une figure récente de *Meganeurula gracilipes* Handl. (Protodonate de Houiller). *Bull. Annls. Soc. r. ent. Belg.*, **89** : 183–184.
[18] CARPENTIER, F. & LEJEUNE-CARPENTIER, M. (1952). Structure de thorax des meganeurides (Protodonates). *Trans. 9th int. Congr. Ent.*, **1** : 161–164.
[19] COMSTOCK, J.H. (1918). *The wings of insects*. Comstock Publishing Co., Ithaca, New York: 430 pp.
[20] CROWSON, R.A., SMART, J. & WOOTTON, R.J. (1967). Insecta. In [31] : 508–528.
[21] EDMUNDS, G.F.Jr. & TRAVER, J.R. (1954). The flight mechanics and evolution of the wings of Ephemeroptera, with notes on the archetype insect wing. *J. Wash. Acad. Sci.*, **44** : 390–400.
[22] EWER, D.W. (1963). On insect flight. *J. ent. Soc. sth Afr.*, **26** : 3–13.
[23] EWER, D.W. & NAYLER, L.S. (1967). The pterothoracic musculature of *Deropeltis erythrocephala*, a cockroach with a wingless female, and the origin of wing movements in insects. *J. ent. Soc. sth Afr.*, **30** : 18–33.
[24] FLOWER, J.W. (1964). On the origin of flight in insects. *J. Insect Physiol.*, **10** : 81–88.
[25] FORBES, W.T.M. (1943). The origin of wings and venational types in insects. *Am. Midl. Nat.*, **29** : 381–405.
[26] FRASER, F.C. (1957). *A reclassification of the Order Odonata*. Royal Zoological Society of New South Wales, Sydney: pp. 1–133.
[27] HAMILTON, K.G.A. (1971). The insect wing, Part I. Origin and development of wings from notal lobes. *J. Kans. ent. Soc.*, **44** : 421–433.

[28] HAMILTON, K.G.A. (1972). The insect wing, Part III. Venation of the orders. *J. Kans. ent. Soc.* 45 : 145–162.

[29] HANDLIRSCH, A. (1908). *Die fossilen Insekten und die Phylogenie der rezenten Formen.* Engelmann, Leipzig: Pp. i–x, 1–1430; Pl. I–LI.

[30] HANDLIRSCH, A. (1925). Palaeontologie. Phylogenie oder Stammesgeschichte. In *Schröder's Handbuch der Entomologie*, 3 : 117–376. Fischer, Jena.

[31] HARLAND, W.B., HOLLAND, C.H., HOUSE, M.R., HUGHES, N.F., REYNOLDS, A.B., RUDWICK, M.J.S., SATTERTHWAITE, G.E., TARLO, L.B.H. & WILLEY, E.C. (eds) (1967). *The fossil record.* Geological Society of London, London : 827 pp.

[32] HATCH, G. (1966). Structure and mechanics of the dragonfly pterothorax. *Ann. ent. Soc. Am.*, 59 : 702–714.

[33] HINTON, H.E. (1963). The origin of flight in insects. *Proc. R. ent. Soc. Lond. (C)*, 28 : 24–25.

[34] HUGHES, N.F. & SMART, J. (1967). Plant-insect relationships in Palaeozoic and later time. In [31] : 107–117.

[35] KUKALOVÁ, J. (1958). Paoliidae HANDLIRSCH (Insecta-Protorthoptera) aus dem Oberschlesischen steinkohlenbecken. *Geologie*, 7 : 935–959.

[36] KUKALOVÁ, J. (1960). New Palaeodictyoptera (Insecta) of the Carboniferous and Permian of Czechoslovakia. *Sb. ústřed. Úst. geol.*, 25 : 239–251.

[37] KUKALOVÁ, J. (1964). To the morphology of the oldest known dragonfly *Erasipteron larischi* Pruvost, 1933 *Vest. ústřed Úst. geol.*, 39 : 463–464, 1 pl.

[38] KUKALOVÁ, J. (1968). Permian mayfly nymphs. *Psyche, Camb.*, 75 : 310–327.

[39] KUKALOVÁ, J. (1969). Revisional study of the Order Palaeodictyoptera in the Upper Carboniferous shales of Commentry, France, Part I. *Psyche, Camb.*, 76 : 163–215.

[40] KUKALOVÁ, J. (1969). Revisional study of the Order Palaeodictoyoptera in the Upper Carboniferous shales of Commentry, France. Part II. *Psyche, Camb.*, 76 : 438–486.

[41] KUKALOVÁ, J. (1970). Revisional study of the Order Palaeodictyoptera in the Upper Carboniferous shales of Commentry, France. Part III. *Psyche, Camb.*, 77 : 1–44.

[42] KUKALOVÁ-PECK, J. (1972). Unusual structures in the Palaeozoic insect Orders Megasecoptera and Palaeodictyoptera, with a description of a new family. *Psyche, Camb.*, 79 : 243–268.

[43] LAMEERE, A. (1923). On the wing-venation of insects. *Psyche, Camb.*, 30 : 123–132. (Trans. A.M. Brues.)

[44] LAURENTIAUX, D. (1952). Découverte d'un rostre chez *Stenodictya lobata* Brgt. (Paléodictyoptère sténodictyide) et le problème des Protohémiptères. *Bull. soc. géol. Fr.* (6), 2 : 233–247.

[45] LAURENTIAUX, D. (1953). Classe des insectes. In Piveteau, J. (ed.) *Traité de Paléontologie:* 397–527. Masson et Cie. Paris.

[46] LEMCHE, H. (1940). The origin of winged insects. *Viddensk. Meddr. dansk naturh. Foren.*, 104 : 127–168.

[47] MACKERRAS, I.M. (1967). Grades in the evolution and classification of insects. *J. Aust. ent. Soc.*, 6 : 3–11.

[48] MARTYNOV, A.V. (1925). Über zwei Grundtypen der Flügel bei den Insekten und ihre Evolution. *Z. Morph. ökol. Tiere*, 4 : 465–501.

[49] MARTYNOV, A.V. (1938). Outlines of the geological history and phylogeny of insectan orders (Pterygota) (In Russian). *Trudy paleont. Inst.*, 7 : 1–149.

[50] MATSUDA, R. (1970). Morphology and evolution of the insect thorax. *Mem. ent. Soc. Can.*, 76 : 3–431.

[51] NEVILLE, A.C. (1960). Aspects of flight mechanics in anisopterous dragonflies. *J. Exp. Biol.*, 37 : 631–656.

[52] RICHARDSON, E.S.Jr. (1956). Pennsylvanian invertebrates from the Mazon Creek area, Illinois. Insects. *Fieldiana, Geol.*, 12 : 15–56.

[53] ROHDENDORF, B.B. (1949). Evolution and classification of the flight apparatuses of insects. (In Russian). *Trudy paleont. Inst.*, 16 : 3–176.

[54] ROHDENDORF, B.B. (1961). The description of the first winged insect from Devonian beds of the Timan (Insecta, Pterygota.) (In Russian). *Ent. Obozr.*, 40 : 485–489.

[55] ROHDENDORF, B.B. (1972). The Devonian Eopterida are not Insecta but Crustacea Eumalacostraca. (In Russian). *Ent. Obozr.*, 51 : 96–97.

[56] SHAROV, A.G. (1961). Order Protoblattodea. In Rohdendorf, B.B. *et al.* Palaeozoic insects of the Kuznetsk Basin. (In Russian). *Trudy paleont. Inst.*, 85 : 157–164.

[57] SHAROV, A.G. (1966). *Basic arthropodan stock*. Pergamon Press, London and New York : 271 pp.

[58] SHAROV, A.G. (1971). Habitat and relationships of Palaeodictyoptera. *Proc. 13th int. Congr. Ent., Moscow* 1968, 1 : 300–303.

[59] SMART, J. (1971). Palaeoecological factors affecting the origin of insects. *Proc. 13th Int. Congr. Ent., Moscow* 1968, 1 : 304–306.

[60] SMART, J. & HUGHES, N.F. (1972). The insect and the plant: progressive palaeoecological integration. In van Emden, H.F. (ed.) *Insect/plant relations. Symp. R. ent. Soc. Lond.*, 6 : 143–155.

[61] SNODGRASS, R.E. (1958). Evolution of arthropod mechanisms. *Smithson. misc. Collns.*, 138 : 1–77.

[62] TANNERT, W. (1958). Die Flügelgelenkung bei Odonaten *Dt. ent. Z.*, 37 : 394–455.

[63] TIEGS, O.W. (1955). The flight muscles of insects—their anatomy and histology; with some observations on the structure of striated muscle in general. *Philos. Trans.* (B), 238 : 221–347.

[64] TILLYARD, R.J. (1932). Kansas Permian Insects. Part 15. The Order Plectoptera. *Am. J. Sci.*, 23 : 97–272.

[65] TILLYARD, R.J. (1936). Kansas Permian Insects. Part 16. The Order Plectoptera (contd). *Am. J. Sci.*, 32 : 435–453.

[66] TSHERNOVA, O.A. (1965). Some fossil mayflies (Ephemeroptera, Misthodotidae) found in Permian deposits in the Ural mountains. *Ent. Rev. Wash.*, 44 : 202–207.

[67] TSHERNOVA, O.A. (1970). On the classification of fossil and Recent Ephemeroptera. *Ent. Rev. Wash.*, 49 : 71–81 (Translation of Entomologicheskoe Obozhrenie).

[68] WEIS-FOGH, T. (1973). Quick estimates of flight fitness in hovering animals, including novel mechanisms for lift production. *J. exp. Biol.*, 59 : 169–230.

[69] WOOTTON, R. J. (1972). Nymphs of Palaeodictyoptera (Insecta) from the Westphalian of England. *Palaeontology*, 15 : 662–675, 1 pl.

12 · The evolution of insect flight

V.B.WIGGLESWORTH

Department of Zoology, University of Cambridge

The evolution of flight throughout the past history of insects is a vast subject; I shall confine myself to two problems: (i) the nature and source of the thoracic appendages which gave rise to wings, and (ii) the selective forces which first led to the evolution of flight. The developments with which we are concerned must have occurred at latest in the mid-Devonian, perhaps much earlier; and we have no fossil record of the insects of those times. Any conclusions to be drawn must be based upon our present knowledge of morphology, physics, physiology and ecology, against the background of what we know about the climatic conditions of the period. This communication is therefore an exercise in uniformitarianism.

HISTORICAL SURVEY

The earliest suggestion for the origin of wings was put forward by Lorenz Oken [33] in his *Naturphilosophie* published in 1811, nearly fifty years before the *Origin of species*. Oken does not discuss questions; he sets down *ex cathedra* statements; he states that the wings are no new or unknown organ, but that in the air-breathing 'Ancyliozoa' the external branchial lamellae, to which the tracheae ran, hardened and were converted into wings. Gegenbaur [14] in his *Comparative Anatomy* likewise took up the idea that the wings of insects evolved from the tracheal gills of an aquatic ancestor, and that the spiracles were first formed as part of this adaptation to an aerial life.

Fritz Müller [32] observed that the first instar larva of *Kalotermes* had prominent wing-like expansions from the terga of the prothorax and the mesothorax, and in later instars from the metathorax, and that these expansions (which appeared to be the precursors of wings) had at first no tracheal supply. From the principle of recapitulation he argued that such structures, devoid of tracheae, could not have been derived from gills and thus claimed to have refuted Gegenbaur's theory!

Opinion was divided as between these two ideas. The description of prothoracic expansions in fossil Palaeodictyoptera by Woodward [57] in a paper which makes no reference at all to the origin of insect wings, swung many authors in favour of Müller's idea. Packard, who had originally accepted the gill theory, conceived the idea that the ancestor of winged insects had legs formed for leaping and that tergal expansions could have served initially as gliders; and this view, put forward in his text-book of 1898 [34], drew more adherents. But Woodworth [58] writing early in this century could still refer to the gill theory as that most widely accepted.

Then in 1916, Crampton [9] wrote a highly judicial paper on the subject aiming at an 'unbiased assessment' of all the evidence with a view to arriving at a 'working basis for further study'. He set out 21 items of evidence in support of the gill theory, and 21 items of evidence supporting what he called the 'paranotal' theory; and he pronounces judgement unequivocally in favour of the paranotal theory. Snodgrass, in all his writings from 1927 to 1963, takes the same view. He dismisses the gill theory out of hand as being 'too roundabout' a process, or on the grounds that there is no evidence for an aquatic ancestor for winged insects.

At the present time there is almost universal support for the paranotal theory. Most authors suppose that the outgrowths were favoured in selection by their ability to serve as planes utilized in gliding from high vegetation to the ground. Hinton [20] starts further back and considers the first step towards flight to have been attitude control of legs and body to ensure that the falling insect landed on its feet and could make a quick escape from predators, notably spiders. Enlargement of lateral outgrowths would be a later step making true gliding possible. Flower [12], who suggests that gliding could have begun in small insects of 1 cm length even before lateral planes appeared, points out that diminutive paranotal lobes at their earliest appearance could have played an important role in attitude control.

Then, as visualized also by Pringle [36], hinges to the planes and muscular control of their inclination would enable large adult insects to exercise control over the direction of gliding; and, finally, relatively slight alterations in the muscular system would provide for slow flapping flight. That is probably the most widely accepted conception of the evolution of wings and flight [13, 16, 43].

RE-EXAMINATION OF THE GILL THEORY

In going through the literature I gradually came to feel that the case for the gill theory had been rather summarily dismissed. I did not find Crampton's assessment of the evidence pro and con the gill theory entirely convincing; and I was impressed by Woodworth's thoughtful paper which took twenty years in the writing. Woodworth [58] points out that gills are soft, permeable, floppy structures, singularly unsuited for conversion into wings operating in dry air. Taking up a point made by Lang [25] he argues that it is more probable that wings should have been formed from some structure accessory to the gill; and he develops a case for the mobile gill-plates on the abdomen of mayfly larvae being homologous with wings.

As Heymons [19], Börner [4] and others have shown, the gills of *Ephemera* and *Chloëon* are clearly derived from the segmental appendages of the abdomen. The abdominal appendages in the Apterygota are represented by the styli (Fig. 12.1A). The styli are not in fact vestigial legs; they are the exopodites arising from rounded limb bases or coxae, the main appendages, or telopodites, of which have been lost. In the larvae of mayflies these abdominal appendages undergo the most varied development. In Palaeozoic mayflies the gills are usually tapering styliform structures. Figure 12.1B, C shows their morphology in *Ephemera*, where they are feathered structures. It is in the free swimming forms (*Chloëon*, *Baëtis* etc.) that they become most wing-like. They migrate laterally and may even come to be articulated to the dorsum of the abdomen (Fig. 12.2).

Indeed as Dürken [10] has shown in *Ephemerella*, at their earliest appearance they may be tergal in position.

The intrinsic muscles attached to the base of the styli of Apterygota become the muscles which move the gill lamellae. The vibrating lamellae in these forms are not as a rule themselves the gills: they may be only weakly tracheated; they are often stiffened by veins, commonly independent of the tracheal supply; the leading margin is often stiff and furnished with hairs and spicules, the posterior margin more flexible. The respiratory components take the form of bundles of fine tracheated filaments; or they may be simply richly tracheated areas of the abdominal surface; and the function of the beating

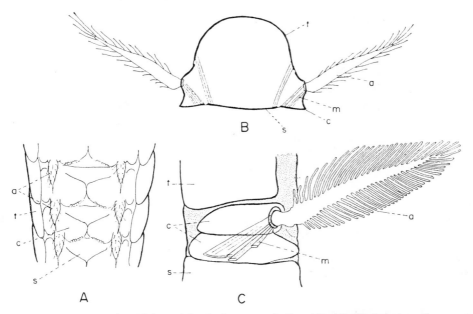

FIG. 12.1. A, ventral view of three abdominal segments in *Petrobius* (after Delaney); B, diagrammatic cross-section through abdomen of *Ephemera* showing articulation of the gill appendages; C, surface view of the same showing gill muscles. (Both after Snodgrass). *a*, styli in A, and homologous gill appendages in B and C; *c*, coxa or 'limb base'; *m*, muscles (homologous with muscles of styli) actuating the gill appendages; *s*, sternum; *t*, tergum

gill-plates is to drive currents of water over these surfaces. The sudden darts through the water that are made by *Chloëon* when disturbed are brought about by rectal jet-propulsion as in dragonfly larvae; but their more gentle swimming is done by the combined action of the gill-plates and the feathered tail filaments [54].

One of the most remarkable wing-like modifications is that seen in *Caenis* in which the functional gills are limited to the dorsum of the third, fourth, fifth and sixth segments of the abdomen (Fig. 12.2A); as Eastham [11] showed they set up transverse currents across the abdomen, from right to left or from left to right. On the second segment gills are wanting, but there is a pair of articulated gill-plates, called 'pseudo-elytra' by Eastham, which extend backwards to cover the gills of the next four segments (Fig. 12.2B). *Caenis* is a bottom-living species, resting in the surface mud; it does not vibrate its pseudo-elytra but raises and lowers them from time to time to regulate the entry of fresh water to the gill chamber. In passing it may be pointed out that in *Baëtisca* larvae the gills on the dorsum of the anterior abdominal segments are covered by an extension of the hind-wing lobes which reach almost to the 4th abdominal segment; and in the strange larva of *Prosopistoma* the fore-wings extend backward over the abdomen to form a closed gill chamber.

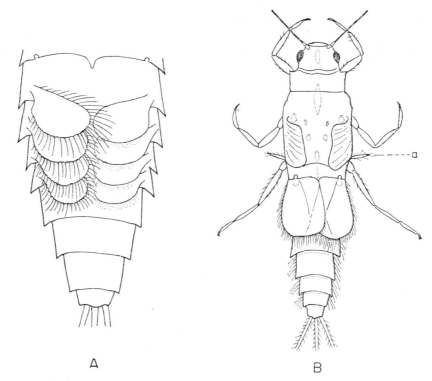

A   B

FIG. 12.2. **Gills in** *Caenis* **larva** (after Eastham). A, abdomen after removal of the 'pseudo-elytra' of segment 2, showing the mobile gill-plates on segments 3–6. B, *Caenis* larva with the wing-like gill covers of segment 2 in situ, concealing the gill-plates. *a*, the styliform vestigial gill of abdominal segment 1.

All these wing-like gill-plates are confined to the abdomen. Mayfly larvae never develop gill-plates on the thorax. In *Oligoneuria* there are well developed bundles of branchial tubes below the head, arising from the basal segment of the maxilla, their origin being normally concealed by the labium; and in *Jolia* larvae, besides the abdominal gills, there is a cluster arising from the basal segment of each maxilla and another cluster on the thorax, arising from the ventral cuticle adjacent to the basal segment of the pro-thoracic legs [53]. The *Caenis* larva has a little tapering structure, resembling an un-changed stylus, projecting laterally below the wing-lobe (Fig. 12.2B). But this is in fact the vestige of the gill belonging to the first abdominal segment and it is articulated to the posterolateral angle of that segment.

It is surprising that the vibratile gill-plates of mayfly larvae should be confined to the abdomen. One would have expected them to be conspicuously developed on the thorax, which is the main locomotor centre, with the major ganglia and with abundant muscles. That is where the gills are always best developed in Crustacea. The explanation, of course, is that the thorax is already committed to the development of wings. Figure 12.3 is a view of the thorax in the *Chloëon* larva based on Börner [4]. The modified limb bases which carry the mobile gill-plates are seen in the abdomen. In the thorax the limb base has provided the pleural stiffening for the thoracic box which extends up to the attachment of the larval wing-lobe; there are no gill-plates to be seen.

If there is any real relation between the wing-like gill-plates of the abdomen and the wings of the thorax it must be sought in the non-existent ancestors of the Pterygote insects. Fossils of these have not yet been found; the best we can do is to look at the Thysanura which are the nearest relatives we have of the early Pterygota. *Petrobius* (*Machilis*) *maritimus* for example, is something like a living fossil; it clearly missed the boat, and has been sitting around on the rocks by the sea-shore since Devonian times.

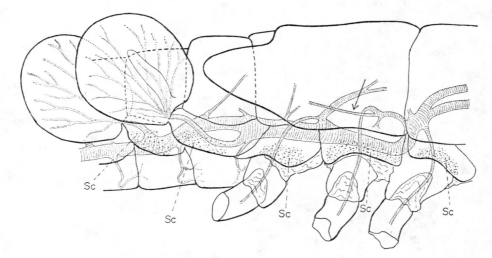

FIG. 12.3. **Lateral view of thorax and anterior abdomen of** *Chloëon* **larva** (greatly simplified after Börner) showing (i) that the sclerites to which the gill-plates are articulated are serially homologous with the subcoxal sclerites of the thorax (*Sc*); (ii) that the tracheae supplying the gills correspond with the tracheae of the legs; (iii) that the single trachea which supplies the wing-lobe (marked by the arrow) arises from the trachea of the mesothoracic leg (metathoracic wings are absent in *Chloëon*).

Figure 12.4A shows *Machilis* in side view. On most of the abdominal segments it has styli, such as we have seen gave rise to the oscillating gill-plates of mayflies; and it also has serially homologous styli of the same character on the meso- and the metathorax— the segments on which the wings of Pterygota develop. Is it possible that we are contemplating in *Machilis* the rudiments of insect wings?

The styli of the thorax in *Machilis* arise from the base of the coxa (Fig. 12.4B). They have played an important part in the arguments of morphologists [17]. If they had not survived in *Machilis* the styli of the abdomen would certainly have been regarded as vestigial limbs (telopodites). As it is, these versatile structures have been identified as the exopodites of serial appendages of which the limb base remains but the main limb, or telopodite, has been lost [17, 48, 49]. Exites and endites of the proximal limb segments of arthropods, from trilobites to insects, have an extraordinary history. They have furnished most of the remarkable tools of the phylum: the mandibles and the other mouth parts, the gills of trilobites and Crustacea, the swimming and grasping appendages, the gill-plates of Ephemeroptera, the styliform gills of Neuroptera, Megaloptera and Coleoptera (*Gyrinus*), the specialized parts of the external genitalia and so on. Their evolutionary potential is comparable with that of the vertebrate limb.

There seems no reason why, in a primary Apterygote adapting to aquatic life, the process of gill formation which has taken place in the abdomen of Ephemeroptera should not have occurred. Such an insect would have suffered no phylogenetic inhibitions about developing gill-plates from the styli of the meso- and metathorax. Although morphologists are still unable to agree about the details, the 'limb base' of Apterygota is regarded as the main source of the lateral sclerites which stiffen the thoracic box to meet the needs of walking and flying. This stiffening would be expected to begin to some extent in an aquatic insect with thoracic gill-plates and thoracic legs—although, of course, the more

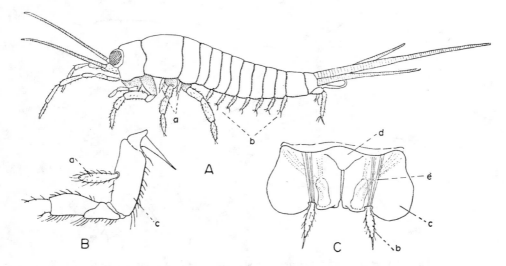

FIG. 12.4. A, *Machilis* in side view showing styli in most abdominal segments and in the meso- and metathorax; B, detail of mesothoracic leg showing stylus arising from the coxa; C, abdominal segment 2 showing the muscles which move the styli (modified after Snodgrass.). *a*, thoracic styli; *b*, abdominal styli; *c*, coxa; *d*, sternum; *e*, muscles of styli.

substantial changes would take place in the flying insect. In the course of this process, if the changes in the abdomen of mayflies are any guide, the gill-plates would have migrated towards the dorsum and their articulated bases would have become approximated to the tergum. It is of interest to note that certain of the Symphylans also carry styli on the base of the second and third pairs of legs, and these styli have become incorporated into the cuticular wall of the segment [46].

Although the pleuron is believed to be derived from the limb base, including (I would suggest) that part of the basal region of the coxa carrying the wing rudiment, there is no sign of the recapitulation of this process of pleuron formation during the embryonic development of insects. It is therefore unlikely that we should get evidence in this way of the coxal origin of the wings. The most we can hope for is to see the point of origin of the wing rudiment. Tower [52] looked carefully into this question and found that in Hemimetabola (Hemiptera and Orthoptera) as well as in Coleoptera and Lepidoptera, the imaginal discs of the wings invariably appear at about the level of the spiracles. He could not find the slightest evidence for the development of wings by extension from the terga. In *Periplaneta* he showed that the epidermal thickenings which form the wing-lobes arise on the sides of the thorax at exactly the same sites as in Coleoptera and Lepidoptera—'but the wings soon migrate dorsally and posteriorly, and when they become external appear as direct backward prolongations of the tergum'.

The rather weak muscles which actuate the abdominal gill-plates of mayfly larvae are derived from the muscles of the styli and they retain their attachment to the coxal or subcoxal limb base. The corresponding muscles would doubtless have been retained by thoracic gill-plates (although the thoracic styli of existing Thysanura, like the styli of Symphylans, do not in fact retain their muscles). Such a muscular supply would explain why the two most important direct flight muscles in primitive insects have their origin in the coxa and run all the way through the thorax to be inserted into the basal sclerites of the wing (Fig. 12.5). Börner [4] attempted to identify the coxal muscles that corresponded to the muscles of the gill-plates. That was in order to establish the homology

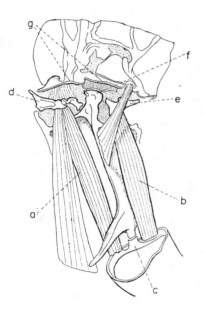

FIG. 12.5. **Flight muscles of a grasshopper** from the mesal aspect (modified after Snodgrass). Two important direct wing muscles, the anterior extensor (*a*) and the posterior extensor (*b*), arise inside the rim of the coxa (*c*) and are inserted respectively into the basalar sclerite (*d*) and the subalar sclerite (*e*), with the pleuron process (*f*) between, and are thus connected to the base of the wing seen in cross-section (*g*).

between the gill bases and the thoracic coxae. For the same purpose he pointed out that the tracheae supplying the gills correspond with the tracheae of the thoracic legs. Börner was opposed to the gill theory of the origin of wings and refers to this scornfully as that former *Schulvergleich*, but it is interesting to note that his figure shows the single trachea which supplies the wing lobe in the nymph of *Chloëon* arising from the leg trachea as it runs towards the coxa (see Fig. 12.4.)

It is likely that thoracic gill-plates would have been supplied by more powerful muscles and could well have been much larger structures than the abdominal gills of existing mayflies. Lubbock [28], who accepted the branchial theory of the origin of wings, did not consider that the branchiae of *Chloëon* are used for swimming (contrary to Wesenberg-Lund), but he claimed that in earlier forms thoracic branchiae might well have been used for locomotion under water. As Lubbock pointed out, the Chalcid wasp *Polynema natans* (*Cataphractus cinctus*), as he himself had discovered [27, 31], rows its way under water by means of its wings; and although he did not for a moment suggest that this had any direct connection with the origin of flight, it did serve to show that the same wings could be used for locomotion in the air or under water.

It may be well to point out again that the branchial theory of the origin of wings as here developed does not require that insects evolved from an aquatic ancestor. It is true that lungs evolved in aquatic vertebrates, but it is generally believed that the tracheal system, which probably evolved independently in arachnids, myriopods and insects, first developed in terrestrial forms; and that the common ancestor of Myriopoda and Insecta was a land animal [18]. Moreover, as Sharov [46] points out, the hexapod structure is characteristic of inhabitants of dry land, and among aquatic arthropods it is found only in insects that have become secondarily aquatic [30].

THE ORIGIN OF FLIGHT

Why should such well adapted insects leave the water? It is reasonable to suppose that their breeding places might dry up. If the drying out of the aquatic habitats were a frequent occurrence in the climatic conditions of the time, an aquatic insect would be faced with the problem of dispersal to new habitats. If it learned to climb up vegetation as high as possible and to entrust itself to the wind (like Tetranychid mites or spiders trailing their gossamer threads) it would be carried away and might be landed by chance in more favourable surroundings; or it could be caught up in rising eddies and carried high by converging winds, as described by Rainey [38] for locusts, and so deposited with the rain in a moist environment.

The extensive collections in the upper air that were made by Glick [15] showed that over the southern United States (Louisiana, Texas) and the Mexican desert there is a vast aerial fauna of insects and other arthropods extending up into the sky to at least 4,200 m. Among the insects it is mainly winged forms which make these aerial migrations: but small hairy caterpillars, such as the gypsy moth, can be carried many miles by the wind [7, 8]; one human flea, *Pulex irritans*, was taken by Glick at a height of 60 m; and wingless Thysanura, Collembola, young stages of Hemiptera and Orthoptera, larvae of Coleoptera, Lepidoptera and Diptera, and great numbers of wingless ants have been captured at heights up to 1,500 m [15].

The possession of large wing-like outgrowths by a stranded aquatic insect would favour this process and would be encouraged by selection. Passive transport would be facilitated if the insect could control the positioning and inclination of its outgrowths, notably during take-off, and there would be further selection pressure for the evolution of controlled flight. Vibratile gill-plates, even if used for locomotion under water, could obviously not be used directly as propellors in the air—but they would be pre-adapted for the development of the necessary skills so that the insect might increase its buoyancy and keep itself afloat. Whether, in the dispersal of Devonian insects to new breeding places, short, direct transport at low altitudes or long-range excursions into the upper air were the more important must be a matter for speculation. But either method would afford opportunities for adapting their movable branchial plates for the purpose of attitude control and for the increase of buoyancy.

DISPERSAL THEORY: PHYSICAL CONSIDERATIONS

I put forward this 'dispersal theory' of the origin of flight some years ago: the proposal being that the insect, carried aloft by winds and thermals, 'learned to fly' while being passively airborne [55, 56]. I described this as a theory of the origin of flight occurring in 'small' insects; not 'tiny' insects as has been asserted for example by Alexander and Brown [1], but small by comparison with the great Palaeodictyoptera which have been pictured as pioneer gliders. In the famous *Lithomantis carbonarius* of Woodward [57] the remnant of the damaged fore-wing is 5·7 cm long; when complete it must have been at least $7\frac{1}{2}$ cm long, giving a wing span exceeding 15 cm.

Obviously, if flight was to be learned in this way, the insect would have to be of the right size. The two examples that I used of the passive transport of winged insects by air-streams were aphids and locusts [55, 56]. Jensen [21] estimated that the sinking speed of gliding *Schistocerca* in the optimal gliding posture was about 0·6 m/s; and Rainey and Waloff [40] had made an estimate of 1 m/s. The terminal falling velocities of aphids are in the range of 1–2 m/s [3]. Rainey and Waloff [40] found that natural up-currents could sometimes exceed 5 m/s and that in some circumstances even locusts with folded wings could be swept upwards by air-streams.

As Hinton [20] has pointed out, the more minute is an insect the more readily is it deflected by the wind; the larger it is the better suited it is to gliding. As several papers in this symposium have made clear the aerodynamics of flight is greatly influenced by overall size. Flower [12] estimated that the first insects to experiment with the aerial biosphere might most conveniently have been about 1 cm long. The winged Paraplecoptera with Thysanura-like early stages as described by Sharov [45] were in the region of 2 cm in length. Existing Thysanura are in the range 0·5–1·5 cm. If the branchiated insects which were forced to take to the air were of this size it would seem to have been a reasonable compromise. But it is hardly necessary to point out that the insect did not learn to fly by studying aerodynamics. It learned by trial and error; and the highly sophisticated outcome of this process of selection is what we endeavour to describe in aerodynamic erms.

PHYSIOLOGICAL CONSIDERATIONS

An aquatic insect that leaves the water must be able to resist desiccation. It must have a cuticle resistant to rapid transpiration. As Beament [2] has shown, the cuticles of many present day aquatic insects have retained most of the permeability properties of their terrestrial relatives.

The airborne insect must also be able to breathe oxygen from the atmosphere. It was suggested by Portier [35] that the wings of butterflies are respiratory organs. They are richly tracheated with gas-filled tubes; the contained air always has a lower partial pressure of oxygen than the atmosphere; and the cuticle is not wholly impermeable to gases. Portier's contention must therefore be true—but the amount of oxygen taken up by this route is probably a negligible fraction of the tracheal respiration through the spiracles. In the aquatic larvae of modern Ephemeroptera the 'spiracles' function only as a means for the ecdysis of the tracheal lining [24], not in respiration, which is effected solely by the gills and other permeable areas. In Odonata the mesothoracic spiracles immediately become functional when the larva leaves the water before eclosion of the adult; the metathoracic and abdominal spiracles seem rarely to be functional in respiration, although they are still patent and will permit the discharge of tracheal air. The difference between the state of affairs in these existing Palaeoptera, and a less specialised condition in which all the spiracles would still function in respiration in an aquatic insect exposed to the air, is so slight that it is not difficult to conceive.

The necessary adaptation of the muscles we have already touched upon (p. 261). But another important element in controlled flight, as stressed by Hinton [20], would be attitude control. As Leston [26] has pointed out, the long slender cerci in Thysanura would have been an asset for flight stability; as indeed, in conjunction with the long fore-legs, they are in existing Ephemeroptera. Even *Calliphora*, a most unstable insect [36], can fly without its halteres if a weighted thread is attached to the tip of the abdomen [23].

ECOLOGICAL CONSIDERATIONS

The theory which I am advocating fits in well with the knowledge we have about the ecological significance of flight in existing insects as set out during this symposium. The central idea which I put forward in 1963 was partly stimulated by Rainey's demonstration of the dispersal of the Desert Locust, carried by converging winds and landed where the rain falls; more recently, Leston [26] and Rainey [39] have favoured the idea of the origin of insect flight in an insect caught in an arid climate. The book by Johnson [22] on the *Migration and dispersal of insects by flight* contains endless examples of the central importance of aerial dispersal.

It is obvious that in very minute arthropods, aided by silken threads, as is seen today in Tetranychid mites, aerial dispersal was possible without the evolution of controlled flight. As Flower [12] has pointed out, such arthropods could well have been selected for diminishing size. But for the branchiated ancestor which we are envisaging, the evolution of controlled flight would not only have been an aid to long range dispersal; it would have brought the innumerable ecological benefits that we see in existing winged insects—for local migration, for mating, for escape from predators, for seeking out oviposition sites etc., and would have opened up the whole glorious future of the Pterygota.

Writing in 1883, Lubbock [28] says: 'It is possible that the principal use of the wings was, primordially, to enable the mature forms to pass from pond to pond thus securing fresh habitats and avoiding in-and-in breeding. If this were so, the development of wings would gradually have been relegated to a late period of life; and by the tendency to the inheritance of characters at corresponding ages, which Mr Darwin has pointed out, the development of wings would have become associated with the maturity of the insect'. We should probably have judged that the moulting of cuticular wings would be impossible—were it not for the fact that existing mayflies regularly make one moult after reaching the flying state. Sharov [45] has shown that the Permian *Atactophlebia termitoides* (Paraplecoptera) had at least two moults in the winged state. The early stages of this insect are strikingly thysanuroid in form, but Sharov regards them as having been terrestrial in habit. If that interpretation is correct it means that in some groups the restriction of functional wings to the final instar was delayed until long after the attainment of effective flight.

CLIMATIC CONSIDERATIONS

Assuming that the evolution of wings took place in Devonian times, were the climatic conditions such as to provide a reasonable background for the theory I have outlined? The combined Lower, Middle and Upper Devonian occupied a long time—some 65 million years. Attempts have been made by Seward [44] to visualize the scene as a back-ground for the evolution of plants; similarly by Romer [42] for vertebrates; as well as by geologists in general [6].

One reads that by the Middle Devonian the two gigantic continents of Laurasia and Gondwana had risen from the sea; that a highly varied flora developed, with vascular plants growing to a height of 20 ft or more and with what appear to be single species having an amazingly widespread distribution, suggesting less diversification of climate than we experience today. We read about evidence of torrential rain, of extensive swamps, of arid conditions and the formation of aeolian sandstones and wind-driven layers of sand between sedimentary deposits: in fact all the elements which, if they occurred over large areas and for long periods of time, would provide us with just the conditions we require for the evolution that I postulate.

Sharov [46] pictures the first insects as subsisting on rotting stems and trunks and plant detritus along the banks of lakes and swamps. As he points out, existing Thysanura are 'the pitiful remains' of thysanuran-like wingless insects belonging to many orders, from one of which the winged insects arose [see also 13, 5]. There must have been intense taxonomic differentiation in the Devonian, for by the middle Carboniferous there was an immense variety of groups in existence. Mamajev [29] considers that the Devonian insects were associated with the vegetation in the swamp forest along the banks of lakes and that they were probably not true terrestrial forms, but followed a semi-aquatic or amphibious mode of life. It would be indeed surprising if, in the course of 50 million years, early hexapods had failed to colonize the inland waters.

Those aquatic forms which could survive drying out, and migrate to new rain-formed habitats, would be favoured by selection. The evolution of wings with a standard tracheal pattern shows that this triumph of adaptation was probably accomplished only by a fairly closely related group. These aquatic ancestors would presumably feed like mayfly larvae today: by scraping up the algae and diatoms which must have abounded on submerged stones and vegetation. It is likely that they built up gigantic populations—like mayflies: so that enormous mortality could be tolerated by the stranded migrants. A single insect entrusting itself to winds and thermals would have a very small chance of passive transport to a suitable habitat; but from a cloud of insects containing hundreds of thousands of individuals, as is seen in present day Ephemeroptera, a small number would be likely to succeed.

The arid conditions which repeatedly supervened have so far left us no fossil record to test this argument. As Rainey [39] points out, the very paucity of the early fossil record would be consistent with the evolution of flight having occurred under arid conditions, in which the likelihood of fossilization would be minimal. Arid and semi-arid areas are characterized by vigorous thermal up-currents developing regularly by day over heated ground and rising at speeds of several metres per second [37]. Such currents could lift any organism with a sinking speed less than these rates of ascent. That would readily lift the the type of insect we are considering. Such insects would be caught up also in whirlwinds and dust-devils and carried up into rain clouds.

It is worth recalling the parallel theory put forward by Romer [41, 42] which ascribes the evolution of terrestrial locomotion in Crossopterygian fishes to their efforts to crawl out of one drying pool during the seasonal droughts of the late Devonian and hobble over land to another pool with water in it. Walking limbs arose, not in an effort to colonize the land, but in order to survive in water. That is the argument for the aquatic ancestor of winged insects. Wings likewise arose to make possible a return to water.

Their subsequent exploitation for purposes of terrestrial life was a later development, which has led to the near domination of the world by insects.

I am well aware that plausible arguments for other theories of the origin of wings and flight in insects can be advanced. But having thought around this subject over the past ten years I have become convinced that the theory I am putting forward, most of the elements of which have appeared at one time or another in the literature of the past century, provides a history that offers the least strain to the imagination—because, as Southwood, Rainey, Johnson and others have pointed out, most of the processes involved can be seen in operation around us to the present day.

SUMMARY

Origin of wings.   After a brief historical survey, the 'gill theory' of the origin of wings is re-examined.

It is argued that the apparent similarity between wings and the vibratile gill-plates of Ephemeroptera is not just a superficial resemblance but may well be based on a true homology.

It is generally agreed that the abdominal gills of mayflies arose from the styli, regarded as the exites of vanished abdominal appendages, articulated to the coxal limb base. It is here suggested that in certain aquatic pre-Pterygotes similar gill-plates were developed from the serially homologous styli of the meso- and metathorax—as seen in *Machilis* as exites from the base of the coxae of these segments.

The sites of articulation of the abdominal gills of many mayflies have migrated dorsally so that they appear superficially to arise from the tergites. It is suggested that the basal parts of the coxal limb bases of the thorax became similarly incorporated into the thoracic wall to form the pleuron and at the same time the sites of articulation of the thoracic gill-plates moved dorsally to the margin of the notum.

It is suggested that such tracheated gill-plates on the thorax would be richly supplied with muscles; in some forms they could well be quite large in size and used for locomotion under water, as well as for driving currents over the respiratory surfaces.

Origin of flight.   The dispersal theory of the origin of flight is re-stated.

It is suggested that the thoracic branchial plates of an aquatic Apterygote (as described above) were pre-adapted to facilitate passive transport, by winds and thermals, of insects stranded by the drying out of their habitats.

In their general form, in the articulation of the base, and in their muscular supply the gill-plates would also be pre-adapted for conversion into wings, first to reduce sinking speed (p. 48) and ultimately for controlled flight.

The climatic conditions of Devonian times would be in conformity with this theory.

Subsequent developments in the evolution of wings are not discussed in this paper.

DISCUSSION

Professor R.S.Scorer: While this is not really my subject, I would like to make three points. One is that trial and error *is* the way we study aerodynamics ourselves, as well as the way the aerodynamics of insects has evolved. The second point is that the atmosphere in the Devonian and previous eras may not have been the same as the air we experience today; it may perhaps have become gradually more hospitable for flight during the Carboniferous era. Finally, the water may have been the best place to learn to 'fly' because there was not so much weight to carry; successful 'flight' in water may have helped in beginning to learn to fly in air.

Pringle: Naturally I find this new theory rather a shock, since I have always accepted the paranotal theory—without very much morphological evidence, but, I think, with a good deal of functional evidence. But perhaps Professor Wigglesworth could say why he thinks insect wings did not, alternatively, evolve from legs. Most insects which swim under water, swim with their legs; and, if such an insect should get carried up into the air and should attempt to delay its descent, it might be expected to use its legs. Legs might thus evolve into wings, in the same kind of way that the Crossopterygian fins evolved into legs.

Wigglesworth: One wants to keep one's hypotheses, one's options, as wide open as possible—obviously. But I do not visualize these appendages as having been initially developed primarily for under-water locomotion—I would think primarily for driving water over respiratory surfaces, as the mayfly does. As I said, there has been considerable argument as to whether *Chloëon* uses its branchial plates for swimming, as Wesenberg-Lund says it does. Its main method of locomotion is jet-propulsion, like the larvae of Odonata, but in gentle swimming it uses its tail filaments and its abdominal branchial plates. Had such respiratory plates occurred also on the thorax of our hypothetical ancestral form, I agree with Lubbock that they might subsequently have become of importance in locomotion.

Gunn: A possible reason why insects did not use their legs in flight is suggested by the penguins, which use their wings in swimming (unlike most birds, which swim with their feet), and as a result cannot fly. Had the insect legs been used in flight they might have become unsuitable for walking.

Dr J.Simpson: While many birds, such as ducks and moorhens, use their legs in swimming under water, others such as the dipper swim under water with their wings, but nevertheless can still fly.

Kristensen: It seems to me that the paranotal theory still provides the most straightforward explanation of the origin of wings, if the term 'paranotum' is taken to mean just a tergal expansion. Huge tergal expansions are present in Thysanura, and their tracheal system (studied by Sulc and Ander) is certainly no less similar to that of wings than that of the gills illustrated. Deriving the wing from a coxal style would create enormous difficulties in the interpretation of the Pterygote coxa, which would correspond to only the distal part of the Thysanuran coxa. Direct muscles from coxa to wing may probably be derived straightforwardly from extrinsic coxal muscles in Thysanura; and the serial homology of coxal and abdominal styles in Machilidae is not universally accepted.

Wigglesworth: All I have tried to show is that there is, I think, a very reasonable degree of support from physiological, ecological and other grounds for the gill theory; all I have done is to provide material for the morphologists of the future to pay attention again to these problems. I can see that Dr Kristensen is going to prove a worthy successor to his compatriot H.J.Hansen, whom I have studied extensively, and to whose works I would refer Dr Kristensen on this question of the serial homology of the styli.

(Communicated): The objection is often raised that the thoracic styli lack muscles, which persist at the base of the abdominal styli; but Hansen points out that in many Diptera the maxillary palps have no muscles, and yet their homology is not in doubt.

REFERENCES

[1] ALEXANDER, R.D. & BROWN, W.L. (1963). Mating behaviour and the origin of insect wings. *Occasional Papers Mus. Zool. Univ. Michigan*, no. 628: 19 pp.

[2] BEAMENT, J.W.L. (1961). The waterproofing mechanism of arthropods. II. The permeability of the cuticle of some aquatic insects. *J. exp. Biol.*, **38** : 277–290.

[3] BERRY, R.E. & TAYLOR, L.R. (1968). High altitude migration of aphids in maritime and continental climates. *J. anim. Ecol.*, **37** : 713–722.

[4] BÖRNER, C. (1909). Die Tracheenkiemen der Ephemeriden. *Zool. Anz.*, **33** : 806–823.

[5] CARPENTER, F.M. (1971). Adaptations among paleozoic insects. *Proc. North Am. Paleontological Convention*, 1969 (Part I) : 1236–1251.

[6] COLBERT, E.H. (1964). Climatic zonation and terrestrial faunas. *In* Nairn, A.E.M. (ed.) *Problems in palaeoclimatology* : 617–642. Wiley, New York.

[7] COLLINS, C.W. (1917). Methods used in determining wind dispersion of the gipsy moth and some other insects. *J. econ. Entomol.*, **10** : 170–177.

[8] COLLINS, C.W. & BAKER, W.L. (1934). Exploring the upper air for wind-borne gipsy moth larvae. *J. econ. Entomol.*, **27** : 320–327.

[9] CRAMPTON, G.C. (1916). The phylogenetic origin and the nature of the wings of insects according to the paranotal theory. *J.N.Y. entom. Soc.*, **24** : 1–38.

[10] DÜRKEN, B. (1923). Die postembryonalen Entwicklung der Tracheenkiemen und ihrer Muskulatur bei *Ephemerella ignita*. *Zool. Jahrb. Abt. Anat.*, **44** : 439–614.

[11] EASTHAM, L.E.S. (1934). Metachronal rhythms and gill movements of the nymph of *Caenis horaria* (Ephemeroptera) in relation to water flow. *Proc. Roy. Soc. Lond.* B, **115** : 30–48.

[12] FLOWER, J.W. (1964). On the origin of flight in insects. *J. Insect Physiol.*, **10** : 81–88.

[13] FORBES, W.T.M. (1943). The origin of wings and venational types in insects. *Amer. Mid. Nat.*, **29** : 381–405.

[14] GEGENBAUR, C. (1870). *Grundzüge der vergleichenden Anatomie*. Leipzig, 1870.

[15] GLICK, P.A. (1939). The distribution of insects, spiders and mites in the air. *Tech. Bull. U.S. Dept. Agric.*, no. 673 : 150 pp.

[16] HAMILTON, K.G.A. (1971). The insect wing. I. Origin and development of wings from notal lobes. *J. Kansas entom. Soc.*, **44** : 421–433.

[17] HANSEN, H.J. (1930). *Studies on Arthropoda III*. Copenhagen : 376 pp.

[18] HENNIG, W. (1969). *Die Stammesgeschichte der Insekten*. Kramer, Frankfort : 436 pp.

[19] HEYMONS, R. (1896). Grundzüge der Entwicklung und des Körperbaues von Odonaten und Ephemeriden. *Abhandl. königl. preuss. Akad. Wiss. Berlin*, 1896 : 66 pp.

[20] HINTON, H.E. (1963). The origin of flight in insects. *Proc. R. entom. Soc. Lond. (C)*, **28** : 23–32.

[21] JENSEN, M. (1956). Aerodynamics of locust flight. *Phil. Trans. R. Soc. (A)*, **239** : 511–552.

[22] JOHNSON, C.G. (1969). *Migration and dispersal of insects by flight*. Methuen, London; 763 pp.

[23] JOUSSET DE BELLESME (1878). *Recherches expérimentales sur les fonctions du balancier chez les insectes diptères*. Paris.

[24] LANDA, V. (1948). Contributions to the anatomy of Ephemerid larvae. I. Topography and anatomy of tracheal system. *Věstník čsl. zool. Spol.*, **12** : 25–82.

[25] LANG, A. (1884). *Lehrbuch der vergleichenden Anatomie*. Jena: 566 pp.

[26] LESTON, D. (1963). The origin of flight in insects. *Proc. R. entom. Soc. Lond. (C)*, **28** : 23–32.

[27] LUBBOCK, J. (1863). On two aquatic Hymenoptera, one of which uses its wings in swimming. *Trans. Linn. Soc. Lond.*, **24** : 135–142.

[28] LUBBOCK, J. (1883). *On the origin and metamorphosis of insects*. Macmillan, London : 108 pp.

[29] MAMAJEV, B.M. (1971). The significance of dead wood as an environment in insect evolution. *Proc. 13th Int. Congr. Ent., Moscow*, 1968, **1** : 269.

[30] MANTON, S.M. (1953). Locomotory habits and the evolution of the larger arthropodan groups. *Symp. Soc. exp. Biol.*, **7** : 339–376.

[31] MATHESON, R. & CROSBY, C.R. (1912). Aquatic Hymenoptera in America. *Ann. ent. Soc. Amer.*, **5** : 65–71.

[32] MÜLLER, F. (1875). Beiträge zur Kenntnis der Termiten. *Jena. Z. Naturwiss.*, **9** : 241–265.

[33] OKEN, L. (1811). *Lehrbuch der Naturphilosophie.* Jena.

[34] PACKARD, A.S. (1898). *A textbook of entomology.* Macmillan, New York : 729 pp.

[35] PORTIER, P. (1930). Respiration pendant le vol chez les Lépidoptères. *C.R. Soc. Biol.*, **105** : 760–764.

[36] PRINGLE, J.W.S. (1957). *Insect Flight.* Cambridge University Press: 133 pp.

[37] RAINEY, R.C. (1958). Some observations on flying locusts and atmospheric turbulence in eastern Africa. *Q. Jl R. met. Soc.*, **84** : 334–354.

[38] RAINEY, R.C. (1963). Meteorology and the migration of Desert Locusts. *Tech. Notes Wld met. Org.*, no. 54 : 115 pp.

[39] RAINEY, R.C. (1965). The origin of insect flight: some implications of recent findings from palaeoclimatology and locust migration. *Proc. 12th Int. Congr. Entom.*, London, 1964 : 134 pp.

[40] RAINEY, R.C. & WALOFF, Z. (1951). Flying locusts and convection currents. *Anti-Locust Bull.*, **9** : 51–72.

[41] ROMER, A.S. (1966). *Vertebrate paleontology* Edn. 3, University of Chicago Press : 468 pp.

[42] ROMER, A.S. (1968). *The procession of life.* Weidenfeld and Nicolson, London : 323 pp.

[43] ROSS, H.H. (1955). The evolution of the insect orders. *Entom. News*, **66** : 197–207.

[44] SEWARD, A.C. (1931). *Plant life through the ages.* Cambridge Univ. Press : 601 pp.

[45] SHAROV, A.G. (1957). Types of insect metamorphosis and their relationship. *Rev. Entom. URSS*, **36** : 569–576. (In Russian with English summary.)

[46] SHAROV, A.G. (1966). *Basic arthropodan stock with special reference to insects.* Pergamon, Oxford: 271 pp.

[47] SNODGRASS, R.E. (1929). How insects fly. *Smithsonian Rep.* , **1929** : 383–421.

[48] SNODGRASS, R.E. (1931). Morphology of the insect abdomen. I. General structure of the abdomen and its appendages. *Smithsonian misc. Coll.*, **85** : no. 6, 128 pp.

[49] SNODGRASS, R.E. (1935). *Principles of insect morphology.* McGraw-Hill, New York : 667 pp.

[50] SNODGRASS, R.E. (1954). Insect metamorphosis. *Smithsonian misc. Coll.*, **122** : no. 9 : 124 pp.

[51] SNODGRASS, R.E. (1958). Evolution of arthropod mechanisms. *Smithsonian misc. Coll.*, **138** : no. 2 : 77 pp.

[52] TOWER, W.L. (1903). The origin and development of the wings of Coleoptera. *Zool. Jahrb. Abt. Anat.*, **17** : 517–572.

[53] VAYSSIÈRE, A. (1882). Recherches sur l'organisation des larves du Éphémèrines. *Ann. Sci. Nat. Zool. Sér.* 6, **13** : 1–137.

[54] WESENBERG-LUND, C. (1943). *Biologie der Süsswasserinsekten.* Springer, Berlin : 682 pp.

[55] WIGGLESWORTH, V.B. (1963). Origin of wings in insects. *Nature*, **197** : 97–98.

[56] WIGGLESWORTH, V.B. (1963). The origin of flight in insects. *Proc. R. entom. Soc. Lond. (C)*, **28** : 23–32.

[57] WOODWARD, H. (1876). On a remarkable fossil Orthopterous insect from the coal measures of Scotland. *Quart. J. Geol. Soc. Lond.*, **32** : 60–65.

[58] WOODWORTH, C.W. (1906). The wing veins of insects. *Univ. Calif. Publ. Agric. Exp. Sta. (Entomology)*, **1** : 1–152.

General discussion

Introducing the concluding discussion, **Professor Southwood** (Chairman) invited further comments on the evolutionary issues raised by the papers of Sir Vincent Wigglesworth and Dr Wootton.

Pringle: A further question to Professor Wigglesworth; one of the general features of Pterygote insects is the suppression of adult characters by the juvenile hormone until the final moult. Is there any evidence from the hormonal situation in the Apterygota which will help to explain this situation in the Pterygota?

Wigglesworth: I believe I anticipated Novák in suggesting that the juvenile hormone secretion of the corpus allatum was originally used primarily in reproduction. That is still the case in the Apterygota, in which it has virtually no morphological effects and is still a reproductive hormone; most Apterygota go through repeated cycles of alternating moults and reproduction. When reproduction became confined to the final instar, the juvenile hormone, as I see it, acquired the new function of regulating the morphological change.

Pringle: That seems to me to be evidence against Professor Wigglesworth's theory of the origin of flight, because the thoracic styli develop only in the last instar in Thysanura.

Wigglesworth: But the early Pterygotes may well have continued to moult in the winged stage; Sharov has described species in the Permian that moulted several times after getting their wings.

Southwood: Can Dr Wootton give us some idea of the range of size of the fossil flying insects? One has the impression (probably erroneous) that they tended to be larger than the average size of insects today, but this might be because of variations in the preservation and the discovery of fossils. From the papers we heard yesterday, size could have important implications for the sort of aerodynamics that could have been used in early flight, for the kind of musculature, and for the degree of independence from the leg muscles.

Wootton: First, the question of preservation: under suitable conditions, even tiny insects can be well preserved. I have seen insects of the size of aphids and small midges in good lake deposits from the Jurassic. Secondly: yes, known Carboniferous insects were on average larger than insects today. The largest were the big Meganeurid Protodonata, whose wings were up to 300 mm long. The largest of the Palaeodictyopterous wings I illustrated (Fig. 11.5) was 190 mm long, implying a wing span of over 400 mm. The smallest were probably also Palaeodictyoptera, of which there were some with a span of no more than 32 mm in the Upper Carboniferous. Later, in the Permian, some of the little Diaphanopterodea were down to mosquito size.

Southwood: This evidence on the size of the Carboniferous insects does of course lend support to Professor Pringle's suggestion yesterday that the evolution of fibrillar muscle may have contributed to a reduction in the size of insects.

Pringle: Yes indeed.

Wootton: It was with this point in mind that I asked my question yesterday of Professor Pringle, because of the tiny size of some of the Diaphanopterodea, particularly of the

Asthenohymenidae. They have very long cerci and do not look in the least like any of the modern insects with fibrillar muscle. My suspicion is that they had a low wing-loading and therefore a relatively low wing-beat frequency.

Southwood: Something like male Coccids?

Wootton: Yes. (Communicated subsequently) Further to my previous reply, a very few smaller Carboniferous insects are in fact known. *Archimioptera carbonaria* Guthorl (Miomoptera) had wings 5·4 mm in length. This is a late Carboniferous form: but the probably related *Metropator pusillus*, which had a 7 mm wing, is Namurian in age, and therefore one of the earliest fossil insects. The presence in the lower Permian of further Miomoptera, and also small Hemiptera and Psocoptera, increases the proportion of small forms—the smallest recorded being the Pscopteran *Cyphoneura permiana* Carpenter with a fore-wing of 1·9 mm. The wings of the smallest described Astheno-hymenid Diaphanopterodea were 4·3 mm long.

Rainey: A further question for Dr Wootton: have not significant changes in the composition of the atmosphere, of the kind Professor Scorer mentioned, been suggested in the lower Palaeozoic?

Wootton: Yes, this is so, with for example a progressive cutting down of the penetration of ultra-violet radiation by increasing ozone, but the whole question of the Palaeozoic atmosphere is still very much a controversial one. [An authoritative recent review (Meadows, A.J., 1972, The atmospheres of the earth and the terrestrial planets; their origin and evolution, *Physics Reports* 5C : pp. 199–238) states that the scheme of Berkner and Marshall (1965) for the growth of oxygen in the atmosphere, as discussed by Smart and Hughes at the Society's previous Symposium, has received quite widespread support, though without being universally accepted. This scheme envisages oxygen building up to about one-tenth of its present level during the Silurian, when land flora began to appear in profusion, with a subsequent rapid rise during the Carboniferous, to values perhaps temporarily in excess of the present level. These suggested changes in oxygen are illustrated in Fig. 2 of Smart and Hughes' paper, with ordinates (accidentally omitted) on a logarithmic scale of units of the present atmospheric level—Ed.]

Kristensen: A further point: some recent aquatic insects do retain primitive features, but, if the ancestor of the Pterygota was aquatic, it is surprising it should have been holopneustic.

Rainey (communicated subsequently): On that unknown and crucial first stage in the evolution of insect flight, the central problem seems to be that of envisaging frequently occurring circumstances in which becoming temporarily airborne could have had an immediate survival value greater than that of not doing so. One suggestion has been for escape from predators; but for this purpose it might be thought that the first wing-rudiments, not yet capable of providing sustained flight, would initially have been a handicap rather than of assistance. I still find it easiest to imagine that the first accidentally-airborne insects were terrestrial, and were carried up in vigorous thermal up-currents like dust-devils, in the kind of semi-desert environment in which down-wind displacement (towards areas of wind-convergence with their rain) could have had an immediate survival value. I find it more difficult to imagine a comparably reliable mechanism for becoming frequently airborne into the much less turbulent air which is likely to have been associated with aquatic habitats sufficiently permanent to have been appropriate to ancestral aquatic insects not yet capable of flight. Transport by air-currents suggests small size (cf. the Namurian *Metropator* with its 7 mm wing), at which wings could have been initially unnecessary, and with sinking-speeds small enough to have been later offset perhaps by modest power outputs, applied possibly to slowly-flapping wings not yet exploiting a resonant thoracic structure.

Southwood: Looking further back through our programme; we talked about the movement of insects in the air, and swarms, and the behaviour of individual insects and their orientation, and Professor Schaefer flung out a number of challenges which people were not able to take up because of pressure of time. Perhaps Dr Rainey would like to open this part of the discussion.

Rainey: The point which I had hoped to emphasize was the evidence of concentration of airborne insects by wind-convergence, down to spacings and up to densities which mean that the individual insects are within a matter of a few metres of each other even at heights of several hundreds of metres. This poses the question that they may be being brought by the wind to within range of mutual perception, which I think is a fascinating possibility from a behavioural point of view.

Gunn: I would like to ask a rather broader question which my position this morning rather inhibited me from asking, particularly since we were running behind itme, and this refers to this convincing evidence of great concentrations, linear concentrations of insects in certain conditions, and the possibility (which may be no more than the gleam in Vernon Joyce's eye) of large-scale control of pests by something like national effort, greatly diminishing the use of insecticide, making control much more efficient and effective. Is this a feasible, a plausible plan arising out of the discoveries of the people we heard this morning, or are those great concentrations—extensive linearly as they are—are they still only a small proportion of the pest populations? This is the kind of thing that I would like to hear more about—what do they think is going to happen in the future?

Southwood: Who has his crystal ball with him, and is willing to project into the future?

Joyce: I will start off, but I am sure any details will be answered much better by Dr Schaefer and perhaps Dr Rainey. I think there is a long way to go yet. The first thing is we often do not know what these concentrations of insects are composed of, what species are involved. The concentrations which it was possible to sample in Sudan did turn out in fact to be largely composed of grasshoppers and predominantly one species —*Aiolopus simulatrix*—which is a serious pest of *Sorghum* in the area. The sampling in Canada of aerial concentrations was less complete than we would have hoped, but I think that the evidence was that these concentrations were largely of the spruce budworm moth. Clearly it would be just as harmful to spray concentrations of flying insects willy-nilly in respect of their composition; but I find it difficult to believe that there will not be some mechanism of segregation of insects in the air, due to the structure of the air, to differences in the flight behaviour (falling-speeds, gliding-speeds and so on) of different insects, and to their different habits in take-off. But this remains to be seen. I myself certainly hope that it will be possible to develop means of insect control through attacking airborne concentrations, not so much in respect of problems on which conventional crop protection techniques have been developed in Europe and North America, but in respect of the very high proportion of the damage suffered by crops throughout the world which is in developing countries and is rarely seen and rarely reported. Thus the only control which has been adopted against armyworm on any large scale has been for the sake of expensive cash crops, while most of the damage is done to peasant farms from which no reports come and where people can lose their entire crops. Again, I have seen hundreds of acres of rice in Pakistan decimated by rice *Hispa*, and starvation resulting from it; and pest outbreaks like these, as far as I can see, do not occur through breeding locally, but occur through aerial concentration, giving rise to a situation which can only be rectified by being in a position to predict it; and the only practical way to my mind of tackling this sort of problem is along the particular lines which we are investigating.

Southwood: Professor Schaefer, in your talk you said you thought the bulk of the biomass was flying at night rather than during the day, though of course your observations in the tropics might well be dealing with different meteorological situations perhaps than occur most of the time in temperate countries. But certainly my remembering of the Rothamsted data (Dr Johnson will correct me if I am wrong) is that suction-traps at night did not seem to produce the sort of biomass that was present during the day. Although there could be waves of concentration which could be missed by suction-traps—and you have got your observations in Canada—I wonder whether this phenomenon of these concentrations is something which differs in its magnitude in different climatic zones of the world, and that it is perhaps not so invariably important in a temperate country like Britain with its disturbed weather systems. Could you comment on that?

Schaefer: Yes, on the two aspects. I was referring to my biomass comparisons in the locations where I have been working and where I have data—radar for the larger insects, and the aircraft netting in the Sudan and in Canada for the smaller insects, with suction-traps; and there the biomass is certainly greater by night than by day on the whole. Further, I think that concentrating mechanisms of strong characteristics occur more often in the tropics than in temperate latitudes, where we are more concerned with fairly mild frontal conditions as opposed to these very strong systems that I was talking about—tropical storms and so on—though that is a very general statement.

Southwood: When we were studying the movement of the fritfly (*Oscinella frit*—Chloropidae) in the air, again at the time of maximum flight in the middle of the day, one did get a profile showing that a very large part of the population was high up in the air. This was from suction-trap data, and some of the density/height profiles for this insect throughout the time of its flight (Johnson, C.G., Taylor, L.R. & Southwood, T.R.E., 1962, *J. Anim. Ecol.*, **31** : 373–383) did fit in with some of the profiles that you showed. So I think that we may be looking at different ends of the spectrum in these two sorts of studies rather than dealing with a completely different phenomenon. Perhaps Dr Johnson would like to comment.

Johnson: Well, I think Professor Schaefer and I know each other's views well enough. But I would not like the rest of you to think that we believe that there is just a simple logarithmic relationship with height—we have always known that this is not so. As far back as 1939 P.A.Glick (*Tech. Bull. U.S. Dep. Agric.*, no. 673) had some aeroplane catches over the Mexican desert where the whole insect population appeared to have lifted off, giving just such a profile as Professor Schaefer showed, with very little below about 100 metres and all the rest above, and a concentration somewhere about halfway up. One ought also perhaps to say that in our night-time catches we did not deal with biomass, we dealt with numbers; and in biomass one large insect could make up for a good many aphids. And suction-traps, as Professor Schaefer quite rightly said, are not very good for sampling the very low densities that you were recording with your radar and could express in terms of biomass. Also, Dr Taylor found that in the warmer conditions in America even aphids which were daytime fliers in Britain were kept up in the air much longer at night-time over there (Berry, R.E. & Taylor, L.R., 1968, *J. Anim. Ecol.*, **37** : 713–722). So what Professor Southwood says about these day and night aerial catches in different parts of the world is very true. With regard to spraying in these concentrations, it does rather worry me, and I do not know what the conservationists would say about it. My experience is that there are vast numbers of species in the air; I can imagine a locust swarm or a concentration of armyworm or spruce budworm moths outnumbering everything else, but in general I would have thought that the concentrations were composed of a large number of species, and that to spray them would be a hazardous enterprize unless you had sorted them out or unless they were disentangled in some way. But it is a possibility that remains for the future, and I am not saying that is not possible.

Schaefer: I think that Dr Johnson has covered himself; that we are as much worried about this as he is; and that the aircraft catches, which Dr Rainey knows far more about than I do, can already tell us a good deal about the number of species in the locations where we worked.

Rainey: I would just add the point that a number of the biggest aircraft catches have turned out to be predominantly of a single species, which was different from occasion to occasion. For example, we once took a tiny Anthicid beetle—*Leptaleus sennarensis* (Pic)—at 1600 metres by night in the Sudan at a density of one per 190 m³ (p. 98) and outnumbering everything else in the catch; this was towards the top of the zone of transition between the two wind-currents at the ITF, with the ground-level front 100 km away to the north-west at the time. So that it does look—as Vernon Joyce was hoping—as though there may be mechanisms, biological as well as meteorological, which can operate to some extent to sort out and separate these concentrations, species from species.

(Communicated subsequently): On Dr Gunn's initial question as to the proportion of a total population of a species which may be represented by one of these airborne concentrations: while no immediate answer can be given, we may be within sight of knowing how to provide one, in circumstances like the Gezira where one can envisage ground-surveys of the non-flying stages adequate in nature and extent to complement the observations of the flying insects. On Professor Southwood's discussion with Dr Schaefer on relative biomass by day and by night in the tropics and in temperate latitudes, one wonders whether simply the relatively cool nights to which we are accustomed e.g. in this country, may be the main factor concerned.

Betts: Picking up Dr Johnson's point and referring to a particular incident, the concentration of *S. exempta* moths which moved across Muguga on the night of 9th/10th March, 1970 (Fig. 5.17) was accompanied by enormous numbers of other moths of a wide range of species. Now next morning when we looked at what was left on the ground and there were vast numbers still settled, one of the most conspicuous was a species of *Hypena*, which is an occasional defoliator of trees in Kenya. Within a very short period after this there was indeed a series of outbreaks of *Hypena* larvae some 20 km to the north-east and south-east of Muguga, but, interestingly enough, substantially to the west of the bulk of the *exempta* larval infestations which appeared at the same time 50 to 100 km from Muguga (Fig. 6.7). So although the two species of moth had moved through together, they laid in fact in two different zones. Whether the *Hypena* were ready to breed earlier, or flew for a shorter number of hours during the night and settled earlier, I do not know, but at some point there appears to have been some sifting mechanism.

19

Johnson: Of course if there were discrete sources, and the sources were in different places, we might expect that they would be segregated in the air.

Betts: In actual fact the peak catch of all the species went up together as the westerlies came in through Muguga, so that at that particular stage they were all flying together.

Onyango-Odiyo: I would like to comment on the suggestion of massive control as suggested by Dr Gunn and Mr Joyce. Some of the really big infestations of armyworm in East Africa have not been on crops but on grazing areas or game parks which are not economically sufficiently important to be worth immediate control. But major outbreaks have sometimes started in these out-of-the-way places and then built up to more serious subsequent infestations on crops. To persuade Governments to provide money to control this kind of infestation in the wilderness is not easy, because we have to balance whether the money should best be spent in this way or on more immediately pressing problems. And because armyworms are just part of the pest populations in the area, it may not be easy to organize campaigns on a regional basis until we come to a stage where armyworm control can be linked up possibly with the control of the Desert Locust as well as of other pests.

Southwood: It is I think very fitting that we should be reminded of the economic realities of insect pest control, in what has to be the last observation in this discussion.

Contributors to discussions

Miss Elizabeth BETTS *Centre for Overseas Pest Research (Ministry of Overseas Development), College House, Wrights Lane, London, W8 5SJ*

Dr J. N. BRADY *Imperial College Field Station, Ashurst Lodge, Silwood Park, Ascot, Berkshire SL5 7DE*

Dr D. O. GREENBANK *Maritimes Forest Research Centre (Canadian Forestry Service), P.O. Box 4000, Fredericton, New Brunswick, Canada*

Dr D. L. GUNN, C.B.E. *Well Cottage, Taylor's Hill, Chilham, Canterbury, Kent CT4 8BZ. (sometime Director, International Red Locust Control Service)*

Dr J. W. HARGROVE *Department of Zoology, University of Rhodesia, P.O. Box MP 167, Mount Pleasant, Salisbury, Rhodesia*

Professor B. HOCKING *Department of Entomology, University of Alberta, Edmonton 7, Canada*

Mr G. J. JACKSON *Department of Applied Zoology, University College of North Wales, Bangor, Caernarvonshire*

Dr C. G. JOHNSON, O.B.E. *Coconut Industry Board, P.O. Box 204, Kingston 10, Jamaica, West Indies. (Sometime Head of Entomology Department, Rothamsted Experimental Station)*

Mr R. J. V. JOYCE *Agricultural Aviation Research Unit (Ciba-Geigy Ltd.), Cranfield College of Aeronautics, Bedfordshire. (sometime Director, Desert Locust Control Organization for Eastern Africa)*

Dr J. A. KEFUSS *Institut für Bienenkunde, Universität Frankfurt am Main, Im Rosengärtchen, 637 Oberursel/Ts., Frankfurt, German Federal Republic*

Dr M. P. KRISTENSEN *Zoologisk Museum, Universitetsparken 15, DK-2100 Copenhagen, Denmark*

Dr C. T. LEWIS *Imperial College Field Station, Ashurst Lodge, Silwood Park, Ascot, Berkshire SL5 7DE*

Professor Sir James LIGHTHILL, F.R.S., *Department of Applied Mechanics and Theoretical Physics, University of Cambridge, Silver Street, Cambridge, CB3 9EW. (sometime Director, Royal Aircraft Establishment, Farnborough)*

Professor M. LINDAUER *Institut für vergleichende Physiologie, Universität Würzburg, Rontgenring 10, 87 Würzburg, German Federal Republic*

Professor B. MULLONEY *Departments of Biology and Psychiatry, University of California—San Diego, La Jolla, California 92037, U.S.A. (now Department of Zoology, University of California–Davis)*

Professor W. NACHTIGALL *Zoologisches Institut der Universität des Saarlandes, 6600 Saarbrücken, German Federal Republic*

Dr A. C. NEVILLE *Department of Zoology, University of Bristol, Woodland Road, Bristol, BS8 1UG*

Mr P. ONYANGO-ODIYO *East African Agriculture and Forestry Research Organization, P.O. Box 30148, Nairobi, Kenya*

Professor J. W. S. PRINGLE, M.B.E., F.R.S., *Department of Zoology, University of Oxford, South Parks Road, Oxford, OX1 3PS*

Dr R. C. RAINEY *Centre for Overseas Pest Research (Ministry of Overseas Development), College House, Wrights Lane, London, W8 5SJ*

Dr C. J. C. REES *Department of Biology, University of York, Heslington, York, YO1 5DD*

Mr M. J. SAMWAYS *Animal Acoustics Unit, City of London Polytechnic, 139–141 Minories, London, E.C.3*

Dr G. W. SCHAEFER *Department of Physics, University of Technology, Loughborough, Leicestershire*

Professor R. S. SCORER *Department of Mathematics, Imperial College of Science & Technology, Exhibition Road, London, S.W.7*

Dr J. SIMPSON *Entomology Department (Bees), Rothamsted Experimental Station, Harpenden, Hertfordshire*

Professor T. R. E. SOUTHWOOD *Department of Zoology and Applied Entomology, Imperial College of Science & Technology, Prince Consort Road, London, S.W.7*

Mr M. V. VENKATESH *Locust Research Station, Nagniji Road, Bikaner, Rajasthan, India*

Professor T. WEIS-FOGH *Department of Zoology, University of Cambridge, Downing Street, Cambridge, CB2 3EJ*

Professor Sir Vincent WIGGLESWORTH, C.B.E., F.R.S., *Department of Zoology, University of Cambridge, CB2 3EJ*

Dr R. J. WOOTTON *Department of Biological Sciences, University of Exeter, Hatherly Laboratories, Prince of Wales Road, Exeter, EX4 4PS*

Index

Acheta domestica; histolysis of flight muscles 225–6

Acrotylus; light-trapping and possible radar records 173, 177

Adaptive dispersal by flight 137, 142, 149, 264

Adenosine triphosphate (ATP), in flight muscles 12

Adult concentration, of locusts; possible role of wind-convergence 101, 106

Aedes aegypti; flight performance 50, 62, 219, 222

communis; histolysis of flight muscles in autogenous race 225–6

nearcticus; relative air-speed of wing-tip 48

sollicitans; use of sugar and glycogen in flight 221

taeniorhynchus; sugar, glycogen, flight and migration 221, 227

Aegeria apiformis; egg-laying before mass flights 223

Aeshna grandis; hovering flight 50, 61

Aeshnidae 60, 64, 67–8

African Rift Convergence Zone (ARCZ) 96–8

Aglais urticae; radar cross-sections 168, 193

Agnoscelis versicolor; crop damage by invading adults 141

Agronomic approach, in field spray-trials; limitations of 136

Agrotis; flight performance 33, 222

exclamationis; radar cross-sections 193

Aiolopus simulatrix (Sudan plague grasshopper) 81, 91, 98, 107, 140, 142, 159, 161–2, 166–8, 171–5, 177–82, 185–6, 189, 194–5, 273

Aircraft, exploration of wind-systems 85, 89–95, 138

observations on locust swarms 77–85, 107, 186

spraying of airborne insects 75, 107, 110, 142, 150–2, 273, 275

trapping of airborne insects 81, 90–7, 99, 138, 162, 168, 172–3, 262, 274

visual navigation of; analogy with foraging bees 28, 201, 214

Air-speed, of beetles 34, 36

grasshoppers 161

locusts 54, 57–60

moths 145, 162

Alatae; variability of flight performance 217

Aleurodidae; synchronous flight muscles despite small size 6–7, 12

Aleyrodes brassicae; autumn migration 140

Alucitidae; lift mechanism 49

Amazilia fimbriata; hovering flight 50, 62

Ambit, of localised population of flying insects 80

Amphimallon solstitialis; hovering flight 50, 62

Amphiprosopia; light-trapping and possible radar records 173, 177

Amyna punctum, and African Rift Convergence Zone 96

Anacridium melanorhodon (Tree Locust) 170–3

Analogue, historical; use in forecasting infestations of migrant pests 115, 123

Anemotaxis, in foraging bees 207

Anispotera 243–4, 250

Anisozygoptera 243

Anomis sabulifera, and African Rift Convergence Zone 96

Antennae; role in flight 10, 11

Antidromic stimulation of flight motor neurones 22–3

Aphididae 6–7, 91, 95, 101, 136, 139, 150, 158, 217, 223, 225–7, 263

Aphis fabae 217–8, 223, 226–7

gossypii 139, 146

Apis dorsata (Indian rock-bee) 201

florea (Indian dwarf honey-bee) 201

indica (Indian house-bee) 201

mellifica (honey-bee) 8, 17–18, 28–9, 31, 35, 38, 41–3, 45, 50, 60, 62, 80, 199–215, 220

carnica (Austrian bee) 200–1, 209–12, 214

ligustica (Italian bee) 200–1, 209–12, 214

Apterygota 256–7, 260, 266, 271

Arachnida and the origin of insects 250, 262

Archimioptera carbonaria; notably small Carboniferous insect 272

Archodonata 237, 244

Assessment, cartographical, of current spatial distribution of mobile pest populations 115, 123

Asthenohymenidae 246, 272

Asynchronous (myogenic) mechanism of flight muscles 1, 5, 12–13, 16–17

Atactophlebia termitoides; three successive winged instars 251, 264

Atmosphere, as environment of airborne insects 75 *et seq.*

Axillary sclerites; role in changes of wing profile 3, 8

Baëtis; wing-like larval gills 256

Baëtisca; larval gills beneath wing lobes 257

Bardohymenidae 246

Bats (Chiroptera) 48–50, 54, 69

Bee-eaters (Meropidae) 50

Behaviour in flight, in relation to:
  gregarious cohesion of travelling locust swarms
    76, 78–9, 109–10
  height of flight of locusts and grasshoppers
    80–2
  mutual perception of sexes after concentration
    by wind-convergence 106
  orientation and navigation of foraging bees
    28–9, 80, 199–215
  orientation of night-flying insects observed by
    radar 160–4, 176, 182, 185–6, 189
  persistence of static populations 78
Belostoma malkini; histolysis of flight muscles 225
Belostomatidae; asynchronous flight muscles
  despite large size 7, 13
Bemisia tabaci (cotton whitefly) 95, 139–40, 146
Bénard convection cells; airborne insects in 186–7
Bernoulli effect, in fluid flow 51
Biomass of airborne insects 181, 196, 274
Birds, flight of 48–50, 54, 57, 59–60, 63, 69
  radar echoes from 157, 164, 168–9, 172–3, 190
  song of 21
Blattinopsidae 240
Blattodea 236–8, 247–51
Bombus, landing approach 61
  terrestris; lift and power in hovering flight 50,
    62
Boundary layer of atmosphere:
  aerodynamic 33 *et seq.*
  biological 80, 137, 139, 164
Budgerigar (*Melopsittacus*); power requirement
  for flight 60

Caenis; wing-like larval gills of 257–8
Calcium; role in flight motor neurones 16
Caliothrips (cotton leaf thrips) 139, 145
  fumipennis 137
Calliphora 23, 264
  erythrocephala 36
  vicina 218–9
  vomitoria 106
Calliphoridae; flight mechanisms 13, 18, 70
Callosobruchus chinensis and *maculatus;*
  individual variations in flight activity 227
Caloneurodea 237, 247
Campaniform sensilla; as strain-gauges 8–9, 11–
  12
Campyloptera 239
Carnus hemapterus; histolysis of flight muscles
  225
Case-studies; use in forecasting infestations of
  mobile pests 115, 123
Cassidinae; paranotal lobes 236
Cataloipus 168, 172–3
Catantops axillaris 91, 162, 168, 172–3, 177
Cataphractus cinctus; underwater use of wings
  261
Catocala; lift and drag measurements 31–2
Catopsilia florella; radar observations in Sahara
  163–4, 166–9
Ceiling of high-flying insects; radar observations
  177–9

Cenocorixa bifida; block in flight muscle
  development 221
Central nervous system (CNS) 19, 22
  ability to produce flight pattern without
    sensory input 16
Cercopidae 6–7, 91, 98
Chemoreception; role in flight 10, 206–7
Chilo suppressalis (rice stem-borer), air-speed of
  145
Chitin; experimental suppression of formation 27
Chloëon; wing-like larval gills 256–7, 259, 261,
  267
Cholinesterase, brain, and the onset of flight
  activity 220
Chordotonal organs; role in flight 8–10
Choristoneura fumiferana (spruce budworm)
  103–4, 138, 141, 158–60, 162–4, 168, 174–5,
  177–81, 188, 191–2, 223
Chortoicetes terminifera (Australian plague locust)
  24, 26, 168, 174, 177–8, 184, 193
Chromosome, Y; absence in male crickets 22
Cicada, wing of Recent fossil 240
Cicadellidae 91, 95, 98, 139, 218
Cicadidae; synchronous wing muscles 6–7
Cicadulina 218–9, 223, 227, 230
  storeyi 218
Circadian rhythms of flight activity 228
Circulation, aerodynamic, around wing 51, 66–7,
  69
  global atmospheric 76
Cixiidae 6–7, 95, 139
Clap; wings brought together at end of stroke 63,
  65, 67
Climbing flight 81, 138, 176–7
Cloud, marking position of convergence zone 100
Coastal front, and insectivorous birds 101–3
Coccidae 6–7, 272
Coccophagus spectabilis 63
Code, of communication between foraging bees
  200–1
Coenagrion dyeri 243
Colias lesbia; line-concentration off Argentine
  coast 108
Collembola, in aircraft catches 262
Computer studies, of redistribution of airborne
  insects at a coastal front 103
Congo Air Boundary and spread of armyworm
  infestations 96, 126
Control plots, unsprayed, as complicated by
  insect flight 136–7
Convective up-currents (thermals) and airborne
  insects 80–1, 109–10, 214, 265–6, 272
Convergence, wind; insects in zones of 75, 83–
  109, 115–9, 126, 128, 138–9, 152, 157, 181–9,
  262, 264, 272, 275
  quantitative estimates of 88–90, 93, 95–6,
    100–4, 181
Cordovan, mutant strain of honey-bee with
  reduced visual pigment 200
Corixa; histolysis of flight muscles 225
Corpora allata 224, 251
Corpora cardiaca 251

Cosmophila flava (cotton semi-looper) and ARCZ 96

Cotton (*Gossypium*) 137, 139–41, 145–50, 174, 176

Crop protection; need for greater appreciation of insect flight activity 135

Crossopterygian fishes; fins as walking limbs 265, 267

Crustacea 21, 238, 258–9

Culex tarsalis; glycogen in flight 222

Culicoides in aircraft catches 90–1

Cyphoneura permiana; very small Permian Psocopteran 272

Danaidae; flight muscles with small fibres 223

Dates, estimation of; inferred egg-laying 115, 123
   potential locust fledging 115
   potential moth emergence 123

Deilephila elpenor; hovering flight 61

Delphacidae, in aircraft and suction-trap catches 90–1, 95, 139

Dendroctonus monticolae; histolysis of flight muscles 226
   pseudotsugae; histolysis and regeneration of flight muscles 226

Density, of insect populations; area 88, 96, 174–81
   volume 90, 93, 96–8, 162, 177–80

Depressions, barometric; effects on insect distribution 76, 104

Dermaptera 6, 225, 237

Deropeltis erythrocephala 248

Detour experiments, on foraging bees 205–6

Diaphanopterodea 237–8, 244–6

Dictyoptera; tubular flight muscles 6–7

Dictyoptilus 244
   peromapteroides 245

Dipper (*Cinclus*); underwater use of wings 267

Ditaxineura anomalostigmata 243

Dorsal light reaction; field observations followed by laboratory evidence 82

Drag, aerodynamic, of insects 31–2, 34, 40, 46, 48 *et seq.*
   coefficient 35, 46
   induced 53, 59
   parasite 58
   profile 53, 58
   use of, in delaying descent of wingless forms 48

Drosophila 11, 13, 19–23, 31, 54, 59, 62–3, 67, 70
   funebris and *melanogaster;* tethered flight performance 218–9
   virilis 50, 62

Ducks (Anatidae) 267

Dura (*Sorghum*) 139, 141, 146, 176, 185

Duration of insect flight; radar evidence 180

Dysdercus intermedius (cotton stainer) 136, 218–19, 225–6

Ecdysone 27

Ecological aspects of insect flight 46, 48–9, 75–110, 114–9, 123–8, 130, 157, 162–4, 174–89, 195–6, 199–215, 217–30, 246, 256, 272–6

Economic threshold of infestation, for control, as complicated by insect flight 136–41

Efficiency, of asynchronous mechanism of flight muscles 12
   dynamic, in hovering flight; contribution of elastic storage system 62–3
   of spraying flying insects, relative to that of crop spraying 150–2

Elytra, of Coleoptera; role in flight 34–6, 46

Empis tessalata; campaniform sensilla at wing-base 9

Empoasca fabae and cold fronts 138
   lybica (cotton jassid) 137, 139, 145

Encarsia formosa, flight of 49–50, 61, 63–7, 70–1, 108

Endogenous bursters; hypothetical flight motor neurones determining wing-beat frequency 22

Endurance, flight 59

Eustheniidae; nymphal wing-pads separated by groove from terga but not movable 250

Energy, and the aerodynamics of insect and bird flight 48 *et seq.*
   consumption as a clue to foraging distance 200
   kinetic, of atmospheric circulation; utilisation by airborne insects 83

Environment, arid, of hypothetical first flying insects 264–5, 272
   of airborne insects 75 *et seq.*
   contamination of, by insecticides; improvement possible by air-to-air spraying 107

Eopteridium; Devonian crustacean, initially identified as insect 238

Eopterum devonicum; Devonian crustacean, initially identified as insect 238

Ephemera 239, 246, 256–7
   vulgata; vertical nuptial dance as possible ancient flight regime 235, 239

Ephemerella; tergal appearance of abdominal gills 257

Ephemeroptera; synchronous flight muscles 6

Erasipteron larischi 242

Erioischia brassicae; nectar and ovarian development 222

Eristalis; antidromic potentials without effect on firing of other neurones 23
   tenax 50, 62

Eudoter delicatulus 240

Eulerian approach, in study of mobile systems 105

Eulophidae; novel aerodynamic mechanisms 63

Eurygaster integriceps (sunn-pest) 141, 221

Euryptilodes horridus, Carboniferous nymph of 249

Eutelia discistriga and ARCZ 96

Exponential height/density gradients, of airborne insects 140, 177–8
   departures from 138–9, 177–9

Eyes; role in flight 10–11, 13, 200 *et seq.*

Fanning, for hive ventilation 41, 45

Flap, of locust fore-wing 54–6

Flatidae; synchronous wing muscles 6–7

Flicker fusion frequency in foraging bees 214

Flight motor 3–9, 12–13, 16, 56–63, 219–28, 266, 273
   neurones 5, 16–28

Fling; wing separation after 'clap' 51, 65, 246
Flip; wing-twist propagated by delayed elasticity 51, 66–9, 243–6
Flower-constancy, of foraging bees 206
Forces on insect wing:
  aerodynamic 1, 6, 9, 12, 31–6, 39–41
  elastic 42
  gyroscopic, on halteres 10, 12
  inertial 12, 42, 57
Forcipomyia; highest recorded wing-beat frequency 5
Forecasting, of armyworm outbreaks 113–14, 123–32
  of locust invasions 113–22, 129–32
Forecasts, verification of 121–2, 128, 131–2
Forest regeneration, role of pest outbreaks in 140
Forficula auricularia; histolysis of flight muscles 225
Formica fusca 225
Formicidae, wingless, in aircraft catches 262
Fossilization; minimal likelihood under arid conditions 265
Frankliniella dampfi (cotton bud thrips) and ITF 96, 98, 136
Frequency of wing-beat 5, 12, 18, 22, 36, 41–3, 45, 165–74
Frontal hairs, as indicators of air-speed and yaw 10–11
Fronts, atmospheric, in relation to airborne insects:
  coastal 101–4, 138
  cold 76, 103–4, 108, 138
  Inter-Tropical 86–96, 138–9, 184–6
  storm-outflow 107, 138, 181–3
  warm 103–4, 108
Fuel, for insect flight:
  fat 222–3
  glycogen 221–22
  sugars 221–22

Geocorisae; tubular subalar muscles 7
Genetics, of action potentials in cricket song 20, 21
Gerris; histolysis of flight muscles 225
  odontogaster; environmental and genetic effects on flight characters 227
Gill theory of origin of wings; review and re-examination 255–272
Gliding, by insects 31, 48, 82, 250, 263
  by sailplane, providing information on thermal and frontal up-currents 81, 101
Glossina (tsetse-fly) 220, 223, 228–9
Glossina morsitans; flight within limited ambit 80
α-glycerophosphate dehydrogenase (GDH) in flight muscle 220
β-glycerophosphatase and the onset of flight activity 220
Goal-finding, by foraging bees 199 *et seq.*
Gomphocerippus; leg muscle innervation for singing and walking 28
Gomphocerinae; learning in sound patterns 27
Gonocephalum simplex; Sudan suction-trapping 95
Groundnuts (*Arachis*), and *Heliothis* 146–7
  radar observations of moth take-off from 174–6

Gryllidae (crickets) 17–19, 21–3, 27, 46
Gryllus 162, 189
Gull, black-headed (*Larus ridibundus*); soaring in thermal up-currents 48
  laughing (*L. atricilla*); minimum aerodynamic power requirement 60
Gyrinus 13, 260

Halteres; role in flight-stability 9–13
Halticinae and ARCZ 97, 99
Height of insect flight 80–2, 90–5, 98, 177–80, 275
Heliocopris; lift and power for hovering flight 50, 62
Heliothis armigera (cotton bollworm) 96, 141, 145–9, 158, 174, 176
  zea; height of flight 149
Heptagenia sulphurea; fore-wing base 239
Hercothrips; important early pest of cotton in Sudan Gezira 145
Hispa; rice damage through possible aerial concentration 274
Histolysis, of flight muscles 136, 225
Homaloneura lehmanni 245
Homoioptera gigantea; wing length 190 mm 245
  woodwardi; Palaeodictyopteran with fore- and hind-wings overlapping 245–6
Hovering flight, of insects and birds 48, 50, 60–4
  power requirements for 62
Hummingbirds (Trochilidae), flight of 50, 57, 59, 61–3, 70
Hyalophora cecropia; JH and lipid transport 222
Hydrocorisae; subalar muscle absent 7
Hydroporus palustris; histolysis of flight muscles 225
Hylobius abietus; histolysis of flight muscles 225
Hypena and ARCZ 96, 275

Ichneumonidae; asynchronous power-producing basalar muscles 7
Idoptilus onisciforme; nymph with precociously incipient axillary plates 250
Ilycoris cimicoides; histolysis of flight muscles 225
Inachis io (peacock butterfly); momentary gliding flight 31
Infra-red radiation; possibility of orientation of flying insects by 195
Inter-Tropical Convergence Zone (ITCZ), in relation to insects 76, 83–96, 107, 109 116–9, 126, 184
Inter-Tropical Front (ITF) 86
  in relation to insects 87–96, 138–9, 184–6, 275
Iphiclides podalirius (scarce swallowtail); gliding flight 31
Ips paraconfusus; degeneration of flight muscles 226
Isoptera (termites); histolysis of flight muscles 225

Jassidae 6–7, 136, 139, 145–6
Johnston's organ; information on relative wind 205, 215
Jolia larva; thoracic branchial tubes 258
Juvenile hormone (JH); possible effects on lipid transport and on flight muscles 222, 226

Kalotermes; wing-like expansions from thoracic terga 255

Kestrel (*Falco tinnunculus*); possible 'flip' effect in hovering 69

Krebs cycle, in 'sprinters' and 'stayers' 223

Lactate dehydrogenase, in jumping muscles 220

Lagrangian approach, in study of mobile systems 79, 105

Lake-breeze, aircraft-trapping in 97

Lamellicornia; wing-loadings and hovering flight 50, 60

Lamproptilia grandeuryi 244–5

Landing (settling) 11, 184

Landmarks, in bee navigation:
 individual 200–1, 203–4, 208–14
 linear 201–3
 complex pattern 28, 208–14

Lasius niger; histolysis of flight muscles 225

Latin square, in large-scale experiment on incidence and control of airborne pests 147

Learning, in insects, by experience 28–9, 199–215
 by genetic mechanism 19–22, 26–7
 generalisation 212–13

Lemmatophora; Protorthoptera with prothoracic paranotal lobes 247

Leptaleus sennarensis in ITCZ 95, 98, 275

Leptinotarsa decemlineata (Colorado beetle) 220–1, 225–6

Leptoblattina exilis; Carboniferous nymph 249

Leptomastix; dorsoventral flight muscle absent 6

Lethocerus maximus; histolysis of flight muscles 225

Lift, aerodynamic, of flying insects 31, 32, 34 *et seq.*
 coefficients 35, 46, 50, 53–4, 60, 66, 68, 71
 generation, by circulation:
  in steady flow 52–6
  in unsteady flow 52–3, 56, 66–7, 71
 from body of insect 34–5, 46

Lift/drag ratio (*l/d*) 31–3

Light intensity and take-off 174–6

Light-trapping, of airborne insects 96, 98, 103, 143, 149
 and absolute insect densities by radar 175, 177

Lipopterna cervi; histolysis of flight muscles 225

Lithomantis carbonarius; paranotal lobes and large size 247, 263

Lithoneura lameeri; Carboniferous Ephemeropteroid with very primitive venation 236, 241–2
 mirifica; known as hind-wing only 241

Locusta migratoria (Migratory Locust) 38, 71, 168, 172–3, 177, 220, 222, 224–5

Longitarsus 97, 99

Longiunguis sacchari (cane aphid) and ITF 139

Lucanus cervus; histolysis of flight muscles 225

Lucilia cuprina; use of carbohydrate for flight 222
 sericata; location of sensilla on halteres 10

Lygaeus kalmii; variability of tethered flight performance 218

Lymantria dispar (gypsy moth); airborne larvae 262

Machilis; coxal styli on meso- and metathorax 259–60, 266

Macroglossum stellatarum; lift and power in hovering flight 50, 62

Madera mamayi; small Palaeodictyopteran 245

Magnus effect, in lift generation 51, 53

Malacosoma pluviale; variability in flight activity and larval behaviour 227

Manduca sexta; lift and power for hovering flight 50, 61–2

Maniola jurtina; flight and persistence of static populations 80

Mapping, of current distribution of pest populations 114–21, 123–30
 of seasonal probabilities of breeding and invasion 115
 of synchronised meteorological observations (synoptic analysis) 116
 stereotaxic, of flight muscles 20

Mechano-receptors involved in flight 10, 11

Mecoptera; synchronous flight muscles 6

Mecynostomata dohrni 245

Megaloptera 260

Megatypus schucherti 243

Melolontha melolontha; estimates of lift from elytra and body 34–5, 46
 vulgaris; lift and power in hovering flight 50, 61–2

Membracidae; synchronous flight muscles 6–7

Metropator pusillus; one of earliest fossil insects, with 7 mm wing 272

Micropterygidae; probably smallest insects with synchronous flight muscles 5

Migration 59, 76, 108, 137–8, 164, 166, 203, 217, 219–24, 226–9

Mischoptera nigra 245, 247

Misthodotes 240

Mochlonyx culiciformis; histolysis of flight muscles 225

Models, theoretical, simulating aircraft spraying of flying insects 107

Monsoons, and distribution of airborne insects 76, 83, 86, 90–1, 185

Moorhen (*Gallinula chloropus*); use of legs in underwater swimming 267

Musca 11, 23
 domestica 219–20

Muscina 11, 46

Muscles, flight 4–6, 16–19, 62, 261

Mymaridae; flight of extremely small insects 51

Myriapoda 262

Mythimna (Pseudaletia) separata (oriental armyworm); prolonged tethered flight 217, 219

Nannophlebia risi 243

Navigation, by foraging bees:
 long-range:
  use of landmarks 28–9, 201–5, 208–12, 214
  use of sun-compass 201–3, 205, 211
 short-range:
  use of olfactory tropotaxis 206–7
  use of visual pattern recognition 207–13

Nectar 28, 176, 200, 207
 constituents, as fuel for flight 221–2

Neighbour, nearest, among flying insects; distances to 98, 105
Nematocera; asynchronous flight muscles 7
Nepa cinerea; histolysis of flight muscles 225-6
Neurones:
  flight motor 16-17, 20, 22
    double-firing 28
    genetically determined pattern of action-potentials 20
    identification of individual 24
    intracellular recording from 25
    number per muscle 27
Neuroptera 6, 69
Night-viewing device; observations of moth take-off 176
Nightjars (Caprimulgidae) 173
Nilaparvata lugens (brown plant-hopper); in frontal systems off Japanese coast 108
Nomadacris septemfasciata (Red Locust); flight in static populations 78, 80
Notonectidae; asynchronous flight muscles 7
Notorachis wolfforum; overlapping wings 245-6

Odonata 6-7, 31, 50, 60, 64, 67, 70-1, 186-7, 229, 235, 237-9, 242-4, 246-7, 249-51, 263, 267
Odour, role in bee navigation:
  colony specific 204
  flower 206
Oedaleus 172-3, 177
Olfactory tropotaxis 206-7
Oligoneuria 258
Oncopeltus: landing response 11
  fasciatus (milkweed bug) 18, 217, 219, 227, 229
Operophtera brumata (winter moth); temperature threshold for flight 105
Optomotor reactions 204
  possible role in uniformity of orientation after concentration by wind-convergence 106
Orientation in flight:
  down-wind, in scattered night-flying locusts 163
  up-wind, in persistent static populations 80
  effectively random, in locust swarms— travelling with velocity of wind 77
  uniform, in groups of locusts within swarms 78
    in other high-flying insects; radar observations 162-4, 176, 182, 185-6, 189
  inwards, at perimeter of travelling swarms, as factor in cohesion 79
  to landmarks 201-5, 208-12
  to sun 201-3, 205, 211
  in relation to spraying 152
Oryctes boas 34
Oscinella frit (frit-fly) 218-19, 225, 274
Ovarian development and flight 233
Oviposition; spatial and temporal patterns, of *Heliothis* on cotton 147-9

Pachydiplosis; main larval damage within hours of hatching 141
Papilio machaon (swallowtail) 31
Palaeodictyoptera 236-8, 241, 243-7, 249-51, 255, 263, 271
Palaeoptera 237-8, 248, 251-2, 263

Paranotal lobes and possible origin of wings 236-7, 256, 267
Paraplecoptera 263
Partridge (*Perdix*); relative speed of wing-tip 48
Passerine birds; wing-loadings 50
Pattern-recognition, in bee navigation 207-13
Penguins (Spheniscidae) 267
Perception, mutual, among flying insects 105-6
Periplaneta 261
Peromaptera filholi 245
Petrobius 257
  maritimus 259
Phasmida; close-packed oblique dorsal flight-muscle 6
Pheasant (*Phasianus*); relative speed of wing-tip 48
Phormia regina 36-7, 39-40, 56, 67, 69, 222
Photoperiod 224, 227
Pieris; both fat and carbohydrate used as fuel for flight 222
  brassicae 31, 67, 70
  napi; lift in hovering flight 50, 63
Pigeon (*Columba*) 48, 58, 70
Pitching in flight (about transverse axis) 10
Plate-organs; site of olfactory receptors on bee antenna 206
Platychirus peltatus; role of alula in flight 68
Plecotus auritus (bat); lift in hovering flight 50
Plusia acuta and ARCZ 96
  gamma; changes in flight activity with age 228
  limberena and *orichalcea* and ARCZ 96
Plutella maculipennis (diamond-back moth) 225
Polymorphism, alary 217, 227-8
Polynema natans; use of wings in underwater swimming 261
Population, density; effects of concentration by wind-convergence 88-109
  discrete mobile; morphometrics of neighbouring locust swarms 77
    numbers in 79, 113
  movements; elucidation by biogeographical analysis 83 *et seq.* 114-32
Power requirements for flight (insect and bird) 52-9
Pressure systems, barometric 76
  surface and upper-air analyses of, in relation to airborne insects 83, 91, 97, 104
Probabilities, seasonal, of locust breeding and occurrence of swarms 115
Prodenia eridania (southern armyworm) 221
Profiles, height/density, of airborne insects 138, 140, 177-9, 183, 185-9, 274-5
Progoneura nobilis 243
Propodeum; involvement in flight-system 3
Prosopistoma; gill chamber formed by larval fore-wings 258
Protereisma permianium; venation and prothoracic paranotal expansions 239-40
Pseudo-elytra; articulated abdominal gill-plates 257, 258
Psocoptera, at ITF 91
Psocus; asynchronous flight muscles 6
Psyllidae; asynchronus flight muscles 6-7

Psylliodes chrysocephala; histolysis of flight muscles 225

Pterophoridae (plume moths); low wing-loading 49

Pterygota 3, 48–9, 235 *et seq.* 259, 264, 267

Ptiliidae; flight of extremely small insects 51

Pulex irritans, in aircraft catch 262

Pyrrhosoma nymphula; terga and wing-bases 246

Pyruvate cycle and development of flight activity 221

Quelea quelea aethiopica (weaver bird); swarming habit and pest status 141

Racial differences in learning, in foraging bees 214

Radar observations of airborne insects 13, 78, 81–2, 91–2, 94–5, 101, 103–4, 107, 113, 138–40, 142, 149, 152, 157–96, 274

Rain and airborne insects 76, 83, 95–8, 107, 116, 141, 149, 181–2, 262, 264–5

Ranatra; histolysis of flight muscles 225

Range, flight, of foraging bees 200

Rayleigh law, in radar reflection 190

Recapitulation, evolutionary 255

Red Sea Convergence Zone 98–101

Reizwechsel (frequency of stimulus–change) 214

Reservations, of pest populations 135, 143, 147

Resilin, storing energy during deformation in flight movements 3

Resonance, mechanical, of thorax-wing system 3, 5, 12, 16, 43, 273

Respiration; abdominal pumping movements inferred from radar 166

Reynolds number 32, 49–51, 58, 60, 70

Rhopalocera 12, 31–4, 50, 61, 63, 69–71, 108, 168, 223

Rice (*Oryza*) 143–5

Richardson number, as criterion for sharpness of wind-shift lines 101–3

Rochdalia parkeri; Palaeodictyopteran nymph 249–50

Rochlingia 247

Rock-dove (*Columba livia*); unsteady-state aerodynamics in initial wing claps 69–70

Rolling in flight (about longitudinal axis) 11, 13

Romalea microptera; flightlessness and enzymes 220

Rook (*Corvus frugilegus*); relative speed of wing-tip 48

Rotation of crops, in relation to airborne pests 145–6, 174, 176

Sampling volume (radar) 160

Sarcophaga 223

Saturniidae; lipases and flight 222

Scales, wing; possible aerodynamic role 33–4, 46

Schistocerca gregaria (Desert Locust) 48, 54–5, 57–9, 70–1, 76–85, 95–6, 98, 101, 104–7, 113–22, 129–32, 137, 140, 142, 150, 158, 162–5, 167–71, 174, 178–81, 184, 186–7, 193–5, 203, 217, 219, 222, 224–5, 229, 263, 276

vaga; GDH and development of flight activity 220

Scolytidae; development and histolysis of flight muscles 220, 225–6

Sea-breeze, in relation to bird and insect flight 101

Servo-mechanisms, of flight control in bees 8

Sesamia 145

Sex, of flying locusts, distinguished by radar 171 linkage in genetic control of cricket song 22

Side-wind compensation, by foraging bees 203–5

Sitona hispidula and *lineatus;* histolysis of flight muscles 225

regensteinensis; development and histolysis of flight muscles 220, 225–6

Sogata furcifera; in frontal systems off Japanese coast 108

Sorghum purpureo-sericeum; abundant wild host-plant of *Heliothis* 148

Sowing date in relation to pest incidence 149

Sphingidae 12, 61, 168–9

Sphinx ligustri; lift and power for hovering flight 50, 62

Spiders (Araneae), airborne 91, 262

Spodoptera exempta (African armyworm) 60, 83, 96–8, 107, 113–4, 123–31, 137, 149, 203, 275–6

exigua (lesser armyworm); reported flight duration 217

littoralis (cotton leafworm) 174, 176

Spraying, insecticidal, synchronized and unsynchronized 145

against flying insects 107, 110, 150–1, 273, 275

Sprinters 223

Stability in flight:

in pitch (fore-and-aft plane) 8–10

in roll (transverse plane) 13

in yaw (horizontal plane) 10

Stalling in flight 35, 41, 54

Staphylinidae and the ITF 95, 139

Stayers 223

Stenodictya; Palaeodictyopteran with haustellate mouthparts 244, 247

fayoli (formerly *agnita*) 246

spinosa 245

Stork (*Ciconia*); principle of flapping flight 55, 57

Strain gauges, in cuticle 8, 11

Stretch-receptors, monitoring wing-beat 8

Stridulation, using muscles and nerves employed in flight 16, 18, 46

Sub-imago; initial flying stage of mayfly, undergoing further moult before maturity 251

Suction-trapping; of airborne insects 91, 93, 95, 97, 139–40, 157

Sun-compass, in bee navigation 201 learning element 203

Sustaia impar; Paoliid fossil with wings in oblique position 248

Swallow (*Hirundo*); wing-loading 50

Swarms, flying:

locust:

cohesion 76–9, 109

direction of displacement 77

ground-speed 77, 186

structure 138, 187, 195

tracks 96

other insects 157 *et seq.*

Swift (*Apus*); in sea-breeze front 101–2
  wing-loading 50
Sylvohymen sibiricus 245
Sympetrum striolatum; age and air-speed 229
Symphyla; thoracic styli 260–1
Symphyta; structure of flight muscles 6
Synchronous, control of pests 107, 143–4, 150
  (neurogenic) mechanism of flight muscles 4,
  12–13, 16–17
Synoptic, analysis of synchronised meteorological
  observations 83, 91, 97, 104
  survey of concurrent pest infestations 107, 141,
  145, 150
Syrphidae; GDH in flight muscles 222
Syrphinae; hovering with horizontal body and
  unsteady-state aerodynamics 60, 63–4, 67–8
Syrphus; evidence for special mechanism of
  hovering 50, 61
  balteatus; proposed 'flip' mechanism of lift
  generation 68

Tabanidae; appreciable lift from insect body 46
Tebanus affinis; relative speed of wing-tip 48
Take-off 10, 104
  radar observations 157, 174–7
Tegmina, asymmetry of 13–14
Teleogryllus commodus (Australian field cricket);
  stridulation and flight 19, 21–2
  oceanicus (Polynesian field cricket); genetics of
  song 21–2
Temperature and flight activity 76, 83, 104–5,
  171, 173, 179, 188, 227
Tenebrio molitor; flight muscles complete at
  emergence, but not enzymes 220
Tern (Sterninae); possible 'flip' mechanism in
  hovering 69
Tethered flight 11, 217 *et seq.*
  unphysiological conditions in absence of
  airflow 41
Tetranychidae; airborne mites on silken threads
  262, 264
Tettigoniidae; stridulation versus flight? 46
Thermals (convective up-currents) and airborne
  insects 80–2, 186–7, 265–6, 272
  possible concentration of flying insects by
  109–10
Thrust; propulsive force developed by wing-
  system 39–40, 45, 55
Thysanoptera 6, 31, 69, 90–1, 95–6, 98, 101, 145,
  237
Thysanura 259–65, 267
Tipula; stability; lift and power in hovering
  flight 13, 50, 62
Trachyostus ghanaensis; histolysis of flight
  muscles 225
Trade-winds and distribution of airborne insects
  76, 86, 90–1, 185
Trialeurodes vaporariorum (greenhouse whitefly)
  64
Trichogrammatidae; flight at very low Reynolds
  numbers 51
Trichoplusia ni; duration of tethered flight 217
Trichoptera; synchronous wing muscles 6

Trilobita 259
Triplosoba; perhaps closest to archetypal insect
  wing 248
  pulchella; oldest known mayfly 241
Trogius; synchronous wing muscles 6
Tropotaxis, olfactory 211
Tryporyza incertulas (rice stemborer) 141, 143–5,
  150
Turbulence, atmospheric 100, 109
  and airborne insects 13, 78, 82, 90, 188–9
Turn-and-bank indicator, halteres as 10
Turning in flight:
  by differential power to wings 9
  by differential wing-warping 8, 18, 28
  by rudder action of abdomen and legs 11, 18,
  28
Typus gracilis 243

Uniformitarianism 255
Unsteady-state aerodynamics; role in insect flight
  12, 41, 48, 64–5, 69, 70
Uropetala 243

Vanessa; single wing-hinge layer 229
Verification, of forecasts 121–2, 128, 131–2
Vespa crabro; lift and power for hovering flight
  50, 62
Viscosity effects, in flight of small insects 53, 71
Visual acuity, in mutual perception of airborne
  insects 105
  reactions, in flight 11, 13, 18, 80, 201 *et seq.*
Vortex, in air circulation of insect wing:
  bound 51–2, 65–6, 68
  pattern in 'fling' mechanism of lift generation
  65
  starting 52, 65
  tip 52, 59, 65

Wafting; wing movements for scent dispersal
  within hive 41
Waggle-dance, in communication between
  foraging bees 197
Wagner effect, delaying development of
  aerodynamic lift 52, 54, 68
  as avoided by 'fling' mechanism 66
Warnings, of armyworm outbreaks 113, 124,
  128–9, 131
  of locust invasions 113, 116–7, 119, 121–2,
  131–2
Wind measurement, by airborne Doppler radar
  85, 88, 138
  by drift-meter 85
  by pilot-balloon 85, 161, 178, 185
  systems 82 *et seq.*
  tunnel, in research on insect flight 19, 31, 36,
  41, 46
Wing-beat frequency; effects of captivity 171
  photographic determinations 36–9, 64, 167–8,
  171
  radar evidence 164–74
  relationship with wing length 168
  coupling 42, 70
  hinge layers 229

Wing-loading 5, 8, 50
Wing, movements, kinematics of:
    Chalcid wasp 64–6
    flies 36
    honey-bee 41
    locust 55, 71
    stork 57
  structure, in relation to function 9, 42, 69
Wood-pigeon (*Columba palumbus*); probably

unsteady-state aerodynamics at take-off 69

Xyela; asynchronous wing muscles 6

Yawing in flight (about vertical axis) 205
Y-chromosome; absence in male crickets 22

Zdenekia grandis; Carboniferous dragonfly with
  very well preserved venation 242